SpringerWienNewYork

Computational Microelectronics

Edited by

Siegfried Selberherr
Technical University Vienna
Vienna, Austria

For further volumes:
www.springer.com/series/1263

Sung-Min Hong
Anh-Tuan Pham
Christoph Jungemann

Deterministic Solvers for the Boltzmann Transport Equation

SpringerWienNewYork

Sung-Min Hong
Device Laboratory
Samsung Information Systems America, Inc.
75 West Plumeria Drive
San Jose, CA 95134
USA
hi2ska2@gmail.com

Anh-Tuan Pham
Institut für Elektronische
Bauelemente und Schaltungstechnik
TU Braunschweig
Postfach 3329
38023 Braunschweig
Germany
pham@nst.ing.tu.bs.de

Christoph Jungemann
Institut für Theoretische Elektrotechnik
RWTH Aachen University
Templergraben 55
52056 Aachen, Germany
christoph.jungemann@ithe.rwth-aachen.de

SpringerWienNewYork is a part of Springer Science+Business Media
springer.at

Cover: WMXDesign GmbH, Heidelberg, Germany
Typesetting: SPi, Chennai, India

Printed on acid-free and chlorine-free bleached paper
SPIN 80062049

With 124 Figures

Library of Congress Control Number: 2011934251

ISBN 978-3-7091-0777-5 e-ISBN 978-3-7091-0778-2
DOI 10.1007/978-3-7091-0778-2
SpringerWienNewYork

Preface

In this book the research of the last decade by the authors regarding deterministic solvers for the Boltzmann transport equation is summarized. The work was started by the third author, who was interested in electronic noise of semiconductor devices. At first glance the use of a deterministic solver seems to be strange, because all groups working at that time on noise in devices solved the Boltzmann transport equation with the stochastic Monte Carlo method, which inherently contains noise and was regarded as the best method to do so. But it soon became clear that the Monte Carlo method, which is also inherently transient, was too slow to calculate noise in the technically relevant frequency range with sufficient accuracy. A deterministic solver on the other hand could yield truly stationary solutions, and exact small-signal analysis could be performed directly in the frequency domain covering the full frequency range. Unfortunately, the deterministic solvers have not only advantages compared to the Monte Carlo method but also disadvantages. The biggest obstacle in the application of deterministic solvers was the huge memory requirement (about 100 GBytes for a realistic 2D device simulation including noise), which is by orders of magnitude larger than the one of the Monte Carlo method. This is the reason, why the interest in deterministic solvers had almost vanished by the end of the last millennium. When this project was started, deterministic solvers still required expensive supercomputers. Fortunately, Moore's law held during the last decade and nowadays computers with sufficient memory are available at a reasonable price. Thus, progress in semiconductor technology solved the biggest problem in applying deterministic solvers to the Boltzmann transport equation.

This work was started in earnest when the third author joined the Institute of Electron Devices and Circuits at the Technical University of Braunschweig in Braunschweig, Germany. A little later the second author joined this institute, which is headed by Prof. Dr. Bernd Meinerzhagen. Both authors are very grateful for his support of their work and his valuable advice. In 2006, the third author became a professor at the Bundeswehr University in Neubiberg, Germany and soon after the first author joined this new group. His position was first funded by the Deutsche Forschungsgemeinschaft (DFG) and later by the Bundeswehr University, for which the first and third author are most grateful. The authors would like to thank all persons who have supported or contributed to this work. Without the very memory efficient linear solver ILUPACK developed by Prof. Dr. Matthias Bollhöfer of the TU Braunschweig and his help the project would have faltered early on. The third author gratefully acknowledges discussions with Prof. Dr. Christian Ringhofer of the Arizona State University about the maximum entropy dissipation scheme. At the Bundeswehr University our work has profited from contributions by Dr. Mindaugas Ramonas, Dr. Tanh Viet Dinh, and Gregor Matz. The authors gratefully acknowledge financial support by the EU through the project DotFive under grant agreement no 216110 (FP7/2007-2013). The fruitful collaboration with the group of Prof. Dr. Michael Schröter of the TU Dresden regarding SiGe HBTs was most helpful, especially the supply of experimental data by Dr. Paulius Sakalas and the feedback by the first user of our code, Gerald Wedel. The second author gratefully thanks his colleagues M. Klawitter and Dr. A. Kuligk of Institute of Electron Devices and Circuits (TU Braunschweig) for helpful discussions. The first author gratefully acknowledges many fruitful discussions with Dr. Seonghoon Jin of Synopsys about the inclusion of full band effects.

The first author is very grateful to his wife (Young Min Yoon), his children (Jin Gi and Lena), and all other family members. Especially, the endless love and support from his parents (Tae Suk Hong and Jin Sook Kim) are greatly appreciated. The first year of the first author's research was partially supported by a Korea Research Foundation grant by the Korean Government (MOEHRD) under contract number KRF-2007-357-D00159.

The second author gratefully thanks his friends, his family including two sons Anh-Tri (Chip) and Sy-Tien (Mos), and his grandparents who had unfortunately gone too soon to see their grandson's achievements.

The authors thank Prof. Dr. Siegfried Selberherr of the TU Vienna and editor of this Springer series for the opportunity to publish this work and his encouragement.

Neubiberg and Braunschweig
June 2011

Sung-Min Hong
Anh-Tuan Pham
Christoph Jungemann

Contents

Notation

Symbols

x	Scalar
$\mathbf{x}$	Vector
$\hat{\mathbf{X}}$	Matrix
$\mathbf{0}$	Zero vector
$\hat{\mathbf{0}}$	Zero matrix
$\hat{\mathbf{I}}$	Identity matrix
$\hat{\mathbf{A}}$	Rotation matrix rotated the Luttinger-Kohn basis states to x, y, z basis states
$a_{\mathbf{k}}^{\mu}$	Expansion coefficient of the Bloch function
a_v	Hydrostatic deformation potential
$a, a_{\mathrm{Si}}, a_{\mathrm{Ge}}$	Lattice constant, Si lattice constant, Ge lattice constant
$a_{l,m,l',m',l'',m''}$	Integral of three spherical harmonics
$\mathbf{a}_{l,m,l',m'}$	Integral of two spherical harmonics
$\mathbf{a}_{l,m,l',m',l'',m''}$	Integral of three spherical harmonics and the radial vector
b	Shear strain deformation potential
$b_{m,m'}^{\mathrm{E},\nu}(x,\tilde{\varepsilon})$	b^{E}-factor on the LHS of the projected BTE
$b_{m,m'}^{\mathrm{B},\nu}(x,\tilde{\varepsilon})$	b^{B}-factor on the LHS of the projected BTE
$\mathbf{B}$	Magnetic field
β	Band index
β_{ι}	Ge-content of layer ι
$\mathbf{b}_{l,m,l',m'}$	Integral of two spherical harmonics
$\mathbf{b}_{l,m,l',m',l'',m''}$	Integral of three spherical harmonics
$c_{\eta,\mp}^{\nu',\nu}(x,\mathbf{k}'\vert\mathbf{k})$	Scattering transition rate
$c_{\eta,\mp,l',l}^{\nu',\nu}(x,\tilde{\varepsilon}^{\nu'}(x,\mathbf{k}'),\tilde{\varepsilon}^{\nu}(x,\mathbf{k}))$	Fourier expansion coefficient of scattering transition rate

$c_{l,m}$	Normalization factor of spherical harmonic $Y_{l,m}(\vartheta, \varphi)$
c_m	Normalization factor of Fourier harmonic $Y_m(\phi)$
$\hat{\mathbf{C}}$	Positive-definite diagonal matrix in the matrix form of the discretized $\mathbf{k} \cdot \mathbf{p}$ EFA SE
$\mathscr{C}$	Subset of monotonic interpolation region $\mathscr{M}$
d	Shear strain deformation potential
$d^{\nu}_{m,m'}(x, \tilde{\varepsilon})$	d-factor on the LHS of the projected BTE
$\hat{\mathbf{D}}_j, \hat{\mathbf{D}}^-_j, \hat{\mathbf{D}}^+_j$	Matrix elements at grid point $z' = z'_j$ in the matrix form of the discretized $\mathbf{k} \cdot \mathbf{p}$ EFA SE
$D_t K$	Average optical deformation potential
Δ	Spin-orbit split-off potential, rms height of the surface roughness
$\delta_{\nu',\nu}$	Kronecker symbol
$\delta\mathbf{E}$	Small lateral electric field applied along the channel direction of a homogenous channel MOS structure
δQ_{G}	Small signal variations of the gate charge
$\langle \lvert\Delta_{\mathbf{q}}\rvert^2 \rangle$	Surface roughness power spectrum
$\mathbf{e}_{x_\mathrm{C}}, \mathbf{e}_{y_\mathrm{C}}, \mathbf{e}_{z_\mathrm{C}}$	Unit vectors of the crystallographic coordinate system $(x_\mathrm{C}, y_\mathrm{C}, z_\mathrm{C})$
$\mathbf{e}_{x'}, \mathbf{e}_{y'}, \mathbf{e}_{z'}$	Unit vectors of the primed coordinate system (x', y', z')
$\mathbf{e}_{\varepsilon}, \mathbf{e}_{\vartheta}, \mathbf{e}_{\varphi}$	Unit vectors of the spherical coordinate system $(\varepsilon, \vartheta, \varphi)$
$\mathbf{e}_{ij}$	Unit vector along a line from the node i to the node j
$\mathbf{E}_{\mathrm{H}}$	Hall field
$\mathbf{E}_{\lvert\lvert}$	Lateral electric field applied along the channel direction of a homogenous channel MOS structure
E	Band edge
E_{c}	Energy of the conduction band edge
E_{v}	Valence band offset
$\varepsilon, \varepsilon_{\mathbf{k}}, \varepsilon^{\lambda}_{\Gamma}, \varepsilon^{\nu}$	Total energy, wave vector dependent eigen energy, eigen energy of the Γ-centered eigen problem, eigen energy of eigen state ν
ε_0	eigen energy of degenerated bands
$\tilde{\varepsilon}, \tilde{\varepsilon}'$	Kinetic energy
$\tilde{\varepsilon}_j$	Kinetic energy of grid point j on the kinetic energy axis
$\varepsilon_{\mathrm{pot}}$	Potential energy
$\varepsilon(k, \phi), \varepsilon_m(k), \varepsilon_{m,l}$	Energy as a function of wave vector $\mathbf{k} = (k, \phi)$, Fourier expansion coefficient of $\varepsilon(k, \phi)$, Fourier expansion coefficient of $\varepsilon(k_l, \phi)$
$\epsilon_{xx}, \epsilon_{yy}, \epsilon_{zz}, \epsilon_{xy}, \epsilon_{yz}, \epsilon_{xz}$	Strain tensor components
η	Scattering mechanism index
$f(x, \mathbf{k}), f(x, \tilde{\varepsilon}, \phi)$	Distribution function

$f_{\mathrm{eq}}(x, \varepsilon(\mathbf{k}))$	Equilibrium distribution function
f_{TE}	Transit frequency
$\mathbf{F}, \mathbf{F}_{\mathrm{E}}, \mathbf{F}_{\mathrm{B}}$	Total force, force due to the gradient in the total subband energy, force due to magnetic field
$F_{\mathrm{E,pot}}$	Force due to gradient in the subband potential energy
$\tilde{F}_{\mathrm{E}}$	Force due to gradient in the subband kinetic energy
Φ'	Polar angle of the growth direction z' in the crystallographic coordinate system
Φ_{ms}	Work function difference
ϕ, ϕ'	Polar angle in the 2D **k**-space of the transport plane
ϕ_n	Discrete polar angle in the 2D **k**-space for which the $\mathbf{k} \cdot \mathbf{p}$ EFA SE is solved
φ	Azimuthal angle in the spherical coordinate system
$\varphi, \phi_{N_{\mathrm{irr}}}$	Lower bound of the irreducible ϕ range, upper bound of the irreducible ϕ range
φ_m	Phase of Fourier harmonic $Y_m(\phi)$
φ_{D}	Angle between x_{D}- and x'-axes
$g(x, \tilde{\varepsilon}, \phi)$	Modified distribution function
$g_{\mathrm{D}}, g_{\mathrm{D}}^{\mathrm{DD}}$	Channel conductance in a short channel MOSFET, g_{D} assuming drift-diffusion model for the transport
$g_{\mathrm{D,h}}$	Channel conductance in the homogenous channel MOS structure
$g_m(x, \tilde{\varepsilon})$	Fourier expansion coefficient of modified distribution function $g(x, \tilde{\varepsilon}, \phi)$
$\mathcal{G}_{\iota}^{\nu',\nu}(x)$	Isotropic overlap factor of phonon scattering
$\gamma_1, \gamma_2, \gamma_3$	Luttinger-Kohn $\mathbf{k} \cdot \mathbf{p}$ parameters
$\gamma^{\nu}(x, \mathbf{k})$	Area generation rate
$\gamma_{\mathrm{s}}^{\nu}(x, \mathbf{k})$	Interface generation rate
Γ	The point $\mathbf{k} = \mathbf{0}$ in the **k** space or the piezoresistance coefficient
$\Gamma_{\mathrm{GPN}}^{\nu',\nu}(x, \mathbf{k}', \mathbf{k})$	Approximated surface roughness scattering matrix elements
$\hbar$	Planck's constant divided by 2π
$\mathcal{H}$	Hamiltonian operator
$\mathcal{H}_{\mathbf{kp}}$	$\mathbf{k} \cdot \mathbf{p}$ Hamiltonian operator
$\mathcal{H}_{\mathbf{kp}}^{\lambda\mu}$	Matrix elements of the $\infty \times \infty$ $\mathbf{k} \cdot \mathbf{p}$ Hamiltonian
$\hat{\mathbf{H}}_{\mathbf{kp}}^{\lambda\mu}$	Matrix elements of the $N \times N$ $\mathbf{k} \cdot \mathbf{p}$ Hamiltonian
$\hat{\mathbf{H}}_{\mathbf{kp}}^{3\times3}$	3×3 spin-degenerate $\mathbf{k} \cdot \mathbf{p}$ Hamiltonian
$\hat{\mathbf{H}}_{\mathbf{kp}}^{\prime 3\times3}$	3×3 spin-degenerate $\mathbf{k} \cdot \mathbf{p}$ Hamiltonian associated with the primed coordinate system
$\hat{\mathbf{H}}_{Ev}^{3\times3}$	3×3 diagonal band-offset Hamiltonian
$\hat{\mathbf{H}}_{\mathrm{strain}}^{3\times3}$	3×3 Pikus-Bir strain Hamiltonian
$\hat{\mathbf{H}}_{\mathbf{kp}}^{\mathrm{LK}}$	6×6 Luttinger-Kohn $\mathbf{k} \cdot \mathbf{p}$ Hamiltonian
$\hat{\mathbf{H}}_{\mathbf{kp}}^{\prime\mathrm{LK}}, \hat{\mathbf{H}}_{\mathbf{kp}}^{\prime}$	6×6 Luttinger-Kohn $\mathbf{k} \cdot \mathbf{p}$ Hamiltonian associated with the primed coordinate system

$\hat{\mathbf{H}}_{\text{so}}^{\text{LK}}$	6×6 Luttinger-Kohn spin-orbit Hamiltonian
$\hat{\mathbf{H}}_{\text{so}}^{\prime\text{LK}}$, $\hat{\mathbf{H}}_{\text{so}}^{\prime}$	6×6 Luttinger-Kohn spin-orbit Hamiltonian associated with the primed coordinate system
$\hat{\mathbf{H}}_{\text{strain}}^{\text{LK}}$	6×6 Luttinger-Kohn strain Hamiltonian
$\hat{\mathbf{H}}_{\text{strain}}^{\prime\text{LK}}$, $\hat{\mathbf{H}}_{\text{strain}}^{\prime}$	6×6 Luttinger-Kohn strain Hamiltonian associated with the primed coordinate system
$\hat{\mathbf{H}}^{\text{LK}}$	6×6 Luttinger-Kohn total Hamiltonian
$\hat{\mathbf{H}}^{\prime\text{LK}}$, $\hat{\mathbf{H}}^{\prime}$	6×6 Luttinger-Kohn total Hamiltonian associated with the primed coordinate system
$\hat{\mathbf{H}}^{\prime(0)}$	6×6 Luttinger-Kohn **k**-independent Hamiltonian
I_{D}, I_{ON}	Drain current, ON state current
$\mathbf{j}$, j_x	Current density vector, $\mathbf{e}_x$ component of current density vector
$\tilde{\mathbf{j}}_{m,m'}^{\nu}(x, \tilde{\varepsilon})$	$\tilde{\mathbf{j}}$-factor on the LHS of the projected BTE
$\mathbf{k}$, $\mathbf{k}'$	Wave vector
k_{B}	Boltzmann's constant
k_l, k_L	Discrete magnitude of wave vector at grid point l for which the $\mathbf{k} \cdot \mathbf{p}$ EFA SE is solved, upper bound of k_l
κ	Dielectric constant
l	Deformation potential due to strain
L	Maximum order of Fourier harmonics up to which the scattering transition rate is expanded, correlation length of the surface roughness
L_{g}	Gate length
$\mathscr{L}$, $\mathscr{L}'$	$\mathbf{k} \cdot \mathbf{p}$ parameter
m	Deformation potential due to strain
m_0	Free electron mass
m_d	Density-of-states mass
M	Maximum order of Fourier harmonics up to which the subband quantiles $X(k, \phi)$ are expanded
M'	Maximum order of Fourier harmonics up to which the modified distribution function $g(x, \tilde{\varepsilon}, \phi)$ is expanded
M''	Maximum order of Fourier harmonics up to which the subband quantiles $X(\tilde{\varepsilon}, \phi)$ are expanded
$M_{\text{GPN}}^{\nu',\nu}(x, \mathbf{k}', \mathbf{k})$	Surface roughness scattering matrix elements
$\mathscr{M}$, $\mathscr{M}'$	$\mathbf{k} \cdot \mathbf{p}$ parameter
$\mathscr{M}$	Region for monotonic interpolation

$\hat{\mathscr{M}}^{(2)}_{\mathbf{kp}}$, $\hat{\mathscr{M}}^{(1)}_{\mathbf{kp}}(k'_x, k'_y)$, $\hat{\mathscr{M}}^{(0)}_{\mathbf{kp}}(k'_x, k'_y)$	6×6 matrices in the expansion of $\hat{\mathbf{H}}'^{\mathrm{LK}}_{\mathbf{kp}}$ in polynomial of k'_z
$\hat{\mathscr{M}}^{xx}_{\mathbf{kp}}$, $\hat{\mathscr{M}}^{yy}_{\mathbf{kp}}$, $\hat{\mathscr{M}}^{zz}_{\mathbf{kp}}$, $\hat{\mathscr{M}}^{xy}_{\mathbf{kp}}$, $\hat{\mathscr{M}}^{yz}_{\mathbf{kp}}$, $\hat{\mathscr{M}}^{zx}_{\mathbf{kp}}$	6×6 matrices in the expansion of $\hat{\mathbf{H}}'^{\mathrm{LK}}_{\mathbf{kp}}$ in polynomial of k'_x, k'_y, k'_z
$\hat{\mathscr{M}}^{\phi'}_{\mathbf{kp}}$	Polar angle ϕ' dependent matrix
μ, μ_0	Mobility scalar
$\hat{\mu}$	Mobility tensor
μ_{11}, μ_{12}, μ_{21}, μ_{22}	Mobility components of the mobility tensor
μ_{phon}, μ_{SR}, μ_{alloy}	Phonon limited mobility, surface roughness limited mobility, alloy disorder limited mobility
$\mu_{\mathrm{Matt.'s\ rule}}$	Matthiessen's rule mobility
μ_{eff}	Effective low electric field homogenous channel mobility without magnetic field
$\mu_{\mathrm{MR}}(B)$, $\mu_{\mathrm{MR,0}}$	Magneto resistance mobility, magneto resistance mobility for a vanished magnetic field
n	Deformation potential due to strain
$\mathbf{n}$	Surface vector pointing into the device
$n(x, z)$	Electron density
$n(\hbar\omega)$	Bose-Einstein distribution
N	Number of bands
N_{A}, N_{D}	Acceptor, donor concentrations
N_{c}	Effective density of state of the conduction band edge
N_{inv}	Hole inversion charge
$\mathscr{N}$, $\mathscr{N}'$	$\mathbf{k} \cdot \mathbf{p}$ parameter
ν	Valley index
$p(x, z)$, $\tilde{p}(x, z)$	Hole density, modified hole density
$\mathbf{p}$	Momentum operator
$P_l(x)$	Legendre polynomial
$P^m_l(x)$	Associated Legendre polynomial
$\mathscr{P}^{\lambda\mu}$	Momentum matrix element
$\mathscr{P}^{\nu}(z, \mathbf{k})$, $\mathscr{P}^{\nu}_m(z, k)$, $\mathscr{P}^{\nu}_m(z, k_l)$	Probability density function, Fourier expansion coefficient of $\mathscr{P}^{\nu}(z, k, \phi)$, Fourier expansion coefficient of $\mathscr{P}^{\nu}(z, k_l, \phi)$
ψ	Wave function
q	Positive elementary charge
$\mathbf{r}$	Real space position vector
$\mathbf{R}$	Direct lattice vector

$\hat{\mathbf{R}}$	Unitary rotation matrix used to rotate $\hat{\mathbf{H}}'^{\mathrm{LK}}_{\mathbf{kp}}$ and $\hat{\mathbf{H}}^{\mathrm{LK}}_{\mathbf{kp}}$
$r_{\mathrm{H}}(B)$, $r_{\mathrm{H},0}$	Hall factor, Hall factor for a vanished magnetic field
ρ	Mass density
$\hat{S}^{\nu}\{f\}$, $\hat{S}^{\nu}_{\mathrm{in}}\{f\}$, $\hat{S}^{\nu}_{\mathrm{out}}\{f\}$	Scattering integral associated with subband ν, in-scattering integral associated with subband ν, out-scattering integral associated with subband ν
$S^{\nu',\nu}_{\mathrm{ac}}(x, \mathbf{k}'\|\mathbf{k})$	Transition rate of acoustic phonon scattering
$S^{\nu',\nu}_{\mathrm{opt}}(x, \mathbf{k}'\|\mathbf{k})$	Transition rate of optical phonon scattering
$S^{\nu',\nu}_{\mathrm{alloy}}(x, \mathbf{k}'\|\mathbf{k})$	Transition rate of alloy disorder scattering
$S^{\nu',\nu}_{\mathrm{SR}}(x, \mathbf{k}'\|\mathbf{k})$	Transition rate of surface roughness scattering
T	Lattice temperature
T_{delay}	Intrinsic delay
ϑ	Polar (zenith) angle in the spherical coordinate system
θ'	Azimuth angle of the growth direction z' in the crystallographic coordinate system
$\|\mathbf{u}\rangle^{\mathrm{LK}}$	Luttinger-Kohn basis states
u_l	Longitudinal sound velocity
$u^{\mu}_{\Gamma}(\mathbf{r})$	Γ-centered Bloch functions associated with band μ
$u_{\mathbf{k}}(\mathbf{r})$	Bloch function
U	Alloy scattering potential
$\hat{\hat{\mathbf{U}}}$	Rotation matrix used to transform the crystallographic coordinate system and the primed coordinate system
$\mathbf{v}$	Group velocity
$\langle v\rangle^{\mathrm{H}}_{1}$	$\mathbf{e}_1$ component of average drift velocity due to Hall field
v_k, v_ϕ	$\mathbf{e}_k$, $\mathbf{e}_\phi$ component of $\mathbf{v}$ in the polar coordinate system (k, ϕ)
$v_k(k,\phi)$, $\{v_k\}_m$, $\{v_k\}_{m,l}$	v_k as a function of wave vector $\mathbf{k} = (k,\phi)$, Fourier expansion coefficient of $v_k(k,\phi)$, Fourier expansion coefficient of $v_k(k_l,\phi)$
v_x, v_y	$\mathbf{e}_x$, $\mathbf{e}_y$ component of $\mathbf{v}$ in the Cartesian coordinate system (x, y)
$v_{x_{\mathrm{D}}}$, $v_{y_{\mathrm{D}}}$	$\mathbf{e}_{x_{\mathrm{D}}}$, $\mathbf{e}_{y_{\mathrm{D}}}$ component of $\mathbf{v}$ in the Cartesian coordinate system $(x_{\mathrm{D}}, y_{\mathrm{D}})$
$V(\mathbf{r})$	Electrostatic potential at position $\mathbf{r}$ in the 3D real space
$V(x_{\mathrm{D}}, z_{\mathrm{D}})$	Electrostatic potential at position $(x_{\mathrm{D}}, z_{\mathrm{D}})$ in the 2D real space
V_{DS}, V_{GS}	Drain-to-source bias, gate-to-source bias
$V_{\mathrm{gate,bias}}$	Applied gate bias
$V_{\mathrm{L}}(\mathbf{r})$	Periodic lattice potential at position $\mathbf{r}$ in the 3D real space
W	Channel width

ω	Optical phonon frequency
$\hbar\omega_\eta$	Energy transfer due to scattering
Ω	Solid angle
Ω_s	System volume
$\lvert x\rangle$	$u_\Gamma^X(\mathbf{r})$
x'	x-axis of the primed coordinate system
x_i	Position of grid point i on the x-axis
x_C	x-axis of the crystallographic coordinate system
x_D	x-axis of the device coordinate system (channel direction)
$X(k,\phi)$, X_m, $X_{m,l}$	X as a function of wave vector $\mathbf{k} = (k,\phi)$, Fourier expansion coefficient of $X(k,\phi)$, Fourier expansion coefficient of $X(k_l,\phi)$
$\overline{X}(x)$	Average over 2D $\mathbf{k}$-space of microscopic quantity $X^\nu(x,\mathbf{k})$
Ξ	Average acoustic deformation potential
$\lvert y\rangle$	$u_\Gamma^Y(\mathbf{r})$
y'	y-axis of the primed coordinate system
y_C	y-axis of the crystallographic coordinate system
y_D	y-axis of the device coordinate system
$Y_{l,m}(\vartheta,\varphi)$	Spherical harmonic function
$Y_m(\phi)$	Fourier harmonic function
$\Upsilon_{m,m',m''}$	Integral of product of three Fourier harmonics
$\{\Upsilon C\}_{m,m',m''}$	Integral of product of three Fourier harmonics and cos function
$\{\Upsilon S\}_{m,m',m''}$	Integral of product of three Fourier harmonics and sin function
$\lvert z\rangle$	$u_\Gamma^Z(\mathbf{r})$
z'	z-axis of the primed coordinate system
z_C	z-axis of the crystallographic coordinate system
z_D	z-axis of the device coordinate system
$Z^\nu(\mathbf{r},\varepsilon,\vartheta,\varphi)$	Generalized density of state for a single spin direction associated with valley ν
$Z^\nu(x,\tilde{\varepsilon},\phi)$	Generalized density of state for a single spin direction associated with subband ν
$\zeta_\mu^\nu(\mathbf{r})$	Envelope function associated with eigen state ν, band μ
$\zeta_\mu^{\prime\nu}(\mathbf{r})$	Envelope function associated with eigen state ν, band μ, and the primed coordinate system
$\boldsymbol{\zeta}^\nu(z,\mathbf{k})$, $\boldsymbol{\zeta}^\nu(x,z,\mathbf{k})$	Envelope function vector

Abbreviations

1D, 2D, 3D	One, two, three dimensional
ac	Acoustic
BTE	Boltzmann transport equation
BZ	Brillouin zone
comp	Compressive
CPU	Central processing unit
DC	Direct current
DD	Drift-diffusion
DG	Double gate
DOS	Density of states
EFA	Envelope function approximation
FET	Field effect transistor
Ge	Germanium
HH, hh	Heavy hole
HOI	Heterostructure on insulator
LH, lh	Light hole
LHS	Left hand side
LK	Luttinger Kohn
MC	Monte-Carlo
MOS	Metal oxide semiconductor
opt	Optical
PB	Pikus Bir
PE	Poisson equation
phon	Phonon
RHS	Right hand side
RT	Relaxation time
RTA	Relaxation time approximation
SE	Schrödinger equation
SHE	Spherical harmonics expansion
Si	Silicon
SiGe	Silicon-germanium
SG	Single gate
so	Spin-orbit, split-off
SOI	Semiconductor on insulator or Si on insulator
SR	Surface roughness
TCAD	Technology computer aided design
tens	Tensile
w, w/o	With, without
w.r.t	With respect to

Part I
Overview

Chapter 1
Introduction

Due to the continuous scaling during the last five decades the feature size of semiconductor devices has reached the deca-nanometer range [1], and scaling is expected to continue for some time [2]. For such scaled devices, transport can no longer be described accurately by momentum based models (drift-diffusion or hydrodynamic models) [3, 4], which fail even in the linear transport regime [5, 6].

In the semiclassical framework the BTE gives the most accurate information about the device operation [7–9]. The BTE is the master equation for the distribution function, which is the probability of the occupancy of the electron states in the phase space [10]. Once the distribution function is known, most quantities of interest can be evaluated. However, it is defined over the six-dimensional phase space (three-dimensional real space and three-dimensional wavevector space). The high dimensionality of the phase space makes solving the BTE a very challenging task. Since an analytical solution is not available for the general case, a computational approach is required.

The usual approach for solving the BTE is the Monte Carlo approach, where the BTE is solved by simulating a stochastic process [11–16]. In this approach advanced physical models (e.g. full bands [17, 18]) can be easily implemented. However, the Monte Carlo method has many disadvantages due to its stochastic nature and that it is inherently transient [19]. This makes it, for example, impossible to simulate with the standard Monte Carlo approach the kink effect in SOI devices with a floating body, because the charging time of the body (e.g. milliseconds) is many orders of magnitude larger than the length of the time step (e.g. femtoseconds) necessary to ensure stability of the self-consistent solution of the BTE and PE [20]. This limits the duration of the simulated time in Monte Carlo device simulations to nanoseconds or less and only processes which are considerably faster than nanoseconds can be analyzed. W.r.t. small-signal and noise analyses this means, that they can be performed for devices only at frequencies in the higher GHz-range or beyond [21]. In addition, rare events or small quantities can entail excessive CPU times. For example, the CPU time to simulate the subthreshold current in MOSFETs is prohibitive. Moreover, the stochastic nature of the Monte Carlo approach makes it

S.-M. Hong et al., *Deterministic Solvers for the Boltzmann Transport Equation*, Computational Microelectronics, DOI 10.1007/978-3-7091-0778-2_1,

impossible to apply standard numerical methods used in classical TCAD tools and Monte Carlo simulators are difficult to apply in a TCAD framework.

A possible alternative is to solve the BTE in a deterministic manner without stochastic errors. Its properties are similar to those of the classical TCAD models. Since the distribution function is expanded with specific basis functions, the number of unknown variables after discretization can be very large. As a result, the deterministic approach requires much more computer memory than the Monte Carlo approach. This is the main reason why the application of deterministic BTE solvers to devices was hampered for a long time. The rapid advance of modern computers makes such an approach more and more feasible.

There are many possible choices for the basis functions. For example, the discretization of the entire wavevector space with a three-dimensional grid can be found in [22, 23]. However, the structure of the scattering integral suggests to write the BTE in spherical coordinates and to use spherical harmonics as basis functions for the angular dependence. Regarding the radial dependence of the distribution function, usually the energy is discretized with a grid. In [24–27], the energy dependence of the distribution function is resolved by an expansion with polynomials.

The idea to expand the distribution with spherical harmonics was among the first methods used to solve the BTE by numerical means (e.g. [28]). Since the SHE was first applied to devices in [29], there have been considerable efforts to improve the physical models. In bulk simulations, some full band effects were included in a lowest-order expansion by Vecchi et al. [30]. Complete inclusion of a full band structure was demonstrated for the valance bands in [31]. Simulations including a magnetic field were reported in [32] for bulk and in [33] for devices. The simulation of a MOSFET with the relevant scattering mechanisms can be found in [34]. Since rare events can be easily calculated in contrast to the Monte Carlo approach, the substrate current for MOSFETs, for example, can be readily calculated [34]. Although restricted to a lowest-order expansion, two-dimensional simulations were already reported in [34, 35]. Two-dimensional simulations with a higher-order expansion appeared only recently in [36]. Simulations including some quantum effects were reported in [37].

The order at which the SHE is truncated has a strong impact on the accuracy of the results. For the bulk case, Hennacy et al. presented higher-order expansions [38, 39]. In the case of one-dimensional devices Rahmat et al. demonstrated a third-order expansion [40]. Later expansions of arbitrary order were reported for such devices in [31]. Box integration for the derivation of the discrete system of equations was introduced and exact current continuity achieved [29,35,40]. Without stabilization simulation of realistic semiconductor devices is not possible. The H-transform was developed to stabilize the equations [29]. An upwind discretization has been proposed in [40]. Later a stabilization was introduced [31], which is based on the maximum entropy dissipation scheme [25].

Small-signal analysis is a vital component of modern device simulation. In contrast to the Monte Carlo method, exact small-signal analysis is possible for devices, even directly in the frequency domain, in the case of the SHE [37,41,42].

Noise can be calculated based either on the BTE [43, 44] or Langevin Boltzmann equation [45].

In part II the development of a device simulator based on a SHE of arbitrary order is described in detail with emphasis on the expansion, numerical stability and band structure mapping.

In Chap. 2 the BTE based on the three-dimensional wave vector space including a magnetic field and the Pauli principle together with the necessary framework for device simulation is introduced. The BTE is expanded with spherical harmonics and important properties of the spherical harmonics and of the corresponding projection coefficients are reviewed in detail. The expansion is based on a generalized coordinate transform which maps the wave vector space onto the energy space. The expansion of the BTE results in transport coefficients, which must satisfy certain relations which are fundamental to the stabilization of the resultant balance equations. At the end of the chapter the SHE of the Langevin Boltzmann equation for noise is discussed.

Within semiconductor devices densities vary by many orders of magnitude and realistic real space grids are rather coarse to keep the CPU time and memory requirement low. Robust device simulations require therefore sophisticated stabilization schemes. In Chap. 3 it is shown how to obtain a very robust implementation of the BTE by combining the H-transform with the maximum entropy dissipation scheme. The resultant Jacobi matrix of the non-linear system of equations derived from the discretized BTE is formulated in such a way that it is a non-singular M-matrix for a first order SHE ensuring non-negativeness of the solution. This enhances the stability very much, even in the case of a higher order expansion. In the case of small-signal analysis the stabilization scheme is modified to improve the smoothness of the solution over the device.

Chapter 4 contains the description of the band structure and scattering models. Different approximations of the numerical full band structure are discussed, since direct inclusion of the numerical full band structure is not possible for all energies. A new model is introduced, which is based on an isotropic approximation of the full band structure after mapping the full band structure with the generalized coordinate transformation. It is shown that this isotropic band model is superior to previous isotropic and anisotropic models while retaining at the same time the same CPU efficiency as the simple isotropic models. The results obtained with the new band model are very close to the results by full band MC simulations without approximations for the band structure.

In Chap. 5 the SHE simulator is applied to various devices and its properties and accuracy are discussed. The convergence of the SHE is investigated and it is shown that the necessary order of SHE depends on the device and bias condition. Full band Monte Carlo results are used as a reference. In the case of the n^+nn^+ structure transport due to electric and magnetic fields is investigated. The impact of the Pauli principle on transport and noise is investigated for a double gate NMOSFET. Convergence of the BTE and PE is demonstrated for a partially depleted SOI devices with a floating body and the kink effect can be simulated. SiGe HBTs are discussed including break down due to impact ionization and profile optimization.

In the case of MOSFETs quantization in the channel is very important and can often not be neglected [46, 47]. Especially in the case of PMOSFETs the confinement of the carriers has a strong impact on the band structure in the 2D wavevector space in addition to strain and surface orientation [48]. Since the wavevector space is 2D, instead of spherical harmonics Fourier harmonics are used for the expansion of the BTE. Otherwise most of the methods developed in part II can be directly applied to the BTE in part III, in which the coupled solution of the SE, BTE and PE for PMOSFETs is described. Special emphasis is placed on the derivation of the Hamiltonian of the SE and efficient discretization of the 2D wavevector space.

In Chap. 6 the three different coordinate systems for the real space are introduced, which are aligned with the crystal lattice, the semiconductor/insulator interface and strain, and the device. In addition to part II the SE is solved together with the BTE for the quasi-2D hole gas and the PE.

The $\mathbf{k} \cdot \mathbf{p}$ method is introduced including arbitrary crystallographic orientation and strain in Chap. 7. Since the confinement orientation can deviate form the crystallographic one, the $\mathbf{k} \cdot \mathbf{p}$ Hamiltonian has to be rotated and an efficient approach for an arbitrary rotation is presented. Finally, the symmetry of the suband structure is discussed.

Since the $\mathbf{k} \cdot \mathbf{p}$ Schrödinger solver consumes the largest part of the CPU time for a transport simulation, an efficient approach to discretize the 2D k-space is required, which is described in Chap. 8. In analogy to the 3D case, polar coordinates are used and the dependence on the angle is expanded with Fourier harmonics. The convergence of this Fourier harmonics expansion is investigated. With respect to energy monotonic cubic splines are used and the number of grid points is strongly reduced. In some cases, certain parts of the dispersion relation are not monotonic and a special algorithm is introduced, which can handle this problem. Since the interpolated dispersion relation is continuous and the derivative is continuous, the group velocity can be calculated without further approximations.

In Chap. 9 the BTE for the quasi-2D hole gas including magnetic fields and the Pauli principle is introduced and its projection onto Fourier harmonics is discussed. Exact particle number conservation is ensured by box integration.

The discrete PE and the corresponding boundary conditions are introduced in Chap. 10. In a quasi-2D hole gas the boundary conditions of an ideal ohmic contact deviate from a bulk contact and the special boundary condition for the potential is shown.

The BTE, the PE and the SE are non-linearly coupled and special iteration methods are presented in Chap. 11 in order to obtain a self-consistent solution of all three equations. The simplest approach is a Gummel-type relaxation loop over all three equations. A Newton–Raphson approach is difficult to apply, because the SE is an eigen value equation. To circumvent this problem a simplified perturbation approach is used to capture the coupling of the PE with the BTE via the SE.

The transport model is calibrated with fundamental mobility data for silicon and germanium, and verified by independent experiments for single and double gate PMOSFETs including SiGe HOI structures. The impact of the channel and surface orientations together with strain on the performance of silicon and germanium double gate PMOSFETs is investigated in Chap. 12 for a large number of configurations. The failure of the drift-diffusion approximation is demonstrated for short channel devices. Magnetotransport effects are discussed for devices and it is found that the usual approximations fail.

In the appendix of part III the not so well-known formulas for the Fourier harmonics and the details of the projection are presented and part of the discretization of the SE is described.

References

1. Thompson, S., Alavi, M., Hussein, M., Jacob, P., Kenyon, C., Moon, P., Prince, M., Sivakumar, S., Tyagi, S., Bohr, M.: 130 nm logic technology featuring 60 nm transistors, low-k dielectrics and Cu interconnects. Intel Technol. J. **6**(2), 5–13 (2002)
2. International Roadmap Committee: The international technology roadmap for semiconductors. public.itrs.net (2009)
3. Nekovee, M., Geurts, B.J., Boots, H.M.J., Schuurmans, M.F.H.: Failure of extended-moment-equation approaches to describe ballistic transport in submicrometer structures. Phys. Rev. B **45**(12), 6643–6651 (1992)
4. Fischetti, M.V., Laux, S.E., Crabbe, E.: Understanding hot-electron transport in silicon devices: Is there a shortcut? J. Appl. Phys. **78**, 1058–1087 (1995)
5. Shur, M.S.: Low ballistic mobility in submicron HEMTs. IEEE Electron Dev. Lett. **23**(9), 511–513 (2002)
6. Jungemann, C., Grasser, T., Neinhüs, B., Meinerzhagen, B.: Failure of moments-based transport models in nanoscale devices near equilibrium. IEEE Trans. Electron Dev. **52**(11), 2404–2408 (2005)
7. Madelung, O.: Introduction to Solid State Theory. Springer, Berlin (1978)
8. Jacoboni, C., Lugli, P.: The Monte Carlo method for semiconductor device simulation. Springer, New York (1989)
9. Price, P.J.: Monte Carlo calculation of electron transport in solids. Semiconduct. Seminet. **14**, 249–309 (1979)
10. van Kampen, N.G.: Stochastic Process in Physics and Chemistry. North-Holland, Amsterdam (1981)
11. Kurosawa, T.: Monte Carlo calculation of hot electron problems. J. Phys. Soc. Jpn. **21**, 424–426 (1966)
12. Fawcett, W., Boardman, A.D., Swain, S.: Monte Carlo determination of electron transport properties in gallium arsenide. J. Phys. Chem. Solids **31**, 1963–1990 (1970)
13. Reklaitis, A.: The calculation of electron transient response in semiconductors by the Monte Carlo technique. Phys. Lett. **13**, 367–370 (1982)
14. Reggiani, L.: Hot-Electron Transport in Semiconductors. Springer, Berlin (1985)
15. Nedjalkov, M., Vitanov, P.: Iteration approach for solving the Boltzmann equation with the Monte Carlo method. Solid State Electron. **32**, 893–896 (1989)
16. Moglestue, C.: Monte Carlo Simulation of Semiconductor Devices. Chapman & Hall, London (1993)
17. Hess, K. (ed.): Monte Carlo Device Simulation: Full Band and Beyond. Kluwer, Boston (1991)
18. Fischetti, M.V., Laux, S.E.: Monte Carlo analysis of electron transport in small semiconductor devices including band-structure and space-charge effects. Phys. Rev. B **38**, 9721–9745 (1988)
19. Jungemann, C., Meinerzhagen, B.: Analysis of the stochastic error of stationary Monte Carlo device simulations. IEEE Trans. Electron Dev. **48**(5), 985–992 (2001)

20. Rambo, P.W., Denavit, J.: Time stability of Monte Carlo device simulations. IEEE Trans. Comp. Aided Des. **12**, 1734–1741 (1993)
21. Jungemann, C., Neinhüs, B., Decker, S., Meinerzhagen, B.: Hierarchical 2–D DD and HD noise simulations of Si and SiGe devices: Part II – Results. IEEE Trans. Electron Dev. **49**(7), 1258–1264 (2002)
22. Banoo, K., Lundstrom, M.S.: Electron transport in a model Si transistor. Solid State Electron. **44**, 1689–1695 (2000)
23. Banoo, K., Lundstrom, M.: Direct solution of the Boltzmann transport equation in nanoscale Si devices. In: Proceedings of SISPAD, pp. 50–53 (2000)
24. Ringhofer, C.: Space–time discretization of series expansion methods for the Boltzmann transport equation. SIAM J. Num. Anal. **38**, 442–465 (2000)
25. Ringhofer, C.: Numerical methods for the semiconductor Boltzmann equation based on spherical harmonics expansions and entropy discretizations. Transport. Theor. Stat. Phys. **31**(4–6), 431–452 (2002)
26. Ringhofer, C.: A mixed spectral-difference method for the steady state boltzmann-poisson system. SIAM J. Num. Anal. **41**(1), 64–89 (2003)
27. Ringhofer, C., Schmeiser, C., Zwirchmayer, A.: Moment methods for the semiconductor Boltzmann equation in bounded position domains. SIAM J. Num. Anal. **39**, 1078–1095 (2001)
28. Baraff, G.A.: Maximum anisotropy approximation for calculating electron distributions; Application to high field transport in semiconductors. Phys. Rev. **133**(1A), A26–A33 (1964)
29. Ventura, D., Gnudi, A., Baccarani, G., Odeh, F.: Multidimensional spherical harmonics expansion of Boltzmann equation for transport in semiconductors. Appl. Math. Lett. **5**, 85 (1992)
30. Vecchi, M.C., Rudan, M.: Modeling electron and hole transport with full-band structure effects by means of the spherical-harmonics expansion of the BTE. IEEE Trans. Electron Dev. **45**(1), 230–238 (1998)
31. Jungemann, C., Pham, A.-T., Meinerzhagen, B., Ringhofer, C., Bollhöfer, M.: Stable discretization of the Boltzmann equation based on spherical harmonics, box integration, and a maximum entropy dissipation principle. J. Appl. Phys. **100**, 024502–1–13 (2006)
32. Smirnov, S., Jungemann, C.: A full band deterministic model for semiclassical carrier transport in semiconductors. J. Appl. Phys. **99**, 063707–1–11 (2006)
33. Hong, S.-M., Jungemann, C.: Simulation of magnetotransport in nanoscale devices. In: International Conference on Solid State and Integrated Circuits Technology, pp. 377–380 (2008)
34. Liang, W., Goldsman, N., Mayergoyz, I., Oldiges, P.J.: 2- D MOSFET modeling including surface effects and impact ionization by self-consistent solution of the Boltzmann, Poisson, and hole-continuity equations. IEEE Trans. Electron Dev. **44**(2), 257–267 (1997)
35. Gnudi, A., Ventura, D., Baccarani, G., Odeh, F.: Two-dimensional MOSFET simulation by means of a multidimensional spherical harmonics expansion of the Boltzmann transport equation. Solid State Electron. **36**(4), 575–581 (1993)
36. Hong, S.-M., Jungemann, C., Bollhöfer, M.: A deterministic Boltzmann equation solver for two-dimensional semiconductor devices. In: Proceedings of SISPAD, pp. 293–296 (2008)
37. Goldsman, N., Lin, C.K., Han, Z., Huang, C.K.: Advances in the spherical Harmonic-Boltzmann-Wigner approach to device simulation. Superlattice Microst. **27**, 159–175 (2000)
38. Hennacy, K.A., Goldsman, N.: A generalized legendre polynimial/sparse matrix approach for determining the distribution function in non-polar semiconductors. Solid State Electron. **36**, 869–877 (1993)
39. Hennacy, K.A., Wu, Y.-J., Goldsman, N., Mayergoyz, I.D.: Deterministic MOSFET simulation using a generalized spherical harmonic expansion of the Boltzmann equation. Solid State Electron. **38**, 1485–1495 (1995)
40. Rahmat, K., White, J., Antoniadis, D.A.: Simulation of semiconductor devices using a Galerkin/spherical harmonic expansion approach to solving the coulped Poisson-Boltzmann system. IEEE Trans. Comp. Aided Des. **15**(10), 1181–1196 (1996)

41. Lin, C.-K., Goldsman, N., Han, Z., Mayergoyz, I., Yu, S., Stettler, M., Singh, S.: Frequency domain analysis of the distribution function by small signal solution of the Boltzmann and Poisson equations. In: Proceedings of SISPAD, pp. 39–42 (1999)
42. Jungemann, C.: A deterministic approach to RF noise in silicon devices based on the Langevin Boltzmann equation. IEEE Trans. Electron Dev. **54**(5), 1185–1192 (2007)
43. Jungemann, C., Graf, P., Zylka, G., Thoma, R., Engl, W.L.: New highly efficient method for the analysis of correlation functions based on a spherical harmonics expansion of the BTE's Green's function. In: Proceedings of IWCE, Portland, Oregon, pp. 45–48, May 1994
44. Korman, C.E., Mayergoyz, I.D.: Semiconductor noise in the framework of semiclassical transport. Phys. Rev. B **54**, 17620–17627 (1996)
45. Jungemann, C., Meinerzhagen, B.: A legendre polynomial solver for the Langevin Boltzmann equation. J. Comput. Electron. **3**, 157–160 (2004)
46. Stern, F., Howard, W.E.: Properties of semiconductor surface inversion layers in the electric quantum limit. Phys. Rev. **163**, 816–835 (1967)
47. Ando, T., Fowler, A.B., Sterns, F.: Electronic properties of two-dimensional systems. Rev. Mod. Phys. **54**, 437–672 (1982)
48. Fischetti, M.V., Ren, Z., Solomon, P.M., Yang, M., Rim, K.: Six-band $k \cdot p$ calculation of the hole mobility in silicon inversion layers: Dependence on surface orientation, strain, and silicon thickness. J. Appl. Phys. **94**, 1079–1095 (2003)

Part II
Electron Transport in the 3D k-Space

Chapter 2
The Boltzmann Transport Equation and Its Projection onto Spherical Harmonics

In the framework of the semiclassical transport theory, the BTE governs the spatiotemporal evolution of the particle gas. In this chapter, the BTE in the three-dimensional wave vector space is introduced in Sect. 2.1. The PE is required for the calculation of the electric field, which enters the BTE. If only one carrier type is simulated, a drift-diffusion model is solved for the other type. The PE and drift-diffusion model are discussed in Sect. 2.2. In Sect. 2.3, basic properties of the spherical harmonics are reviewed. A generalized coordinate transform from the wavevector space to the energy space is introduced, and some important relations between transport coefficients in the energy space are explicitly derived. The spherical harmonics expansion of the BTE is shown in Sect. 2.5. Finally, noise analysis within the Langevin-Boltzmann framework is discussed in Sect. 2.6.

2.1 The Boltzmann Transport Equation

Our goal is to describe the particle kinetics based on a position-dependent band structure, which contains a few conduction (or valence) bands. However, the electronic band, which is defined in the first Brillouin zone following the reduced zone scheme [1], might not be necessarily suitable for a spherical harmonics expansion. For example, in the case of the first Si conduction band, the origin of the Brillouin zone (the Γ-point) does not coincide with the minimum of the energy.[1] Therefore, in this work, instead of the electronic bands themselves, a mathematical model called "valley" is regarded as the basis of the analysis. From its construction, the energy minimum of the valley is located at its origin in the wave vector space. The valleys are labeled with ν throughout part II. Of course, in order to obtain a meaningful physical description, a sound relationship between the original band

[1] In the case of the Si valence bands, the minimum of the band energy is found at this point.

S.-M. Hong et al., *Deterministic Solvers for the Boltzmann Transport Equation*, Computational Microelectronics, DOI 10.1007/978-3-7091-0778-2_2,

structure and the approximated valley model is required. Specific approaches to construct the band models based on the valley description are explained in Chap. 4. For the νth valley, the relative energy at the position $\mathbf{r}$ in the real space and the wave vector $\mathbf{k}$, which is measured from the valley minimum energy, is described by the dispersion relation, $\varepsilon^\nu(\mathbf{r},\mathbf{k})$. Since the band structure can depend on the real space, the dispersion relation of the valley also can be position dependent.

In the framework of the semiclassical transport theory, the position $\mathbf{r}$ in the real space and the momentum $\hbar\mathbf{k}$ ($\hbar$ denotes Planck's constant divided by 2π) of a particle can be measured simultaneously. Therefore, a state $(\mathbf{r},\mathbf{k})$ in the six-dimensional phase space is required in order to specify the state of a particle. Ignoring many-particle effects except the electric field, screening and the Pauli exclusion principle, the particle ensemble can be completely characterized by the so-called one-particle one-time distribution function $f^\nu(\mathbf{r},\mathbf{k},t)$ for the νth valley defined on the six-dimensional phase space [1]. The number of particles in an infinitesimal volume of the six-dimensional phase space $(\mathrm{d}^3k\mathrm{d}^3r)$ at time t is given by

$$\mathrm{d}N = \frac{2}{(2\pi)^3} f^\nu(\mathbf{r},\mathbf{k},t)\,\mathrm{d}^3k\mathrm{d}^3r, \tag{2.1}$$

where the prefactor $\frac{2}{(2\pi)^3}$ takes account of the spin degeneracy and the minimum phase space volume. The Pauli exclusion principle necessitates a distribution function smaller than one. Thus, the distribution function itself is the probability that the state $(\mathbf{r},\mathbf{k})$ of the νth valley is occupied [2].

A macroscopic quantity $x(\mathbf{r},t)$ is an average of a microscopic quantity $X^\nu(\mathbf{r},\mathbf{k})$ over the $\mathbf{k}$ space. It is an expectation of the form

$$x(\mathbf{r},t) = \frac{2}{(2\pi)^3}\sum_\nu \int X^\nu(\mathbf{r},\mathbf{k}) f^\nu(\mathbf{r},\mathbf{k},t)\,\mathrm{d}^3k. \tag{2.2}$$

When the microscopic quantity is 1 ($X = 1$) and the valleys of the conduction bands (C) are considered, the corresponding macroscopic quantity is the electron density:

$$n(\mathbf{r},t) = \frac{2}{(2\pi)^3}\sum_C \int f^C(\mathbf{r},\mathbf{k},t)\,\mathrm{d}^3k, \tag{2.3}$$

whereas the hole density $p(\mathbf{r},t)$ is obtained by summing over the valence band valleys (V):

$$p(\mathbf{r},t) = \frac{2}{(2\pi)^3}\sum_V \int f^V(\mathbf{r},\mathbf{k},t)\,\mathrm{d}^3k. \tag{2.4}$$

For the group velocity ($X = \mathbf{v}^\nu$), which is given by

$$\mathbf{v}^\nu(\mathbf{r},\mathbf{k}) = \frac{1}{\hbar}\nabla_{\mathbf{k}}\varepsilon^\nu(\mathbf{r},\mathbf{k}), \tag{2.5}$$

the corresponding macroscopic quantity is the particle current density:

$$\mathbf{j}(\mathbf{r},t) = \frac{2}{(2\pi)^3}\sum_{\nu}\int \mathbf{v}^{\nu}(\mathbf{r},\mathbf{k}) f^{\nu}(\mathbf{r},\mathbf{k},t)\,\mathrm{d}^3k. \tag{2.6}$$

Once the particle distribution function is known, any macroscopic quantity of interest can be calculated.

In the framework of the semiclassical transport theory, the spatiotemporal evolution of a particle gas is described by the BTE [1–4]

$$\frac{\partial f^{\nu}}{\partial t} + \frac{1}{\hbar}\mathbf{F}^{\nu}\cdot\nabla_{\mathbf{k}} f^{\nu} + \mathbf{v}^{\nu}\cdot\nabla_{\mathbf{r}} f^{\nu} = \hat{S}^{\nu}\{f\}, \tag{2.7}$$

where $\hat{S}^{\nu}\{f\}$ is the single-particle scattering integral including the Pauli exclusion principle

$$\hat{S}^{\nu}\{f\} = \frac{\Omega_{\mathrm{s}}}{(2\pi)^3}\sum_{\nu'}\int (1 - f^{\nu}(\mathbf{r},\mathbf{k},t)) S^{\nu,\nu'}(\mathbf{r},\mathbf{k}|\mathbf{k}') f^{\nu'}(\mathbf{r},\mathbf{k}',t)$$
$$-(1 - f^{\nu'}(\mathbf{r},\mathbf{k}',t)) S^{\nu',\nu}(\mathbf{r},\mathbf{k}'|\mathbf{k}) f^{\nu}(\mathbf{r},\mathbf{k},t)\,\mathrm{d}^3k'. \tag{2.8}$$

Ω_{s} is the system volume and $S^{\nu,\nu'}(\mathbf{r},\mathbf{k}|\mathbf{k}')$ the transition rate from the initial state $(\nu',\mathbf{k}')$ into the final $(\nu,\mathbf{k})$ [4]. Specific expressions for the transition rates of several scattering mechanisms are shown in Chap. 4. The total force $\mathbf{F}^{\nu}$ due to a gradient in the total energy and a magnetic field is given by

$$\mathbf{F}^{\nu}(\mathbf{r},\mathbf{k},t) = -\nabla_{\mathbf{r}}(\mp qV(\mathbf{r},t) \pm E^{\nu}(\mathbf{r}) + \varepsilon^{\nu}(\mathbf{r},\mathbf{k})) \mp q\mathbf{v}^{\nu}(\mathbf{r},\mathbf{k})\times\mathbf{B}, \tag{2.9}$$

where q is the positive electron charge, V the electrostatic potential, E the valley minimum in the electron picture, $\mathbf{B}$ the magnetic field, and the upper and lower signs are for electrons and holes, respectively. It is assumed that the magnetic field is time-independent and constant over the simulation domain. Components of the total force are denoted as

$$\mathbf{F}^{\nu}_{\mathrm{E,pot}}(\mathbf{r},t) = -\nabla_{\mathbf{r}}(\mp qV(\mathbf{r},t) \pm E^{\nu}(\mathbf{r})), \tag{2.10}$$

$$\tilde{\mathbf{F}}^{\nu}_{\mathrm{E}}(\mathbf{r},\mathbf{k}) = -\nabla_{\mathbf{r}}\varepsilon^{\nu}(\mathbf{r},\mathbf{k}), \tag{2.11}$$

$$\mathbf{F}^{\nu}_{\mathrm{E}}(\mathbf{r},\mathbf{k},t) = \mathbf{F}^{\nu}_{\mathrm{E,pot}}(\mathbf{r},t) + \tilde{\mathbf{F}}^{\nu}_{\mathrm{E}}(\mathbf{r},\mathbf{k}), \tag{2.12}$$

$$\mathbf{F}^{\nu}_{\mathrm{B}}(\mathbf{r},\mathbf{k}) = \mp q\mathbf{v}^{\nu}(\mathbf{r},\mathbf{k})\times\mathbf{B}. \tag{2.13}$$

Note that when the band structure does not depend on the position, $\tilde{\mathbf{F}}^{\nu}_{\mathrm{E}}$ vanishes. A useful relation is shown here for later use $(\nabla_{\mathbf{k}}\cdot(\mathbf{v}^{\nu}\times\mathbf{B}) = (\nabla_{\mathbf{k}}\times\frac{1}{\hbar}\nabla_{\mathbf{k}}\varepsilon^{\nu})\cdot\mathbf{B} = 0)$

$$\frac{1}{\hbar}\nabla_{\mathbf{k}}\cdot\mathbf{F}^{\nu} = \frac{1}{\hbar}\nabla_{\mathbf{k}}\cdot\mathbf{F}^{\nu}_{\mathrm{E}} = \frac{1}{\hbar}\nabla_{\mathbf{k}}\cdot\tilde{\mathbf{F}}^{\nu}_{\mathrm{E}} = -\nabla_{\mathbf{r}}\cdot\mathbf{v}^{\nu}. \tag{2.14}$$

For device simulations, suitable boundary conditions at the device boundaries are required. The Neumann boundary condition is imposed on a non-contact boundary. For a contact boundary, instead of the Dirichlet boundary condition we use Neumann boundary conditions together with an interface generation rate [5, 6]

$$\gamma_s^\nu(\mathbf{r},\mathbf{k}) = [f^{\mathrm{eq}}(\mathbf{r},\mathbf{k})\Theta(\mathbf{v}^\nu\cdot\mathbf{n}) + f^\nu(\mathbf{r},\mathbf{k})\Theta(-\mathbf{v}^\nu\cdot\mathbf{n})]\mathbf{v}^\nu\cdot\mathbf{n}, \tag{2.15}$$

where $\mathbf{n}$ is a surface vector pointing into the device, $\Theta(x)$ the step function, and f^{eq} the equilibrium distribution specified by the particle quasi-Fermi level of the contact. When the Pauli principle is considered in the simulation, f^{eq} is given by the Fermi-Dirac distribution, otherwise by the Maxwell–Boltzmann distribution. This boundary condition corresponds to a thermal bath contact similar to the ones used in Monte Carlo simulations [7]. The injected particle flux (the first term on the RHS) is the result of an equilibrium distribution enforced by the externally applied bias, whereas the extracted particle flux (the second term) is due to the distribution within the device.

2.2 Poisson Equation and Drift-Diffusion Model

Since the driving force for particles is a function of the electrostatic potential, which is again a function of the particle density, it should be calculated in a self-consistent manner. The electrostatic potential $V(\mathbf{r},t)$ is the solution of the Poisson equation

$$\nabla_{\mathbf{r}}\cdot(\kappa(\mathbf{r})\nabla_{\mathbf{r}}V(\mathbf{r},t)) = -q(N_{\mathrm{D}}(\mathbf{r}) - N_{\mathrm{A}}(\mathbf{r}) - n(\mathbf{r},t) + p(\mathbf{r},t)), \tag{2.16}$$

where κ is the dielectric constant, N_{A} and N_{D} are the acceptor and donor concentrations, respectively.

The electron density $n(\mathbf{r},t)$ and the hole density $p(\mathbf{r},t)$ are given by (2.3) and (2.4), respectively, if the BTE is solved for the respective particle type. In many applications, only one type of particles (either electrons or holes) is treated with the BTE. In such situations, the macroscopic carrier density of the other type of particles is directly calculated based on the drift-diffusion model, where the mobility does not depend on the driving force and is calculated at equilibrium [8].

For both, the Poisson equation and the drift-diffusion model the Neumann boundary condition is imposed on a non-contact boundary and the Dirichlet boundary condition on a contact boundary.

2.3 Basic Properties of Spherical Harmonics

Since the phase space has six dimensions, it is very inconvenient to discretize it in a straightforward manner with finite differences [9]. While it is possible to reduce the number of dimensions in the real space by assuming translational symmetry in one of more dimensions, this is not possible in the **k** space. Instead of using finite differences in conjunction with cartesian coordinates, spherical coordinates are used and the dependence on the two angles is expanded with spherical harmonics. The expansion is truncated at a maximum order leading to a finite number of unknowns for the angular dependence. In the case that the SHE converges rapidly, this leads to a considerable lower number of unknowns than a finite difference approach. In this section, some useful properties of the spherical harmonics are briefly discussed and the SHE is introduced.

The spherical harmonics can be defined as either complex-valued or real-valued functions. Since the complex computation produces a lot of overhead real spherical harmonics will be used in this work

$$Y_{l,m}(\vartheta,\varphi) = \begin{cases} c_{l,m} P_l^m(\cos\vartheta)\cos(m\varphi) & \text{for } m \geq 0 \\ c_{l,m} P_l^{|m|}(\cos\vartheta)\sin(|m|\varphi) & \text{for } m < 0 \end{cases}, \tag{2.17}$$

where ϑ and φ are angles in the spherical coordinate system. The indices l and m determine the order and the sub-order of the spherical harmonics[2] and they have the following range:

$$l \geq 0, \tag{2.18}$$

$$-l \leq m \leq l. \tag{2.19}$$

The normalization factor $c_{l,m}$ is given as:

$$c_{l,m} = \begin{cases} \sqrt{\dfrac{(2l+1)}{4\pi}} & \text{for } m = 0 \\ \sqrt{\dfrac{(2l+1)(l-|m|)!}{2\pi(l+|m|)!}} & \text{for } m \neq 0 \end{cases}. \tag{2.20}$$

$P_l^m(x)$ is the associated Legendre polynomial

$$P_l^m(x) = (1-x^2)^{\frac{m}{2}} \frac{\mathrm{d}^m}{\mathrm{d}x^m} P_l(x), \tag{2.21}$$

where $P_l(x)$ is the Legendre polynomial

$$P_l(x) = \frac{1}{2^l l!} \frac{\mathrm{d}^l}{\mathrm{d}x^l}(x^2-1)^l. \tag{2.22}$$

The spherical harmonics up to the first order are

$$Y_{0,0}(\vartheta,\varphi) = \frac{1}{\sqrt{4\pi}}, \tag{2.23}$$

[2] In the literature, the order l and the sub-order m are usually called the degree l and the order m, respectively.

$$Y_{1,-1}(\vartheta,\varphi) = \sqrt{\frac{3}{4\pi}}\sin\vartheta\sin\varphi, \tag{2.24}$$

$$Y_{1,0}(\vartheta,\varphi) = \sqrt{\frac{3}{4\pi}}\cos\vartheta, \tag{2.25}$$

$$Y_{1,1}(\vartheta,\varphi) = \sqrt{\frac{3}{4\pi}}\sin\vartheta\cos\varphi. \tag{2.26}$$

A quantity $X(\vartheta,\varphi)$ can be expanded with spherical harmonics by

$$X_{l,m} = \oint X(\vartheta,\varphi)Y_{l,m}(\vartheta,\varphi)\mathrm{d}\Omega, \tag{2.27}$$

where Ω is the solid angle, $\mathrm{d}\Omega = \sin\vartheta\mathrm{d}\vartheta\mathrm{d}\varphi$ and the integral extends over the whole unit sphere. The original form is recovered by

$$X(\vartheta,\varphi) = \sum_{l=0}^{\infty}\sum_{m=-l}^{l} X_{l,m}Y_{l,m}(\vartheta,\varphi). \tag{2.28}$$

The sums are often abbreviated by $\sum_{l,m} = \sum_{l=0}^{\infty}\sum_{m=-l}^{l}$.

The product of two spherical harmonics is normalized when integrated over the unit sphere

$$a_{l,m,l',m'} = \oint Y_{l,m}(\vartheta,\varphi)Y_{l',m'}(\vartheta,\varphi)\mathrm{d}\Omega = \delta_{l,l'}\delta_{m,m'}, \tag{2.29}$$

The integral yields zero if the two spherical harmonics have different indices. In this sense the spherical harmonics are orthogonal and they form a complete set of basis functions. Other integrals related to a product of two spherical harmonics can be defined. Among many possible forms, only two frequently used quantities are shown here

$$\mathbf{a}_{l,m,l',m'} = \oint \mathbf{e}_{\varepsilon}Y_{l,m}Y_{l',m'}\mathrm{d}\Omega \tag{2.30}$$

and

$$\mathbf{b}_{l,m,l',m'} = \oint \left(\mathbf{e}_{\vartheta}\frac{\partial Y_{l,m}}{\partial\vartheta} + \mathbf{e}_{\varphi}\frac{1}{\sin\vartheta}\frac{\partial Y_{l,m}}{\partial\varphi}\right)Y_{l',m'}\mathrm{d}\Omega, \tag{2.31}$$

where $\mathbf{e}_{\varepsilon}$, $\mathbf{e}_{\vartheta}$ and $\mathbf{e}_{\varphi}$ are the unit vectors in the spherical coordinate system. Since $Y_{0,0}$ is constant over the solid angle, $\mathbf{b}_{l,m,l',m'}$ vanishes when the first index pair is (0,0),

$$\mathbf{b}_{0,0,l',m'} = 0. \tag{2.32}$$

In the derivation and the implementation of the balance equations obtained from the projection of the BTE, products of three spherical harmonics are frequently used. They can be defined as direct extensions of the double products.

A scalar triple product $a_{l,m,l',m',l'',m''}$ is given by

$$a_{l,m,l',m',l'',m''} = \oint Y_{l,m}(\vartheta,\varphi) Y_{l',m'}(\vartheta,\varphi) Y_{l'',m''}(\vartheta,\varphi) \mathrm{d}\Omega. \tag{2.33}$$

From its definition, $a_{l,m,l',m',l'',m''}$ is invariant under the six permutations of the three index pairs

$$\begin{aligned} a_{l,m,l',m',l'',m''} &= a_{l'',m'',l,m,l',m'} = a_{l',m',l'',m'',l,m} \\ &= a_{l,m,l'',m'',l',m'} = a_{l',m',l,m,l'',m''} = a_{l'',m'',l',m',l,m}. \end{aligned} \tag{2.34}$$

The vector triple product $\mathbf{a}_{l,m,l',m',l'',m''}$ is given by

$$\mathbf{a}_{l,m,l',m',l'',m''} = \oint \mathbf{e}_\varepsilon Y_{l,m} Y_{l',m'} Y_{l'',m''} \mathrm{d}\Omega. \tag{2.35}$$

It is also invariant under six permutations of index pairs.

The vector triple product $\mathbf{b}_{l,m,l',m',l'',m''}$ is defined as

$$\mathbf{b}_{l,m,l',m',l'',m''} = \oint \left(\mathbf{e}_\vartheta \frac{\partial Y_{l,m}}{\partial \vartheta} + \mathbf{e}_\varphi \frac{1}{\sin\vartheta} \frac{\partial Y_{l,m}}{\partial \varphi} \right) Y_{l',m'} Y_{l'',m''} \mathrm{d}\Omega. \tag{2.36}$$

Due to its definition, $\mathbf{b}_{l,m,l',m',l'',m''}$ is invariant only under two permutations of the index pairs

$$\mathbf{b}_{l,m,l',m',l'',m''} = \mathbf{b}_{l,m,l'',m'',l',m'}. \tag{2.37}$$

Since $Y_{0,0}$ is constant over the solid angle, $\mathbf{b}_{l,m,l',m',l'',m''}$ vanishes, when the first index pair is (0,0),

$$\mathbf{b}_{0,0,l',m',l'',m''} = 0. \tag{2.38}$$

All three triple products defined above reduce to the corresponding double products, when the second (or third) index pair is (0,0),

$$a_{l,m,l',m',0,0} = Y_{0,0} a_{l,m,l',m'} = \frac{1}{\sqrt{4\pi}} \delta_{l,l'} \delta_{m,m'}, \tag{2.39}$$

$$\mathbf{a}_{l,m,l',m',0,0} = Y_{0,0} \mathbf{a}_{l,m,l',m'}, \tag{2.40}$$

$$\mathbf{b}_{l,m,l',m',0,0} = Y_{0,0} \mathbf{b}_{l,m,l',m'}. \tag{2.41}$$

The following relation holds for an arbitrary analytic function $X(\vartheta,\varphi)$ defined on the solid angle,

$$\oint \left[\mathbf{e}_\varepsilon 2 - \mathbf{e}_\vartheta \frac{\partial}{\partial \vartheta} - \mathbf{e}_\varphi \frac{1}{\sin\vartheta} \frac{\partial}{\partial \varphi} \right] X(\vartheta,\varphi) \mathrm{d}\Omega = 0. \tag{2.42}$$

This relation can be checked by inserting the following two equations into the original one:

$$\frac{\partial}{\partial\vartheta}\mathbf{e}_\vartheta = -\mathbf{e}_\varepsilon \tag{2.43}$$

and

$$\frac{\partial}{\partial\varphi}\mathbf{e}_\varphi = -\cos\vartheta\,\mathbf{e}_\vartheta - \sin\vartheta\,\mathbf{e}_\varepsilon. \tag{2.44}$$

This relation is quite useful to derive a relation between the two vector products, **a** and **b**. By inserting $Y_{l,m}Y_{l',m'}$ or $Y_{l,m}Y_{l',m'}Y_{l'',m''}$ as $X(\vartheta,\varphi)$, it is found that

$$2\mathbf{a}_{l,m,l',m'} = \mathbf{b}_{l,m,l',m'} + \mathbf{b}_{l',m',l,m}, \tag{2.45}$$

$$2\mathbf{a}_{l,m,l',m',l'',m''} = \mathbf{b}_{l,m,l',m',l'',m''} + \mathbf{b}_{l',m',l,m,l'',m''} + \mathbf{b}_{l'',m'',l,m,l',m'}. \tag{2.46}$$

A spherical harmonic $Y_{l,m}(\vartheta,\varphi)$ has the following property

$$Y_{l,m}(\pi-\vartheta,\pi+\varphi) = (-1)^l Y_{l,m}(\vartheta,\varphi). \tag{2.47}$$

Therefore, spherical harmonics with even l's are inversion-symmetric, while spherical harmonics with odd l's are anti-symmetric w.r.t. inversion. In this work, we explicitly assume that the dispersion relation of the band structure is inversion-symmetric. Therefore, only the expansion coefficients with even order do not vanish. On the other hand, the group velocity of such a band structure has non-vanishing coefficients only for odd l's.

Sometimes the spherical coordinate system for the band structure and the one for the device can be different. In order to minimize the number of the nonzero terms used for describing the band structure, each valley is usually described in its internal coordinate system [6]. However, in a simulation a global device coordinate system is used, which is shared by all valleys. Note that the device coordinate system can change according to the purpose of the simulation. For this, an algorithm for rotating the spherical harmonics with arbitrary Euler angles [10] is implemented. Up to the first (or second) order, an analytical expression can be easily used. However, since we are dealing with spherical harmonics of arbitrary order, a general scheme for the rotation is required. Rotation of spherical harmonics is performed using the "Wigner matrices" relative to the chosen spherical harmonics basis. Invariance of the lth spherical harmonics subspace holds. A spherical harmonic keeps its l value after the rotation. For a given l, the Wigner matrix is a $(2l+1)\times(2l+1)$ orthogonal matrix, which relates the spherical harmonics of order l,

$$R\cdot Y_{l,m} = \sum_n D^l_{n,m}(R)Y_{l,n}, \tag{2.48}$$

where $R\cdot Y_{l,m}$ is the rotated spherical harmonic. Let an arbitrary function x have the spherical harmonics coefficients $x_{l,m}$, then, if $R\cdot x$ has spherical harmonics coefficients denoted as $x'_{l,m}$, the Wigner matrix connects the two coefficients via

$$x'_{l,m} = \sum_{n=-l}^{l} D^l_{m,n}(R) x_{l,n}. \tag{2.49}$$

Using the rotation from the global device coordinate system to the internal coordinate system, quantities calculated in the internal coordinate system can be written in the global device coordinate system.

2.4 Coordinate Transform

The particle distribution function is expanded with spherical harmonics. Before the dependence of the distribution function on the angles ϑ and φ can be expanded with spherical harmonics, it has to be decided whether the distribution function is expanded for constant energy or constant modulus of the wave vector. An expansion of the distribution function on equienergy surfaces has many advantages over an expansion w.r.t. to the modulus of the wave vector. For example, at equilibrium the distribution function is isotropic on equienergy surfaces. In many scattering models the scattering rate is a function of energy and the energy transfer during scattering is constant. Thus, the scattering integral can be easily formulated on equienergy surfaces. In addition, if a magnetic force is included in the BTE, this formulation ensures that the magnetic force term conserves energy.

In order to expand the distribution function on an equienergy surface, a coordinate transformation is required. For example, in [11], a coordinate transformation, in which the angle variables are the same as in the original **k** space, has been proposed. In this work, a generalized coordinate transform which does not necessarily preserve the angle variables is introduced. This allows us more freedom to choose the coordinate transformation. Also from this study, some relations between coefficients, which are important not only for deriving the balance equations but also for the stabilization scheme, can be obtained in a natural way. The scheme keeping the angle variables [11] is easily obtained as a special case of this generalized transformation.

Due to the position dependent band structure considered in this work, the partial derivative w.r.t. the real space is also effected by the coordinate transformation. In order to avoid confusion, subscripts **k** and ε are used for denoting the real space variables in the original space and the transformed space, respectively. All equations shown in Sect. 2.1 should be interpreted in the original space. The following mapping of the original space $(x_\mathbf{k}, y_\mathbf{k}, z_\mathbf{k}, k_x, k_y, k_z)$ onto the transformed space $(x_\varepsilon, y_\varepsilon, z_\varepsilon, \varepsilon, \vartheta, \varphi)$ is used

$$x_\varepsilon = x_\mathbf{k}, \tag{2.50}$$

$$y_\varepsilon = y_\mathbf{k}, \tag{2.51}$$

$$z_\varepsilon = z_\mathbf{k}, \tag{2.52}$$

$$\varepsilon = \varepsilon^{\nu}(\mathbf{r_k}, \mathbf{k}), \tag{2.53}$$

$$\vartheta = \vartheta^{\nu}(\mathbf{r_k}, \mathbf{k}), \tag{2.54}$$

$$\varphi = \varphi^{\nu}(\mathbf{r_k}, \mathbf{k}), \tag{2.55}$$

where $\varepsilon^{\nu}(\mathbf{r_k}, \mathbf{k})$ is the dispersion relation for the νth valley, and the appropriate choice of $\vartheta^{\nu}(\mathbf{r_k}, \mathbf{k})$ and $\varphi^{\nu}(\mathbf{r_k}, \mathbf{k})$ depends on the underlying band structure. Note that they can be dependent on the real space up to this point. In the following section (Sect. 2.5), it is found that a simple form of balance equations can be obtained when ϑ^{ν} and φ^{ν} do not depend on the real space. We assume that the Jacobian determinant of the transform is nonzero. The angle-preserving transform [11] can be obtained trivially by setting ϑ and φ as follows:

$$\cos\vartheta = \frac{k_z}{k}, \tag{2.56}$$

$$\sin\vartheta\cos\varphi = \frac{k_x}{k}, \tag{2.57}$$

$$\sin\vartheta\sin\varphi = \frac{k_y}{k}, \tag{2.58}$$

where the zenith direction is aligned with the z-axis.

The del operator in the wave vector space is given by

$$\nabla_{\mathbf{k}} = \hbar\mathbf{v}^{\nu}\frac{\partial}{\partial\varepsilon} + (\nabla_{\mathbf{k}}\vartheta^{\nu})\frac{\partial}{\partial\vartheta} + (\nabla_{\mathbf{k}}\varphi^{\nu})\frac{\partial}{\partial\varphi}, \tag{2.59}$$

where $\mathbf{v}^{\nu}$, $\nabla_{\mathbf{k}}\vartheta^{\nu}$, and $\nabla_{\mathbf{k}}\varphi^{\nu}$ are calculated in the original space. Also the del operator in the real space is given by

$$\nabla_{\mathbf{r_k}} = \nabla_{\mathbf{r}_\varepsilon} + (\nabla_{\mathbf{r_k}}\varepsilon^{\nu})\frac{\partial}{\partial\varepsilon} + (\nabla_{\mathbf{r_k}}\vartheta^{\nu})\frac{\partial}{\partial\vartheta} + (\nabla_{\mathbf{r_k}}\varphi^{\nu})\frac{\partial}{\partial\varphi}. \tag{2.60}$$

Note that the last three terms originate from the position dependent band structure.

The infinitesimal volume in the original $\mathbf{k}$ space is given by

$$\mathrm{d}^3k = |\det(\hat{\mathbf{J}}^{-1})|\mathrm{d}\varepsilon\mathrm{d}\vartheta\mathrm{d}\varphi, \tag{2.61}$$

where the Jacobian matrix of the transform $\hat{\mathbf{J}}$ is[3]

[3] Since the coordinate transform is applied to the six-dimensional phase space, this is only a part of the total Jacobian matrix, corresponding to the $\mathbf{k}$ space. Because the transform for the real space is just an identity transform, the entire 6×6 Jacobian matrix can be inverted blockwise.

$$\hat{\mathbf{J}} = \begin{pmatrix} \frac{\partial \varepsilon^\nu}{\partial k_x} & \frac{\partial \varepsilon^\nu}{\partial k_y} & \frac{\partial \varepsilon^\nu}{\partial k_z} \\ \frac{\partial \vartheta^\nu}{\partial k_x} & \frac{\partial \vartheta^\nu}{\partial k_y} & \frac{\partial \vartheta^\nu}{\partial k_z} \\ \frac{\partial \varphi^\nu}{\partial k_x} & \frac{\partial \varphi^\nu}{\partial k_y} & \frac{\partial \varphi^\nu}{\partial k_z} \end{pmatrix} = \begin{pmatrix} \nabla_{\mathbf{k}} \varepsilon^\nu \\ \nabla_{\mathbf{k}} \vartheta^\nu \\ \nabla_{\mathbf{k}} \varphi^\nu \end{pmatrix}, \tag{2.62}$$

where the gradients on the RHS should be understood as row vectors. Since we are working in the energy space $(\varepsilon, \vartheta, \varphi)$, it is useful to explicitly express it in terms of the Jacobian matrix of the inverse transform,

$$\hat{\mathbf{J}} = (\hat{\mathbf{J}}^{-1})^{-1} = \begin{pmatrix} \frac{\partial k_x^\nu}{\partial \varepsilon} & \frac{\partial k_x^\nu}{\partial \vartheta} & \frac{\partial k_x^\nu}{\partial \varphi} \\ \frac{\partial k_y^\nu}{\partial \varepsilon} & \frac{\partial k_y^\nu}{\partial \vartheta} & \frac{\partial k_y^\nu}{\partial \varphi} \\ \frac{\partial k_z^\nu}{\partial \varepsilon} & \frac{\partial k_z^\nu}{\partial \vartheta} & \frac{\partial k_z^\nu}{\partial \varphi} \end{pmatrix}^{-1} = \frac{1}{\det(\hat{\mathbf{J}}^{-1})} \begin{pmatrix} \frac{\partial \mathbf{k}^\nu}{\partial \vartheta} \times \frac{\partial \mathbf{k}^\nu}{\partial \varphi} \\ \frac{\partial \mathbf{k}^\nu}{\partial \varphi} \times \frac{\partial \mathbf{k}^\nu}{\partial \varepsilon} \\ \frac{\partial \mathbf{k}^\nu}{\partial \varepsilon} \times \frac{\partial \mathbf{k}^\nu}{\partial \vartheta} \end{pmatrix}, \tag{2.63}$$

where the cross products in the RHS should be understood as row vectors. The Jacobian determinant $\det(\hat{\mathbf{J}}^{-1})$ is given by

$$\det(\hat{\mathbf{J}}^{-1}) = \frac{\partial \mathbf{k}^\nu}{\partial \varepsilon} \cdot \left(\frac{\partial \mathbf{k}^\nu}{\partial \vartheta} \times \frac{\partial \mathbf{k}^\nu}{\partial \varphi} \right). \tag{2.64}$$

Therefore, we have the following relations:

$$\nabla_{\mathbf{k}} \varepsilon^\nu = \frac{1}{\det(\hat{\mathbf{J}}^{-1})} \left(\frac{\partial \mathbf{k}^\nu}{\partial \vartheta} \times \frac{\partial \mathbf{k}^\nu}{\partial \varphi} \right), \tag{2.65}$$

$$\nabla_{\mathbf{k}} \vartheta^\nu = \frac{1}{\det(\hat{\mathbf{J}}^{-1})} \left(\frac{\partial \mathbf{k}^\nu}{\partial \varphi} \times \frac{\partial \mathbf{k}^\nu}{\partial \varepsilon} \right), \tag{2.66}$$

$$\nabla_{\mathbf{k}} \varphi^\nu = \frac{1}{\det(\hat{\mathbf{J}}^{-1})} \left(\frac{\partial \mathbf{k}^\nu}{\partial \varepsilon} \times \frac{\partial \mathbf{k}^\nu}{\partial \vartheta} \right). \tag{2.67}$$

A similar procedure can be applied to the partial derivative w.r.t. the real space. By inspecting the inverted total Jacobian matrix, the following relation can be found

$$\begin{pmatrix} \frac{\partial k_x^\nu}{\partial x_\varepsilon} & \frac{\partial k_x^\nu}{\partial y_\varepsilon} & \frac{\partial k_x^\nu}{\partial z_\varepsilon} \\ \frac{\partial k_y^\nu}{\partial x_\varepsilon} & \frac{\partial k_y^\nu}{\partial y_\varepsilon} & \frac{\partial k_y^\nu}{\partial z_\varepsilon} \\ \frac{\partial k_z^\nu}{\partial x_\varepsilon} & \frac{\partial k_z^\nu}{\partial y_\varepsilon} & \frac{\partial k_z^\nu}{\partial z_\varepsilon} \end{pmatrix} = -\hat{\mathbf{J}}^{-1} \begin{pmatrix} \frac{\partial \varepsilon^\nu}{\partial x_{\mathbf{k}}} & \frac{\partial \varepsilon^\nu}{\partial y_{\mathbf{k}}} & \frac{\partial \varepsilon^\nu}{\partial z_{\mathbf{k}}} \\ \frac{\partial \vartheta^\nu}{\partial x_{\mathbf{k}}} & \frac{\partial \vartheta^\nu}{\partial y_{\mathbf{k}}} & \frac{\partial \vartheta^\nu}{\partial z_{\mathbf{k}}} \\ \frac{\partial \varphi^\nu}{\partial x_{\mathbf{k}}} & \frac{\partial \varphi^\nu}{\partial y_{\mathbf{k}}} & \frac{\partial \varphi^\nu}{\partial z_{\mathbf{k}}} \end{pmatrix}. \tag{2.68}$$

It can be written in a component-wise fashion,

$$\frac{\partial k^{\nu}_{d_1}}{\partial d_{\varepsilon}} = - \sum_{\alpha=\varepsilon,\vartheta,\varphi} \frac{\partial k^{\nu}_{d_1}}{\partial \alpha} \frac{\partial \alpha^{\nu}}{\partial d_{\mathbf{k}}}, \tag{2.69}$$

where d and d_1 are arbitrary directions, which can be x, y, z, and α represents $\varepsilon, \vartheta, \varphi$. Multiplying $\hat{\mathbf{J}}$ and using (2.63), we obtain

$$\begin{pmatrix} \nabla_{\mathbf{r_k}} \varepsilon^{\nu} \\ \nabla_{\mathbf{r_k}} \vartheta^{\nu} \\ \nabla_{\mathbf{r_k}} \varphi^{\nu} \end{pmatrix} = -\frac{1}{\det(\hat{\mathbf{J}}^{-1})} \begin{pmatrix} \frac{\partial \mathbf{k}^{\nu}}{\partial \vartheta} \times \frac{\partial \mathbf{k}^{\nu}}{\partial \varphi} \\ \frac{\partial \mathbf{k}^{\nu}}{\partial \varphi} \times \frac{\partial \mathbf{k}^{\nu}}{\partial \varepsilon} \\ \frac{\partial \mathbf{k}^{\nu}}{\partial \varepsilon} \times \frac{\partial \mathbf{k}^{\nu}}{\partial \vartheta} \end{pmatrix} \begin{pmatrix} \frac{\partial k^{\nu}_x}{\partial x_{\varepsilon}} & \frac{\partial k^{\nu}_x}{\partial y_{\varepsilon}} & \frac{\partial k^{\nu}_x}{\partial z_{\varepsilon}} \\ \frac{\partial k^{\nu}_y}{\partial x_{\varepsilon}} & \frac{\partial k^{\nu}_y}{\partial y_{\varepsilon}} & \frac{\partial k^{\nu}_y}{\partial z_{\varepsilon}} \\ \frac{\partial k^{\nu}_z}{\partial x_{\varepsilon}} & \frac{\partial k^{\nu}_z}{\partial y_{\varepsilon}} & \frac{\partial k^{\nu}_z}{\partial z_{\varepsilon}} \end{pmatrix}, \tag{2.70}$$

where the cross products and the gradients should be considered as row vectors. When an arbitrary direction d is considered, three relations are derived:

$$(\nabla_{\mathbf{r_k}} \varepsilon^{\nu}) \cdot \mathbf{e}_d = -\frac{1}{\det(\hat{\mathbf{J}}^{-1})} \left(\frac{\partial \mathbf{k}^{\nu}}{\partial \vartheta} \times \frac{\partial \mathbf{k}^{\nu}}{\partial \varphi} \right) \cdot \frac{\partial \mathbf{k}^{\nu}}{\partial d_{\varepsilon}}, \tag{2.71}$$

$$(\nabla_{\mathbf{r_k}} \vartheta^{\nu}) \cdot \mathbf{e}_d = -\frac{1}{\det(\hat{\mathbf{J}}^{-1})} \left(\frac{\partial \mathbf{k}^{\nu}}{\partial \varphi} \times \frac{\partial \mathbf{k}^{\nu}}{\partial \varepsilon} \right) \cdot \frac{\partial \mathbf{k}^{\nu}}{\partial d_{\varepsilon}}, \tag{2.72}$$

$$(\nabla_{\mathbf{r_k}} \varphi^{\nu}) \cdot \mathbf{e}_d = -\frac{1}{\det(\hat{\mathbf{J}}^{-1})} \left(\frac{\partial \mathbf{k}^{\nu}}{\partial \varepsilon} \times \frac{\partial \mathbf{k}^{\nu}}{\partial \vartheta} \right) \cdot \frac{\partial \mathbf{k}^{\nu}}{\partial d_{\varepsilon}}, \tag{2.73}$$

where $\mathbf{e}_d$ is the d-directional unit vector.

The generalized density-of-states for a single spin direction [11][4] is defined as

$$Z^{\nu}(\mathbf{r}, \varepsilon, \vartheta, \varphi) = \frac{1}{(2\pi)^3} |\det(\hat{\mathbf{J}}^{-1})| \frac{1}{\sin \vartheta}, \tag{2.74}$$

in order for the following conventional relation to hold

$$\frac{1}{(2\pi)^3} \mathrm{d}^3 k = Z^{\nu} \mathrm{d}\varepsilon \mathrm{d}\Omega. \tag{2.75}$$

The Jacobian determinant of the inverse transform $\det(\hat{\mathbf{J}}^{-1})$ in the above equations can be eliminated by multiplying a gradient ((2.65), (2.66), or (2.67)) and the DOS:

[4]This definition does not include the integration over the solid angle. The generalized DOS is therefore for the case that it does not depend on the angles by a factor of 4π smaller than the conventional expression.

$$(\nabla_{\mathbf{k}}\varepsilon^{\nu})Z^{\nu} = \pm\left(\frac{\partial\mathbf{k}^{\nu}}{\partial\vartheta}\times\frac{\partial\mathbf{k}^{\nu}}{\partial\varphi}\right)\frac{1}{(2\pi)^3}\frac{1}{\sin\vartheta}, \tag{2.76}$$

$$(\nabla_{\mathbf{k}}\vartheta^{\nu})Z^{\nu} = \pm\left(\frac{\partial\mathbf{k}^{\nu}}{\partial\varphi}\times\frac{\partial\mathbf{k}^{\nu}}{\partial\varepsilon}\right)\frac{1}{(2\pi)^3}\frac{1}{\sin\vartheta}, \tag{2.77}$$

$$(\nabla_{\mathbf{k}}\varphi^{\nu})Z^{\nu} = \pm\left(\frac{\partial\mathbf{k}^{\nu}}{\partial\varepsilon}\times\frac{\partial\mathbf{k}^{\nu}}{\partial\vartheta}\right)\frac{1}{(2\pi)^3}\frac{1}{\sin\vartheta}, \tag{2.78}$$

where $\pm$ appears due to the sign of the Jacobian determinant.

There might be many possible relations between the quantities shown above. However, since the density-of-states Z^{ν} and the group velocity $\mathbf{v}^{\nu}$ are the most important transport coefficients, it is expected that partial derivatives of these quantities are relevant in the derivation of the balance equations. In the remaining part of this section, only three relations will be explicitly derived, because they play important roles in the derivation of balance equations and their stabilization. The first one is related to the partial derivative w.r.t. the energy space, while the second and third ones are related to the partial derivative w.r.t. the real space. Note that the second and third ones originate from the position-dependent band structure. Therefore, in the case of a position-independent one they vanish completely.

Since the second partial derivative does not depend on the order, in which the two derivatives are taken, the following relation results

$$\frac{\partial}{\partial\varepsilon}\left(\frac{\partial\mathbf{k}^{\nu}}{\partial\vartheta}\times\frac{\partial\mathbf{k}^{\nu}}{\partial\varphi}\right)+\frac{\partial}{\partial\vartheta}\left(\frac{\partial\mathbf{k}^{\nu}}{\partial\varphi}\times\frac{\partial\mathbf{k}^{\nu}}{\partial\varepsilon}\right)+\frac{\partial}{\partial\varphi}\left(\frac{\partial\mathbf{k}^{\nu}}{\partial\varepsilon}\times\frac{\partial\mathbf{k}^{\nu}}{\partial\vartheta}\right)=0. \tag{2.79}$$

As an immediate result, we obtain the first relation

$$\frac{\partial}{\partial\varepsilon}\left(\mathbf{v}^{\nu}Z^{\nu}\sin\vartheta\right)+\frac{\partial}{\partial\vartheta}\left((\nabla_{\mathbf{k}}\vartheta^{\nu})\frac{Z^{\nu}}{\hbar}\sin\vartheta\right)+\frac{\partial}{\partial\varphi}\left((\nabla_{\mathbf{k}}\varphi^{\nu})\frac{Z^{\nu}}{\hbar}\sin\vartheta\right)=0. \tag{2.80}$$

Expansion of the above relation with a spherical harmonic $Y_{l,m}(\vartheta,\varphi)$ yields

$$\frac{\partial}{\partial\varepsilon}\oint\mathbf{v}^{\nu}Z^{\nu}Y_{l,m}\mathrm{d}\Omega=\oint(\nabla_{\mathbf{k}}\vartheta^{\nu})\frac{Z^{\nu}}{\hbar}\frac{\partial Y_{l,m}}{\partial\vartheta}\mathrm{d}\Omega+\oint(\nabla_{\mathbf{k}}\varphi^{\nu})\frac{Z^{\nu}}{\hbar}\frac{\partial Y_{l,m}}{\partial\varphi}\mathrm{d}\Omega, \tag{2.81}$$

which is a generalized form of a previous result [5]. In the following, this is explicitly shown.

When the angle variables are kept the same in both spaces, the gradients of the two angle variables are given by

$$\nabla_{\mathbf{k}}\vartheta^{\nu} = \mathbf{e}_{\vartheta}\frac{1}{k^{\nu}}, \tag{2.82}$$

$$\nabla_{\mathbf{k}}\varphi^{\nu} = \mathbf{e}_{\varphi}\frac{1}{k^{\nu}\sin\vartheta}, \tag{2.83}$$

which yields

$$\frac{\partial}{\partial \varepsilon} \oint \mathbf{v}^{\nu} Z^{\nu} Y_{l,m} \mathrm{d}\Omega = \oint \frac{Z^{\nu}}{\hbar k^{\nu}} \left(\mathbf{e}_{\vartheta} \frac{\partial Y_{l,m}}{\partial \vartheta} + \mathbf{e}_{\varphi} \frac{1}{\sin \vartheta} \frac{\partial Y_{l,m}}{\partial \varphi} \right) \mathrm{d}\Omega. \tag{2.84}$$

Expansion of $\mathbf{v}Z$ and $Z/\hbar k$ with a spherical harmonic $Y_{l',m'}(\vartheta, \varphi)$ by (2.27) [11]

$$\mathbf{v}^{\nu} Z^{\nu} = \sum_{l',m'} \{\mathbf{v}Z\}^{\nu}_{l',m'} Y_{l',m'}, \tag{2.85}$$

$$\frac{Z^{\nu}}{\hbar k^{\nu}} = \sum_{l',m'} \left\{ \frac{Z}{\hbar k} \right\}^{\nu}_{l',m'} Y_{l',m'}, \tag{2.86}$$

yields

$$\frac{\partial}{\partial \varepsilon} \{\mathbf{v}Z\}^{\nu}_{l,m} = \sum_{l',m'} \left\{ \frac{Z}{\hbar k} \right\}^{\nu}_{l',m'} \mathbf{b}_{l,m,l',m'}, \tag{2.87}$$

where the orthonormality of spherical harmonics was used. Similarly, we can also show that

$$\sum_{l'',m''} \frac{\partial}{\partial \varepsilon} \{\mathbf{v}Z\}^{\nu}_{l'',m''} a_{l,m,l',m',l'',m''} = \sum_{l'',m''} \left\{ \frac{Z}{\hbar k} \right\}^{\nu}_{l'',m''} (\mathbf{b}_{l,m,l',m',l'',m''} + \mathbf{b}_{l',m',l,m,l'',m''}) . \tag{2.88}$$

If an isotropic band structure with only one valley is considered, further simplifications are possible. In this case the group velocity is aligned with the radial direction,

$$\mathbf{v}Z = \mathbf{e}_{\varepsilon} v(\varepsilon) Z(\varepsilon), \tag{2.89}$$

and Z and k do not depend on the angle variables,

$$\frac{Z}{\hbar k} = \frac{Z(\varepsilon)}{\hbar k(\varepsilon)}. \tag{2.90}$$

From these properties and (2.80), we obtain a relation between the magnitude of the group velocity and the DOS which does not involve any spherical harmonics [5]

$$\frac{\partial v(\varepsilon) Z(\varepsilon)}{\partial \varepsilon} = 2 \frac{Z(\varepsilon)}{\hbar k(\varepsilon)}. \tag{2.91}$$

The second relation involves a gradient of the DOS. The d-directional component of the gradient of $\det(\hat{\mathbf{J}}^{-1})$, which is taken in the transformed space, evaluates to

$$\frac{\partial}{\partial d_{\varepsilon}} \det(\hat{\mathbf{J}}^{-1}) = \frac{\partial^2 \mathbf{k}^{\nu}}{\partial \varepsilon \partial d_{\varepsilon}} \cdot \left(\frac{\partial \mathbf{k}^{\nu}}{\partial \vartheta} \times \frac{\partial \mathbf{k}^{\nu}}{\partial \varphi} \right)$$

$$+\frac{\partial^2 \mathbf{k}^\nu}{\partial\vartheta\partial d_\varepsilon}\cdot\left(\frac{\partial \mathbf{k}^\nu}{\partial\varphi}\times\frac{\partial \mathbf{k}^\nu}{\partial\varepsilon}\right)$$
$$+\frac{\partial^2 \mathbf{k}^\nu}{\partial\varphi\partial d_\varepsilon}\cdot\left(\frac{\partial \mathbf{k}^\nu}{\partial\varepsilon}\times\frac{\partial \mathbf{k}^\nu}{\partial\vartheta}\right). \tag{2.92}$$

With (2.71)–(2.73) and (2.79) we obtain

$$\begin{aligned}\frac{\partial}{\partial d_\varepsilon}\det(\hat{\mathbf{J}}^{-1}) = &-\frac{\partial}{\partial\varepsilon}\left[(\nabla_{\mathbf{r_k}}\varepsilon^\nu)\cdot\mathbf{e}_d\det(\hat{\mathbf{J}}^{-1})\right]\\ &-\frac{\partial}{\partial\vartheta}\left[(\nabla_{\mathbf{r_k}}\vartheta^\nu)\cdot\mathbf{e}_d\det(\hat{\mathbf{J}}^{-1})\right]\\ &-\frac{\partial}{\partial\varphi}\left[(\nabla_{\mathbf{r_k}}\varphi^\nu)\cdot\mathbf{e}_d\det(\hat{\mathbf{J}}^{-1})\right],\end{aligned} \tag{2.93}$$

which can be expressed as a gradient of the DOS (2.74)

$$\begin{aligned}&\nabla_{\mathbf{r}_\varepsilon}(Z^\nu\sin\vartheta)+\frac{\partial}{\partial\varepsilon}\Big((\nabla_{\mathbf{r_k}}\varepsilon^\nu)Z^\nu\sin\vartheta\Big)\\ &+\frac{\partial}{\partial\vartheta}\Big((\nabla_{\mathbf{r_k}}\vartheta^\nu)Z^\nu\sin\vartheta\Big)+\frac{\partial}{\partial\varphi}\Big((\nabla_{\mathbf{r_k}}\varphi^\nu)Z^\nu\sin\vartheta\Big)=0.\end{aligned} \tag{2.94}$$

This is the second relation. When ϑ^ν and φ^ν are position-independent, the last two terms vanish and the term $\sin\vartheta$ can be removed.

The third relation is obtained by the following calculation. Using (2.69) and (2.79) and taking the gradient in the real space in the transformed energy space, we can find that

$$\begin{aligned}\nabla_{\mathbf{r}_\varepsilon}\cdot\left(\frac{\partial \mathbf{k}^\nu}{\partial\vartheta}\times\frac{\partial \mathbf{k}^\nu}{\partial\varphi}\right) = &\sum_\alpha\frac{\partial}{\partial\vartheta}\left((\nabla_{\mathbf{r_k}}\alpha^\nu)\cdot\left(\frac{\partial \mathbf{k}^\nu}{\partial\varphi}\times\frac{\partial \mathbf{k}^\nu}{\partial\alpha}\right)\right)\\ &+\frac{\partial}{\partial\varphi}\left((\nabla_{\mathbf{r_k}}\alpha^\nu)\cdot\left(\frac{\partial \mathbf{k}^\nu}{\partial\alpha}\times\frac{\partial \mathbf{k}^\nu}{\partial\vartheta}\right)\right).\end{aligned} \tag{2.95}$$

If ϑ^ν and φ^ν are position-independent, we obtain with (2.76)–(2.78)

$$\begin{aligned}\nabla_{\mathbf{r}_\varepsilon}\cdot(\mathbf{v}^\nu Z^\nu\sin\vartheta) = &-\frac{\partial}{\partial\vartheta}\left(\tilde{\mathbf{F}}_\mathrm{E}^\nu\cdot(\nabla_\mathbf{k}\vartheta^\nu)\frac{Z^\nu}{\hbar}\sin\vartheta\right)\\ &-\frac{\partial}{\partial\varphi}\left(\tilde{\mathbf{F}}_\mathrm{E}^\nu\cdot(\nabla_\mathbf{k}\varphi^\nu)\frac{Z^\nu}{\hbar}\sin\vartheta\right),\end{aligned} \tag{2.96}$$

where the definition of $\tilde{\mathbf{F}}_\mathrm{E}$ (2.11) was used. Expansion of this equation with spherical harmonics, multiplication with two more spherical harmonics and integration over the unit sphere yields

$$\nabla_{\mathbf{r}_\varepsilon} \cdot \sum_{l'',m''} \{\mathbf{v}Z\}^{\nu}_{l'',m''} a_{l,m,l',m',l'',m''}$$
$$= \sum_{l'',m''} \left\{ \tilde{\mathbf{F}}_E \frac{Z}{\hbar k} \right\}^{\nu}_{l'',m''} (\mathbf{b}_{l,m,l',m',l'',m''} + \mathbf{b}_{l',m',l,m,l'',m''}) \,. \quad (2.97)$$

Thus, under the coordinate transformation defined by (2.50)–(2.55), three important relations between transport coefficients, (2.80), (2.94), and (2.96), hold.

2.5 Spherical Harmonics Expansion of the Boltzmann Transport Equation

In order to obtain a projection of the BTE, which conserves the particle number, the projection onto spherical harmonics (2.27) is modified and an additional projection onto energy is performed, where both projections can be combined by integration over the **k** space [11, 12]. This projection reads for a microscopic quantity $X^{\nu}(\mathbf{r}, \mathbf{k})$

$$\{XZ\}^{\nu}_{l,m}(\mathbf{r}, \varepsilon) = \frac{1}{(2\pi)^3} \int \delta[\varepsilon - \varepsilon^{\nu}(\mathbf{r}, \mathbf{k})] Y_{l,m}(\vartheta^{\nu}(\mathbf{r}, \mathbf{k}), \varphi^{\nu}(\mathbf{r}, \mathbf{k})) X^{\nu}(\mathbf{r}, \mathbf{k}) \, \mathrm{d}^3 k$$
$$= \oint Y_{l,m}(\vartheta, \varphi) X^{\nu}(\mathbf{r}, \varepsilon, \vartheta, \varphi) Z^{\nu}(\mathbf{r}, \varepsilon, \vartheta, \varphi) \mathrm{d}\Omega \,. \quad (2.98)$$

The multiplication with delta function leads to the projection onto energy. Thus, the microscopic quantity $X^{\nu}(\mathbf{r}, \varepsilon, \vartheta, \varphi)$ has to be multiplied with the DOS $Z^{\nu}(\mathbf{r}, \varepsilon, \vartheta, \varphi)$ (2.74) before projection onto spherical harmonics by (2.27).

For $X^{\nu}(\mathbf{r}, \mathbf{k}) = 1$, the projection of the DOS for the νth valley is obtained

$$Z^{\nu}_{l,m}(\mathbf{r}, \varepsilon) = \frac{1}{(2\pi)^3} \int \delta[\varepsilon - \varepsilon^{\nu}(\mathbf{r}, \mathbf{k})] Y_{l,m}(\vartheta^{\nu}(\mathbf{r}, \mathbf{k}), \varphi^{\nu}(\mathbf{r}, \mathbf{k})) \, \mathrm{d}^3 k. \quad (2.99)$$

The conventional DOS is given by

$$\frac{1}{(2\pi)^3} \int \delta[\varepsilon - \varepsilon^{\nu}(\mathbf{r}, \mathbf{k})] \mathrm{d}^3 k = \oint Z^{\nu}(\mathbf{r}, \varepsilon, \vartheta, \varphi) \mathrm{d}\Omega = Z^{\nu}_{0,0}(\mathbf{r}, \varepsilon) \frac{1}{Y_{0,0}} \quad (2.100)$$

With (2.33) a DOS with two index pairs can be defined

$$Z^{\nu}_{l,m,l',m'}(\mathbf{r}, \varepsilon) = \sum_{l'',m''} Z^{\nu}_{l'',m''}(\mathbf{r}, \varepsilon) a_{l,m,l',m',l'',m''}. \quad (2.101)$$

and the projection of the $X^{\nu}(\mathbf{r}, \varepsilon, \vartheta, \varphi) Z^{\nu}(\mathbf{r}, \varepsilon, \vartheta, \varphi)$ can be directly calculated with the individually projected quantities

$$\{XZ\}^{\nu}_{l,m}(\mathbf{r}, \varepsilon) = \sum_{l',m'} X^{\nu}_{l',m'}(\mathbf{r}, \varepsilon) Z^{\nu}_{l,m,l',m'}(\mathbf{r}, \varepsilon). \quad (2.102)$$

With $X^\nu(\mathbf{r},\mathbf{k},t) = f^\nu(\mathbf{r},\mathbf{k},t)$, the generalized energy distribution is obtained

$$g^\nu_{l,m}(\mathbf{r},\varepsilon,t) = \frac{1}{(2\pi)^3}\sum_\nu \int_{BZ^\nu} \delta[\varepsilon - \varepsilon^\nu(\mathbf{r},\mathbf{k})] Y_{l,m}(\vartheta^\nu(\mathbf{r},\mathbf{k}),\varphi^\nu(\mathbf{r},\mathbf{k})) f^\nu(\mathbf{r},\mathbf{k},t)\,\mathrm{d}^3k$$
$$= \oint Y_{l,m}(\vartheta,\varphi) g^\nu(\mathbf{r},\varepsilon,\vartheta,\varphi,t)\mathrm{d}\Omega, \qquad (2.103)$$

where the generalized energy distribution is defined as

$$g^\nu(\mathbf{r},\varepsilon,\vartheta,\varphi,t) = Z^\nu(\mathbf{r},\varepsilon,\vartheta,\varphi) f^\nu(\mathbf{r},\varepsilon,\vartheta,\varphi,t). \qquad (2.104)$$

Similar to (2.102) one obtains

$$g^\nu_{l,m}(\mathbf{r},\varepsilon,t) = \sum_{l',m'} Z^\nu_{l,m,l',m'}(\mathbf{r},\varepsilon) f^\nu_{l',m'}(\mathbf{r},\varepsilon,t). \qquad (2.105)$$

The BTE is projected in the same way as the microscopic quantity (2.98) [11]

$$\frac{1}{(2\pi)^3}\int \delta[\varepsilon - \varepsilon^\nu(\mathbf{r},\mathbf{k})] Y_{l,m}(\vartheta^\nu(\mathbf{r},\mathbf{k}),\varphi^\nu(\mathbf{r},\mathbf{k}))\{BTE\}\mathrm{d}^3k. \qquad (2.106)$$

In the next subsections, the spherical harmonics expansion of the three individual terms of the BTE – the time derivative, free-streaming operator, and scattering integral – is discussed.

2.5.1 Time Derivative

The partial derivative w.r.t. time yields

$$\frac{\partial}{\partial t} f^\nu(\mathbf{r},\mathbf{k},t) \to \frac{1}{(2\pi)^3}\sum_\nu \int_{BZ^\nu} \delta[\varepsilon - \varepsilon^\nu(\mathbf{r},\mathbf{k})] Y_{l,m}(\vartheta^\nu(\mathbf{r},\mathbf{k}),\varphi^\nu(\mathbf{r},\mathbf{k})) \frac{\partial f^\nu(\mathbf{r},\mathbf{k},t)}{\partial t}\mathrm{d}^3k$$
$$= \frac{\partial}{\partial t} g^\nu_{l,m}(\mathbf{r},\varepsilon,t) = \frac{\partial}{\partial t}\sum_{l',m'} Z^\nu_{l,m,l',m'}(\mathbf{r},\varepsilon) f^\nu_{l',m'}(\mathbf{r},\varepsilon,t). \qquad (2.107)$$

by interchanging integration and differentiation.

2.5.2 Free-Streaming Operator

Two terms of the free-streaming operator can be recognized. They are the drift term by the total force and the diffusion term.

The drift term evaluates with (2.59) and integration by parts to

$$\begin{aligned}
\frac{1}{\hbar}\mathbf{F}^{\nu}\cdot\nabla_{\mathbf{k}}f^{\nu} \to & \oint Y_{l,m}\mathbf{F}^{\nu}\cdot\left[\mathbf{v}^{\nu}Z^{\nu}\frac{\partial}{\partial\varepsilon}+\frac{Z^{\nu}}{\hbar}(\nabla_{\mathbf{k}}\vartheta^{\nu})\frac{\partial}{\partial\vartheta}+\frac{Z^{\nu}}{\hbar}(\nabla_{\mathbf{k}}\varphi^{\nu})\frac{\partial}{\partial\varphi}\right]f^{\nu}\mathrm{d}\Omega \\
= & \oint Y_{l,m}\mathbf{F}^{\nu}\cdot\mathbf{v}^{\nu}Z^{\nu}\frac{\partial f^{\nu}}{\partial\varepsilon}\mathrm{d}\Omega \\
& -\int_0^{2\pi}\int_0^{\pi}\frac{\partial}{\partial\vartheta}\left[Y_{l,m}\mathbf{F}^{\nu}\cdot\frac{Z^{\nu}}{\hbar}(\nabla_{\mathbf{k}}\vartheta^{\nu})\sin\vartheta\right]f^{\nu}\mathrm{d}\vartheta\mathrm{d}\varphi \\
& -\int_0^{2\pi}\int_0^{\pi}\frac{\partial}{\partial\varphi}\left[Y_{l,m}\mathbf{F}^{\nu}\cdot\frac{Z^{\nu}}{\hbar}(\nabla_{\mathbf{k}}\varphi^{\nu})\sin\vartheta\right]f^{\nu}\mathrm{d}\vartheta\mathrm{d}\varphi \\
= & \oint \mathbf{F}^{\nu}\cdot\frac{\partial}{\partial\varepsilon}\left[Y_{l,m}\mathbf{v}^{\nu}Z^{\nu}f^{\nu}\right]\mathrm{d}\Omega \\
& -\oint\frac{\partial}{\partial\vartheta}\left[Y_{l,m}\mathbf{F}^{\nu}\right]\cdot\frac{Z^{\nu}}{\hbar}(\nabla_{\mathbf{k}}\vartheta^{\nu})f^{\nu}\mathrm{d}\Omega \\
& -\oint\frac{\partial}{\partial\varphi}\left[Y_{l,m}\mathbf{F}^{\nu}\right]\cdot\frac{Z^{\nu}}{\hbar}(\nabla_{\mathbf{k}}\varphi^{\nu})f^{\nu}\mathrm{d}\Omega,
\end{aligned} \tag{2.108}$$

where the first relation (2.80) was used. With (2.59) and (2.14), this yields

$$\begin{aligned}
\frac{1}{\hbar}\mathbf{F}^{\nu}\cdot\nabla_{\mathbf{k}}f^{\nu} \to & \oint\frac{\partial}{\partial\varepsilon}\left[\mathbf{F}^{\nu}\cdot Y_{l,m}\mathbf{v}^{\nu}Z^{\nu}f^{\nu}\right]\mathrm{d}\Omega \\
& -\oint\frac{\partial Y_{l,m}}{\partial\vartheta}\mathbf{F}^{\nu}\cdot\frac{Z^{\nu}}{\hbar}(\nabla_{\mathbf{k}}\vartheta^{\nu})f^{\nu}\mathrm{d}\Omega \\
& -\oint\frac{\partial Y_{l,m}}{\partial\varphi}\mathbf{F}^{\nu}\cdot\frac{Z^{\nu}}{\hbar}(\nabla_{\mathbf{k}}\varphi^{\nu})f^{\nu}\mathrm{d}\Omega \\
& +\oint Y_{l,m}(\nabla_{\mathbf{r_k}}\cdot\mathbf{v}^{\nu})Z^{\nu}f^{\nu}\mathrm{d}\Omega.
\end{aligned} \tag{2.109}$$

Note that the gradient in the above equation is taken in the original space ($\nabla_{\mathbf{r_k}}$).

The diffusion term of the free-streaming operator evaluates to

$$\mathbf{v}^{\nu}\cdot\nabla_{\mathbf{r_k}}f^{\nu} \to \oint Y_{l,m}\mathbf{v}^{\nu}Z^{\nu}\cdot\nabla_{\mathbf{r_k}}f^{\nu}\mathrm{d}\Omega, \tag{2.110}$$

where the gradient in the real space is also taken in the original space.

The total free-streaming operator can be split into two parts w.r.t. the derivatives in the $\mathbf{k}$- and real space

$$\frac{1}{\hbar}\mathbf{F}^{\nu}\cdot\nabla_{\mathbf{k}}f^{\nu}+\mathbf{v}\cdot\nabla_{\mathbf{r_k}}f^{\nu} \to D_{\mathbf{k}}+D_{\mathbf{r}}, \tag{2.111}$$

where

$$D_{\mathbf{k}} = \oint \frac{\partial}{\partial \varepsilon}\left[(\mathbf{F}^{\nu}_{\mathrm{E,pot}} + \tilde{\mathbf{F}}^{\nu}_{\mathrm{E}}) \cdot Y_{l,m}\mathbf{v}^{\nu} Z^{\nu} f^{\nu}\right] \mathrm{d}\Omega$$
$$- \oint \frac{\partial Y_{l,m}}{\partial \vartheta} \mathbf{F}^{\nu} \cdot \frac{Z^{\nu}}{\hbar} (\nabla_{\mathbf{k}} \vartheta^{\nu}) f^{\nu} \mathrm{d}\Omega$$
$$- \oint \frac{\partial Y_{l,m}}{\partial \varphi} \mathbf{F}^{\nu} \cdot \frac{Z^{\nu}}{\hbar} (\nabla_{\mathbf{k}} \varphi^{\nu}) f^{\nu} \mathrm{d}\Omega, \tag{2.112}$$

and

$$D_{\mathbf{r}} = \oint Y_{l,m} Z^{\nu} \cdot \nabla_{\mathbf{r_k}} (\mathbf{v}^{\nu} f^{\nu})\, \mathrm{d}\Omega. \tag{2.113}$$

Using the expression for the gradient in the real space (2.60), $D_{\mathbf{r}}$ can be rearranged,

$$D_{\mathbf{r}} = \nabla_{\mathbf{r}_{\varepsilon}} \cdot \oint Y_{l,m} Z^{\nu} \mathbf{v}^{\nu} f^{\nu} \mathrm{d}\Omega - \oint \nabla_{\mathbf{r}_{\varepsilon}} [Y_{l,m} Z^{\nu}] \cdot \mathbf{v}^{\nu} f^{\nu} \mathrm{d}\Omega$$
$$+ \frac{\partial}{\partial \varepsilon} \oint Y_{l,m} Z^{\nu} (\nabla_{\mathbf{r_k}} \varepsilon^{\nu}) \cdot (\mathbf{v}^{\nu} f^{\nu}) \mathrm{d}\Omega - \oint \frac{\partial}{\partial \varepsilon} [(\nabla_{\mathbf{r_k}} \varepsilon^{\nu}) Y_{l,m} Z^{\nu}] \cdot \mathbf{v}^{\nu} f^{\nu} \mathrm{d}\Omega$$
$$- \int_0^{2\pi} \int_0^{\pi} \frac{\partial}{\partial \vartheta} [(\nabla_{\mathbf{r_k}} \vartheta^{\nu}) Y_{l,m} Z^{\nu} \sin \vartheta] \cdot \mathbf{v}^{\nu} f^{\nu} \mathrm{d}\vartheta \mathrm{d}\varphi$$
$$- \int_0^{2\pi} \int_0^{\pi} \frac{\partial}{\partial \varphi} [(\nabla_{\mathbf{r_k}} \vartheta^{\nu}) Y_{l,m} Z^{\nu} \sin \vartheta] \cdot \mathbf{v}^{\nu} f^{\nu} \mathrm{d}\vartheta \mathrm{d}\varphi$$
$$= \nabla_{\mathbf{r}_{\varepsilon}} \cdot \oint Y_{l,m} Z^{\nu} \mathbf{v}^{\nu} f^{\nu} \mathrm{d}\Omega + \frac{\partial}{\partial \varepsilon} \oint Y_{l,m} Z^{\nu} (\nabla_{\mathbf{r_k}} \varepsilon^{\nu}) \cdot (\mathbf{v}^{\nu} f^{\nu}) \mathrm{d}\Omega$$
$$- \oint \frac{\partial Y_{l,m}}{\partial \vartheta} (\nabla_{\mathbf{r_k}} \vartheta^{\nu}) Z^{\nu} \cdot \mathbf{v}^{\nu} f^{\nu} \mathrm{d}\Omega - \oint \frac{\partial Y_{l,m}}{\partial \varphi} (\nabla_{\mathbf{r_k}} \vartheta^{\nu}) Z^{\nu} \cdot \mathbf{v}^{\nu} f^{\nu} \mathrm{d}\Omega, \tag{2.114}$$

where the second relation (2.94) was used.

When ϑ^{ν} and φ^{ν} are position-independent, the last two terms in the above equation vanish. Then, the projected free-streaming operator can be written as

$$D_{\mathbf{k}} + D_{\mathbf{r}} = \oint \frac{\partial}{\partial \varepsilon}\left[\mathbf{F}^{\nu}_{\mathrm{E,pot}} \cdot Y_{l,m}\mathbf{v}^{\nu} Z^{\nu} f^{\nu}\right] \mathrm{d}\Omega$$
$$- \oint \frac{\partial Y_{l,m}}{\partial \vartheta} \mathbf{F}^{\nu} \cdot \frac{Z^{\nu}}{\hbar} (\nabla_{\mathbf{k}} \vartheta^{\nu}) f^{\nu} \mathrm{d}\Omega$$
$$- \oint \frac{\partial Y_{l,m}}{\partial \varphi} \mathbf{F}^{\nu} \cdot \frac{Z^{\nu}}{\hbar} (\nabla_{\mathbf{k}} \varphi^{\nu}) f^{\nu} \mathrm{d}\Omega$$
$$+ \nabla_{\mathbf{r}_{\varepsilon}} \cdot \oint Y_{l,m} \mathbf{v}^{\nu} Z^{\nu} f^{\nu} \mathrm{d}\Omega. \tag{2.115}$$

With the coupling coefficients

$$\mathbf{A}^{\nu}_{l,m,l',m'} = \oint Y_{l,m}\mathbf{v}^{\nu} Z^{\nu} Y_{l',m'} \mathrm{d}\Omega, \tag{2.116}$$

$$B^{\nu}_{\mathrm{E},l,m,l',m'} = \oint \mathbf{F}^{\nu}_{\mathrm{E}} \cdot \left[\frac{\partial Y_{l,m}}{\partial \vartheta} \frac{Z^{\nu}}{\hbar} (\nabla_{\mathbf{k}} \vartheta^{\nu}) + \frac{\partial Y_{l,m}}{\partial \varphi} \frac{Z^{\nu}}{\hbar} (\nabla_{\mathbf{k}} \varphi^{\nu}) \right] Y_{l',m'} \mathrm{d}\Omega, \tag{2.117}$$

$$B^{\nu}_{\mathrm{B},l,m,l',m'} = \oint \mathbf{F}^{\nu}_{\mathrm{B}} \cdot \left[\frac{\partial Y_{l,m}}{\partial \vartheta} \frac{Z^{\nu}}{\hbar} (\nabla_{\mathbf{k}} \vartheta^{\nu}) + \frac{\partial Y_{l,m}}{\partial \varphi} \frac{Z^{\nu}}{\hbar} (\nabla_{\mathbf{k}} \varphi^{\nu}) \right] Y_{l',m'} \mathrm{d}\Omega, \tag{2.118}$$

$D_{\mathbf{k}} + D_{\mathbf{r}}$ finally reads

$$\begin{aligned} D_{\mathbf{k}} + D_{\mathbf{r}} = \sum_{l',m'} & \frac{\partial}{\partial \varepsilon} \left[\mathbf{F}^{\nu}_{\mathrm{E,pot}} \cdot \mathbf{A}^{\nu}_{l,m,l',m'} f^{\nu}_{l',m'} \right] + \nabla_{\mathbf{r}_{\varepsilon}} \cdot \left[\mathbf{A}^{\nu}_{l,m,l',m'} f^{\nu}_{l',m'} \right] \\ & - B^{\nu}_{\mathrm{E},l,m,l',m'} f^{\nu}_{l',m'} - B^{\nu}_{\mathrm{B},l,m,l',m'} f^{\nu}_{l',m'}. \end{aligned} \tag{2.119}$$

Actual values of the coupling coefficients $\mathbf{A}^{\nu}_{l,m,l',m'}$, $B^{\nu}_{\mathrm{E},l,m,l',m'}$, and $B^{\nu}_{\mathrm{B},l,m,l',m'}$ depend on the underlying band structure. However, from the definitions, (2.116) and (2.117), and the first and third relations, (2.80) and (2.96), we can readily find the following relations valid for the general case with a position-dependent band structure and the generalized coordinate transform:

$$\mathbf{A}^{\nu}_{l,m,l',m'} = \mathbf{A}^{\nu}_{l',m',l,m}, \tag{2.120}$$

$$\nabla_{\mathbf{r}_{\varepsilon}} \cdot \mathbf{A}^{\nu}_{l,m,l',m'} + \mathbf{F}^{\nu}_{\mathrm{E,pot}} \cdot \frac{\partial}{\partial \varepsilon} \mathbf{A}^{\nu}_{l,m,l',m'} = B^{\nu}_{\mathrm{E},l,m,l',m'} + B^{\nu}_{\mathrm{E},l',m',l,m}, \tag{2.121}$$

$$B^{\nu}_{\mathrm{E},0,0,l',m'} = B^{\nu}_{\mathrm{B},0,0,l',m'} = 0. \tag{2.122}$$

Note that the above (2.120)–(2.122) are generalized versions of (15)–(17) in [6], respectively, which were derived under the assumption of a position-independent band structure and an angle-preserving coordinate transform.

2.5.3 *Scattering Operator*

The scattering integral also has to be expanded with spherical harmonics. A constant energy transfer is considered. The transition rate can then be expressed as

$$S^{\nu,\nu'}_{\eta}(\mathbf{r}, \mathbf{k}|\mathbf{k}') = \frac{1}{\Omega_{\mathrm{s}}} c^{\nu,\nu'}_{\eta}[\mathbf{r}, \mathbf{k}|\mathbf{k}'] \delta(\varepsilon^{\nu}(\mathbf{k}) - \varepsilon^{\nu'}(\mathbf{k}') - \hbar\omega_{\eta}), \tag{2.123}$$

where $\mathbf{k}$ and $\mathbf{k}'$ are the initial and final wave vectors, respectively, $\hbar\omega_{\eta}$ the constant energy transfer, and Ω_{s} the system volume. (Emission and absorption are treated as separate processes.) The scattering integral is split into in-scattering and out-scattering terms,

$$\hat{S}\{f\} = \hat{S}^{\text{in}}\{f\} - \hat{S}^{\text{out}}\{f\}, \tag{2.124}$$

with

$$\begin{aligned}\hat{S}^{\text{in}}\{f\} &= \sum_{\eta} \hat{S}^{\text{in}}_{\eta}\{f\} \\ &= \sum_{\eta} \frac{\Omega_{\text{s}}}{(2\pi)^3} \sum_{\nu'} \int_{BZ} (1 - f^{\nu}(\mathbf{r},\mathbf{k})) S^{\nu,\nu'}_{\eta}(\mathbf{r},\mathbf{k}|\mathbf{k}') f^{\nu'}(\mathbf{r},\mathbf{k}')\mathrm{d}^3k', \end{aligned} \tag{2.125}$$

$$\begin{aligned}\hat{S}^{\text{out}}\{f\} &= \sum_{\eta} \hat{S}^{\text{out}}_{\eta}\{f\} \\ &= \sum_{\eta} \frac{\Omega_{\text{s}}}{(2\pi)^3} \sum_{\nu'} \int_{BZ} (1 - f^{\nu'}(\mathbf{r},\mathbf{k}')) S^{\nu',\nu}_{\eta}(\mathbf{r},\mathbf{k}'|\mathbf{k}) f^{\nu}(\mathbf{r},\mathbf{k})\mathrm{d}^3k'. \end{aligned} \tag{2.126}$$

Projection of the scattering term is given by

$$\hat{S}_{\eta}\{f\} \rightarrow \hat{S}^{\nu}_{\eta,l,m}(\mathbf{r},\varepsilon) = \hat{S}^{\nu,\text{in}}_{\eta,l,m}(\mathbf{r},\varepsilon) - \hat{S}^{\nu,\text{out}}_{\eta,l,m}(\mathbf{r},\varepsilon). \tag{2.127}$$

Projection of the in-scattering term yields

$$\begin{aligned}\hat{S}^{\nu,\text{in}}_{\eta,l,m}(\mathbf{r},\varepsilon) = \sum_{\nu'} \sum_{l',m'} \sum_{l'',m''} &[Z^{\nu}_{l'',m''}(\mathbf{r},\varepsilon) - g^{\nu}_{l'',m''}(\mathbf{r},\varepsilon)] g^{\nu'}_{l',m'}(\mathbf{r},\varepsilon - \hbar\omega_{\eta}) \\ &\Big\{ \oint\oint Y_{l,m}(\vartheta,\varphi) Y_{l'',m''}(\vartheta,\varphi) \\ &\times c^{\nu,\nu'}_{\eta}(\mathbf{r},\varepsilon,\vartheta,\varphi|\varepsilon - \hbar\omega_{\eta},\vartheta',\varphi') \\ &\times Y_{l',m'}(\vartheta',\varphi')\mathrm{d}\Omega'\mathrm{d}\Omega \Big\}. \end{aligned} \tag{2.128}$$

Projection of the out-scattering term yields

$$\begin{aligned}\hat{S}^{\nu,\text{out}}_{\eta,l,m}(\mathbf{r},\varepsilon) = \sum_{\nu'} \sum_{l',m'} \sum_{l'',m''} &[Z^{\nu'}_{l'',m''}(\mathbf{r},\varepsilon + \hbar\omega_{\eta}) - g^{\nu'}_{l'',m''}(\mathbf{r},\varepsilon + \hbar\omega_{\eta})] g^{\nu}_{l',m'}(\mathbf{r},\varepsilon) \\ &\Big\{ \oint\oint Y_{l,m}(\vartheta,\varphi) Y_{l'',m''}(\vartheta',\varphi') \\ &\times c^{\nu',\nu}_{\eta}(\mathbf{r},\varepsilon + \hbar\omega_{\eta},\vartheta',\varphi'|\varepsilon,\vartheta,\varphi) \\ &\times Y_{l',m'}(\vartheta,\varphi)\mathrm{d}\Omega'\mathrm{d}\Omega \Big\}. \end{aligned} \tag{2.129}$$

When the transition rate depends only on the relative angle between the initial and final wave vectors, it can be easily projected onto spherical harmonics with the addition theorem [13]

$$\begin{aligned} c_{\eta}^{\nu,\nu'}&(\mathbf{r},\varepsilon,\vartheta,\varphi|\varepsilon-\hbar\omega_{\eta},\vartheta',\varphi') \\ &= \sum_{l'''=0}^{\infty} c_{\eta,l'''}^{\nu,\nu'}(\mathbf{r},\varepsilon^{\nu}(\mathbf{k})) \sum_{m'''=-l'''}^{l'''} Y_{l''',m'''}(\vartheta,\varphi)Y_{l''',m'''}(\vartheta',\varphi') \end{aligned} \quad (2.130)$$

Projection of the in-scattering term yields

$$\begin{aligned} \hat{S}_{\eta,l,m}^{\nu,\text{in}}(\mathbf{r},\varepsilon) = \sum_{\nu'}\sum_{l',m'}\sum_{l'',m''}&[Z_{l'',m''}^{\nu}(\mathbf{r},\varepsilon) - g_{l'',m''}^{\nu}(\mathbf{r},\varepsilon)]g_{l',m'}^{\nu'}(\mathbf{r},\varepsilon-\hbar\omega_{\eta}) \\ &\times c_{\eta,l'}^{\nu,\nu'}(\mathbf{r},\varepsilon)a_{l,m,l',m',l'',m''}. \end{aligned} \quad (2.131)$$

Projection of the out-scattering term yields

$$\begin{aligned} \hat{S}_{\eta,l,m}^{\nu,\text{out}}(\mathbf{r},\varepsilon) = \sum_{\nu'}\sum_{l',m'}\sum_{l'',m''}&[Z_{l'',m''}^{\nu'}(\mathbf{r},\varepsilon+\hbar\omega_{\eta}) - g_{l'',m''}^{\nu'}(\mathbf{r},\varepsilon+\hbar\omega_{\eta})]g_{l',m'}^{\nu}(\mathbf{r},\varepsilon) \\ &\times c_{\eta,l''}^{\nu',\nu}(\mathbf{r},\varepsilon+\hbar\omega_{\eta})a_{l,m,l',m',l'',m''}. \end{aligned} \quad (2.132)$$

Significant simplification can be obtained when the transition rate does not depend on the scattering angle at all

$$c_{\eta,l}^{\nu,\nu'}(\mathbf{r},\varepsilon) = 4\pi c_{\eta}^{\nu,\nu'}(\mathbf{r},\varepsilon)\delta_{l,0}. \quad (2.133)$$

Projection of the in-scattering term yields

$$\begin{aligned} \hat{S}_{\eta,l,m}^{\nu,\text{in}}(\mathbf{r},\varepsilon) = \sum_{\nu'}&[Z_{l,m}^{\nu}(\mathbf{r},\varepsilon) - g_{l,m}^{\nu}(\mathbf{r},\varepsilon)]g_{0,0}^{\nu'}(\mathbf{r},\varepsilon-\hbar\omega_{\eta}) \\ &\times c_{\eta,0}^{\nu,\nu'}(\mathbf{r},\varepsilon)Y_{0,0}. \end{aligned} \quad (2.134)$$

Projection of the out-scattering term yields

$$\begin{aligned} \hat{S}_{\eta,l,m}^{\nu,\text{out}}(\mathbf{r},\varepsilon) = \sum_{\nu'}&[Z_{0,0}^{\nu'}(\mathbf{r},\varepsilon+\hbar\omega_{\eta}) - g_{0,0}^{\nu'}(\mathbf{r},\varepsilon+\hbar\omega_{\eta})]g_{l,m}^{\nu}(\mathbf{r},\varepsilon) \\ &\times c_{\eta,0}^{\nu',\nu}(\mathbf{r},\varepsilon+\hbar\omega_{\eta})Y_{0,0}. \end{aligned} \quad (2.135)$$

Note that this is a velocity-randomizing scattering mechanism (for example, phonon scattering with constant matrix element). Moreover, in the case of impurity scattering, an approximation by an isotropic-elastic process [7, 14] can be used.[5]

More information about the scattering mechanisms used in this work can be found in Chap. 4.

2.5.4 Boundary Condition

The interface generation rate shown in (2.15) is projected onto spherical harmonics

$$\gamma_{\mathrm{s}}^{\nu} \rightarrow \gamma_{\mathrm{s},l,m}^{\nu}(\mathbf{r},\varepsilon) = \sum_{l',m'} D_{l,m,l',m'}^{\nu} f_{l',m'}^{\nu} + E_{l,m,0,0}^{\nu} f_{0,0}^{\mathrm{eq}}, \tag{2.136}$$

with

$$D_{l,m,l',m'}^{\nu} = \oint \Theta(-\mathbf{v}^{\nu} Z^{\nu} \cdot \mathbf{n}) \mathbf{v}^{\nu} Z^{\nu} \cdot \mathbf{n} Y_{l,m} Y_{l',m'} \mathrm{d}\Omega, \tag{2.137}$$

and

$$E_{l,m,l',m'}^{\nu} = \oint \Theta(\mathbf{v}^{\nu} Z^{\nu} \cdot \mathbf{n}) \mathbf{v}^{\nu} Z^{\nu} \cdot \mathbf{n} Y_{l,m} Y_{l',m'} \mathrm{d}\Omega. \tag{2.138}$$

The interface generation rate is implemented as a volume generation rate within the box of the terminal [5].

2.5.5 Balance Equations

Collecting the previous results, the balance equation for $f_{l,m}^{\nu}$, which is the projection of the BTE onto spherical harmonics, can be formulated

$$\begin{aligned}
\sum_{l',m'} \frac{\partial}{\partial t} \left[Z_{l,m,l',m'}^{\nu} f_{l',m'}^{\nu} \right] &+ \frac{\partial}{\partial \varepsilon} \left[\mathbf{F}_{\mathrm{E}}^{\nu} \cdot \mathbf{A}_{l,m,l',m'}^{\nu} f_{l',m'}^{\nu} \right] \\
&+ \nabla_{\mathbf{r}_{\varepsilon}} \cdot \left[\mathbf{A}_{l,m,l',m'}^{\nu} f_{l',m'}^{\nu} \right] \\
&- B_{\mathrm{E},l,m,l',m'}^{\nu} f_{l',m'}^{\nu} - B_{\mathrm{B},l,m,l',m'}^{\nu} f_{l',m'}^{\nu} \\
&= \sum_{\eta} \hat{S}_{\eta,l,m}^{\nu,\mathrm{in}} - \hat{S}_{\eta,l,m}^{\nu,\mathrm{out}},
\end{aligned} \tag{2.139}$$

[5]The polar optical phonon, which is important for III–V materials, cannot be properly treated as a velocity-randomizing one. In this case, higher-order expansions of the transition rate should be considered, as shown in [15].

where $\gamma^{\nu}_{s,l,m}$ has to be considered on a contact surface.

In order to simulate realistic semiconductor devices, a stabilization scheme is required, which is discussed in Chap. 3.

2.6 Noise Analysis

In this work, only noise analysis under small-signal operation [16] is treated, and we assume that only DC signals are applied to the contacts of the semiconductor device.[6]

Realization of the stochastic process underlying the BTE yields stochastic variables which fluctuate around their expected value [16, 18, 19]. Under stationary conditions the fluctuation of a stochastic variable is defined as the difference between the variable and its expected value.[7]

The fluctuations are characterized by their spectral intensities. The cross-spectrum of two fluctuations $\delta X(t)$ and $\delta Y(t)$ is given by Fourier transformation of the corresponding correlation function [19, 20]

$$S_{XY}(\omega) = 2\int_{-\infty}^{\infty} E\{\delta X(t)\delta Y(0)\}\exp(-\mathrm{i}\omega t)\mathrm{d}t, \tag{2.140}$$

where $E\{\}$ is the expectation. Instead of evaluating the correlation function and Fourier transform it, the spectral intensity can be directly evaluated based on a stochastic differential equation, the Langevin-Boltzmann equation, which can be formulated directly in the frequency domain.

Scattering results in fluctuations of the distribution function, which in turn lead to fluctuations of the expected values. When stationary bias conditions are assumed, noise can be calculated directly in the frequency domain. The deviation of the particle distribution function at angular frequency ω is denoted as $\delta f^{\nu}(\mathbf{r},\mathbf{k})\mathrm{e}^{\mathrm{i}\omega t}$. The fluctuations are assumed to be so small that they can be calculated based on the linearization of the BTE. The linearized BTE reads in the frequency domain

$$L_{\mathrm{BTE}}(\delta f,\delta V) = \mathrm{i}\omega\delta f^{\nu} + \frac{1}{\hbar}\mathbf{F}^{\nu}_{\mathrm{s}}\cdot\nabla_{\mathbf{k}}\delta f^{\nu} + \frac{1}{\hbar}\delta\mathbf{F}^{\nu}\cdot\nabla_{\mathbf{k}} f^{\nu}_{\mathrm{s}} + \mathbf{v}^{\nu}\cdot\nabla_{\mathbf{r}}\delta f - \delta\hat{S}\{\delta f\}, \tag{2.141}$$

where i is the imaginary unit number, ω the angular frequency, $\mathbf{F}^{\nu}_{\mathrm{s}}$ the steady-state solution of the total force, δf^{ν} a fluctuation of the distribution function, $\delta\mathbf{F}^{\nu}$ a fluctuation of the total force defined in (2.9)

$$\delta\mathbf{F}^{\nu} = -\nabla_{\mathbf{r}}(\mp q\delta V), \tag{2.142}$$

[6]Noise analysis under large-signal conditions has been demonstrated for bulk simulations within the framework of the deterministic BTE solver in [17].

[7]From this definition, the expectation of a fluctuation is always zero.

$\delta\hat{S}\{\delta f\}$ the linearized scattering integral

$$\delta\hat{S}\{\delta f\} = \frac{\Omega_s}{(2\pi)^3}\sum_{\nu'}\int_{BZ}(1 - f_s^{\nu}(\mathbf{r},\mathbf{k}))S^{\nu,\nu'}(\mathbf{r},\mathbf{k}|\mathbf{k}')\delta f^{\nu'}(\mathbf{r},\mathbf{k}',\omega)$$
$$-\,\delta f^{\nu}(\mathbf{r},\mathbf{k},\omega)S^{\nu,\nu'}(\mathbf{r},\mathbf{k}|\mathbf{k}')f_s^{\nu'}(\mathbf{r},\mathbf{k}')$$
$$-\,(1 - f_s^{\nu'}(\mathbf{r},\mathbf{k}')S^{\nu',\nu}(\mathbf{r},\mathbf{k}'|\mathbf{k})\delta f^{\nu}(\mathbf{r},\mathbf{k},\omega)$$
$$+\,\delta f^{\nu'}(\mathbf{r},\mathbf{k}',\omega)S^{\nu',\nu}(\mathbf{r},\mathbf{k}'|\mathbf{k})f_s^{\nu}(\mathbf{r},\mathbf{k})\,\mathrm{d}^3k'\,. \tag{2.143}$$

f_s^{ν} is the noiseless solution of the stationary BTE. When the transition rate depends on the particle distribution function or the electrostatic potential, such a dependency should be also considered.

For the calculation of noise, Green's functions are required, which are the solution of the linearized equations for a single particle source term [19]. When the electron is generated in the conduction band, the corresponding Green's functions are defined by

$$L_{\mathrm{BTE}}(G_f, G_V) = (2\pi)^3\delta(\mathbf{r}-\mathbf{r}')\delta(\mathbf{k}-\mathbf{k}')\delta_{\nu,\nu'}. \tag{2.144}$$

$G_f^{\nu,\nu'}(\mathbf{r},\mathbf{r}',\mathbf{k},\mathbf{k}',\omega)$ is the linear response in the frequency domain of the particle distribution function to the generation of a single particle of arbitrary spin.[8] $G_V^{\nu'}(\mathbf{r},\mathbf{r}',\mathbf{k}',\omega)$ is the response of the potential.

With the Green's functions the response of an expectation to the generation of a particle can be calculated. The response of an expectation to the generation of a particle is given by

$$G_X^{\nu'}(\mathbf{r},\mathbf{r}',\mathbf{k}',\omega) = \frac{1}{(2\pi)^3}\sum_{\nu}\int_{BZ} X^{\nu}(\mathbf{r},\mathbf{k})G_f^{\nu,\nu'}(\mathbf{r},\mathbf{r}',\mathbf{k},\mathbf{k}',\omega)\,\mathrm{d}^3k. \tag{2.145}$$

For example, the current response of the kth terminal to the source term is given by the integral of the total current response over the contact area (∂G_k)

$$G_{I_k}^{\nu'}(\mathbf{r}',\mathbf{k}',\omega) = \int_{\partial G_k}[\mp q G_{\mathbf{j}}^{\nu'}(\mathbf{r},\mathbf{r}',\mathbf{k}',\omega) - \mathrm{i}\omega\kappa\nabla_{\mathbf{r}}G_V^{\nu'}(\mathbf{r},\mathbf{r}',\mathbf{k}',\omega)]\mathrm{d}\mathbf{A}, \tag{2.146}$$

where $G_{\mathbf{j}}^{\nu'}$ is the response of the particle flux and the total current response is the sum of the particle and displacement current responses [22]. Only a single carrier type is assumed. If electrons and holes are presented, particle currents for both carrier types have to be included.

[8] A formulation which keeps the spin variable can be found in [21].

Noise is quantified by the corresponding power spectral densities (PSD) of the fluctuations [20]. The fundamental PSD is the one of the distribution function, with which the PSDs of all expectations of the distribution function can be calculated [19,23]

$$S_{ff}^{\nu_1,\nu_2}(\mathbf{r}_1,\mathbf{r}_2,\mathbf{k}_1,\mathbf{k}_2,\omega) = \int K_{ff}^{\nu_1,\nu_2}(\mathbf{r}_1,\mathbf{r}_2,\mathbf{r},\mathbf{k}_1,\mathbf{k}_2,\omega)\mathrm{d}^3r, \tag{2.147}$$

with the local noise contribution

$$\begin{aligned}
&K_{ff}^{\nu_1,\nu_2}(\mathbf{r}_1,\mathbf{r}_2,\mathbf{r},\mathbf{k}_1,\mathbf{k}_2,\omega)\\
&\quad= \frac{4\Omega_\mathrm{s}}{(2\pi)^3}\sum_{\nu}\sum_{\nu'}\int_{BZ}\int_{BZ}\left[G_f^{\nu_1,\nu}(\mathbf{r}_1,\mathbf{r},\mathbf{k}_1,\mathbf{k},\omega) - G_f^{\nu_1,\nu'}(\mathbf{r}_1,\mathbf{r},\mathbf{k}_1,\mathbf{k}',\omega)\right]\\
&\quad\times\left[G_f^{\nu_2,\nu}(\mathbf{r}_2,\mathbf{r},\mathbf{k}_2,\mathbf{k},\omega) - G_f^{\nu_2,\nu'}(\mathbf{r}_2,\mathbf{r},\mathbf{k}_2,\mathbf{k}',\omega)\right]^*\\
&\quad\times\left[1 - f_\mathrm{s}^{\nu}(\mathbf{r},\mathbf{k})\right]S^{\nu,\nu'}(\mathbf{r},\mathbf{k}|\mathbf{k}')f_\mathrm{s}^{\nu'}(\mathbf{r},\mathbf{k}')\mathrm{d}^3k\mathrm{d}^3k'.
\end{aligned} \tag{2.148}$$

The factor of four stems from spin degeneracy and the fact that particle scattering is a Poisson process. Its single-sided white PSD is given by two times the transition rate probability [20]. The PSD is white, because scattering is treated as instantaneous in the framework of the BTE [4]. For the same reason scattering is local in the real space. The scattering process can be viewed as the annihilation of a particle in the initial state and generation of a particle in the final state. The response of the distribution function to the generation of a particle is described by the corresponding Green's function. Annihilation simply requires a negative sign. Thus, the response of the distribution function at $(\mathbf{r},\mathbf{k})$ is given by the difference of the Green's functions for the initial and final states. Correlation with a fluctuation for the state 2, integration over all initial and final wave vectors and summation over all initial and final bands yields (2.148).

The PSD for expectations of two microscopic variables X and Y can be calculated

$$\begin{aligned}
S_{XY}(\mathbf{r}_1,\mathbf{r}_2,\omega) = \frac{1}{(2\pi)^6}\sum_{\nu_1}\sum_{\nu_2}\int_{BZ}\int_{BZ} &X^{\nu_1}(\mathbf{r}_1,\mathbf{k}_1)\\
&\times S_{ff}^{\nu_1,\nu_2}(\mathbf{r}_1,\mathbf{r}_2,\mathbf{k}_1,\mathbf{k}_2,\omega)Y^{\nu_2}(\mathbf{r}_2,\mathbf{k}_2)\,\mathrm{d}^3k_1\mathrm{d}^3k_2.
\end{aligned} \tag{2.149}$$

For device simulations, in addition to the linearized BTE, the linearized PE should be considered in order to obtain the fluctuations of electrostatic potential,

$$L_{\mathrm{PE}}(\delta f,\delta V) = \nabla_{\mathbf{k}}\cdot(\kappa\nabla_{\mathbf{k}}\delta V) + q(-\delta n + \delta p) = 0, \tag{2.150}$$

where no inhomogeneous term for the Poisson equation is imposed. Expressions for δn and δp can be easily obtained with the corresponding Green's functions.

Although evaluation of (2.149) is straightforward in principle, it is extremely CPU intensive. Therefore, (2.147)–(2.149) are replaced by

$$S_{XY}(\mathbf{r}_1, \mathbf{r}_2, \omega) = \int K_{XY}(\mathbf{r}_1, \mathbf{r}_2, \mathbf{r}, \omega) \mathrm{d}^3 r, \tag{2.151}$$

with the local noise contribution

$$\begin{aligned} K_{XY}(\mathbf{r}_1, \mathbf{r}_2, \mathbf{r}, \omega) = \frac{4\Omega_\mathrm{s}}{(2\pi)^6} \sum_\nu \sum_{\nu'} \int \int \left[G_X^\nu(\mathbf{r}_1, \mathbf{r}, \mathbf{k}, \omega) - G_X^{\nu'}(\mathbf{r}_1, \mathbf{r}, \mathbf{k}', \omega) \right] \\ \times \left[G_Y^\nu(\mathbf{r}_1, \mathbf{r}, \mathbf{k}, \omega) - G_Y^{\nu'}(\mathbf{r}_1, \mathbf{r}, \mathbf{k}', \omega) \right]^* \\ \times \left[1 - f^\nu(\mathbf{r}, \mathbf{k})\right] S^{\nu,\nu'}(\mathbf{r}, \mathbf{k}|\mathbf{k}') f^{\nu'} \mathrm{d}^3 k \mathrm{d}^3 k' \end{aligned} \tag{2.152}$$

The Green's function of the expectation for the microscopic quantity X is expanded with spherical harmonics

$$G_{X,l,m}^\nu(\mathbf{r}, \mathbf{r}', \varepsilon', \omega) = \oint G_X^\nu(\mathbf{r}, \mathbf{r}', \mathbf{k}(\varepsilon', \vartheta', \varphi'), \omega) Y_{l,m}(\vartheta', \varphi') \mathrm{d}\Omega'. \tag{2.153}$$

The Green's function $G_{X,l,m}$ can be directly calculated with the generalized adjoint method [24]. This is standard procedure in numerical noise calculations [25]. Due to the stabilization scheme required for reliable device simulation, the actual construction of L_{BTE} needs more efforts. This point is explained in detail in Chap. 3.

The spherical harmonics expansion of the local contribution to the noise yields an expression related with an integral over the product of four spherical harmonics [26]. Again, significant simplification is obtained when the transition rate does not depend on the scattering angle. In this case, we obtain

$$\begin{aligned} K_{XY}(\mathbf{r}_1, \mathbf{r}_2, \mathbf{r}, \omega) = \sum_\nu \sum_{\nu'} \sum_\eta \sum_{l_1,m_1} \sum_{l_2,m_2} \sum_{l_3,m_3} \int_0^\infty G_{X,l_1,m_1}^\nu G_{Y,l_2,m_2}^{\nu *} a_{l_1,m_1,l_2,m_2,l_3,m_3} \\ \times (Z_{l_3,m_3}(\mathbf{r}, \varepsilon) - g_{l_3,m_3}(\mathbf{r}, \varepsilon)) g_{0,0}(\mathbf{r}, \varepsilon - \hbar\omega_\eta) \\ - G_{X,l_1,m_1}^\nu G_{Y,l_2,m_2}^{\nu' *} \delta_{l_1,l_3} \delta_{m_1,m_3} Y_{0,0} \\ \times (Z_{l_1,m_1}(\mathbf{r}, \varepsilon) - g_{l_1,m_1}(\mathbf{r}, \varepsilon)) g_{l_2,m_2}(\mathbf{r}, \varepsilon - \hbar\omega_\eta) \\ - G_{X,l_1,m_1}^{\nu'} G_{Y,l_2,m_2}^{\nu *} \delta_{l_2,l_3} \delta_{m_2,m_3} Y_{0,0} \\ \times (Z_{l_2,m_2}(\mathbf{r}, \varepsilon) - g_{l_2,m_2}(\mathbf{r}, \varepsilon)) g_{l_1,m_1}(\mathbf{r}, \varepsilon - \hbar\omega_\eta) \\ + G_{X,l_1,m_1}^{\nu'} G_{Y,l_2,m_2}^{\nu' *} a_{l_1,m_1,l_2,m_2,l_3,m_3} \\ \times (Z_{0,0}(\mathbf{r}, \varepsilon) - g_{0,0}(\mathbf{r}, \varepsilon)) g_{l_3,m_3}(\mathbf{r}, \varepsilon - \hbar\omega_\eta) \\ Y_{0,0} c_{\eta,0}[\mathbf{r}, \varepsilon] \mathrm{d}\varepsilon. \end{aligned} \tag{2.154}$$

References

1. Madelung, O.: Introduction to Solid State Theory. Springer, Berlin (1978)
2. van Kampen, N.G.: Stochastic Process in Physics and Chemistry. North-Holland, Amsterdam (1981)
3. Price, P.J.: Monte Carlo calculation of electron transport in solids. Semiconduct. Semimet. **14**, 249–309 (1979)
4. Jacoboni, C., Lugli, P.: The Monte Carlo Method for Semiconductor Device Simulation. Springer, New York (1989)
5. Hong, S.-M., Jungemann, C.: A fully coupled scheme for a Boltzmann-Poisson equation solver based on a spherical harmonics expansion. J. Comput. Electron **8**(3), 225–241 (2009)
6. Hong, S.-M., Matz, G., Jungemann, C.: A deterministic Boltzmann equation solver based on a higher-order spherical harmonics expansion with full-band effects. IEEE Trans. Electron Dev. **57**, 2390–2397 (2010)
7. Jungemann, C., Meinerzhagen, B.: Hierarchical Device Simulation: The Monte-Carlo Perspective, Computational Microelectronics. Springer, New York (2003)
8. Selberherr, S.: Analysis and Simulation of Semiconductor Devices. Springer, New York (1984)
9. Banoo, K., Lundstrom, M.: Direct solution of the Boltzmann transport equation in nanoscale Si devices. In: Proceedings of SISPAD, pp. 50–53 (2000)
10. Lisle, I.G., Huang, S.-L.T.: Algorithms for spherical harmonic lighting. In: GRAPHITE '07: Proceedings of the 5th international conference on Computer graphics and interactive techniques in Australia and Southeast Asia, pp. 235–238. ACM, New York (2007)
11. Jungemann, C., Pham, A.-T., Meinerzhagen, B., Ringhofer, C., Bollhöfer, M.: Stable discretization of the Boltzmann equation based on spherical harmonics, box integration, and a maximum entropy dissipation principle. J. Appl. Phys. **100**, 024502–1–13 (2006)
12. Gnudi, A., Ventura, D., Baccarani, G., Odeh, F.: Two-dimensional MOSFET simulation by means of a multidimensional spherical harmonics expansion of the Boltzmann transport equation. Solid State Electron **36**(4), 575–581 (1993)
13. Liang, W., Goldsman, N., Mayergoyz, I., Oldiges, P.J.: 2- D MOSFET modeling including surface effects and impact ionization by self-consistent solution of the Boltzmann, Poisson, and hole-continuity equations. IEEE Trans. Electron Dev. **44**(2), 257–267 (1997)
14. Kosina, H.: A method to reduce small-angle scattering in Monte Carlo device analysis. IEEE Trans. Electron Dev. **46**(6), 1196–1200 (1999)
15. Bieder, J., Hong, S.-M., Jungemann, C.: A deterministic Boltzmann solver for GaAs devices based on the spherical harmonics expansion. In: Proceedings of IWCE International Workshop on Computational Electronics, pp. 31–34 (2010)
16. Bonani, F., Ghione, G.: Noise in Semiconductor Devices, Modeling and Simulation, Advanced Microelectronics. Springer, Heidelberg (2001)
17. Jungemann, C., Meinerzhagen, B.: A frequency domain spherical harmonics solver for the Langevin Boltzmann equation. In: International Conference on Noise in Physical Systems and 1/f Fluctuations, AIP Conference Proceedings vol. 780, pp. 777–782 (2005)
18. Gardiner, C.W.: Handbook of Stochastic Methods. Springer, Berlin (1985)
19. Kogan, Sh.: Electronic Noise and Fluctuations in Solids. Cambridge University Press, Cambridge (1996)
20. Papoulis, A.: Probability, Random Variables, and Stochastic Processes, 3rd edn. McGraw–Hill, NY (1991)
21. Jungemann, C.: A deterministic solver for the Langevin Boltzmann equation including the Pauli principle. In: SPIE: Fluctuations and Noise, vol. 6600, pp. 660007–1–660007–12 (2007)
22. Jackson, J.D.: Classical Electrodynamics, 2nd edn. Wiley, New York (1975)
23. Gantsevich, S.V., Gurevich, V.L., Katilius, R.: Theory of fluctuations in nonequilibrium electron gas. Nuovo Cimento **2**, 5 (1979)
24. Branin, F.H.: Network sensitivity and noise analysis simplified. IEEE Trans. Circuit Theor. **20**, 285–288 (1973)

25. Bonani, F., Ghione, G., Pinto, M.R., Smith, R.K.: An efficient approach to noise analysis through multidimentional physics-based models. IEEE Trans. Electron Dev. **45**(1), 261–269 (1998)
26. Jungemann, C.: Transport and noise calculations for nanoscale Si devices based on the Langevin Boltzmann equation expanded with spherical harmonics. J. Comput. Theor. Nanosci. **5**(6), 1152–1169 (2008)

Chapter 3
Device Simulation

In this chapter, various numerical techniques for stable device simulations are discussed in detail.

The H-transformation and the maximum entropy dissipation scheme, which are two key ingredients for a stable higher-order SHE simulation, are expounded. Moreover, in the special case of the lowest order expansion, it is explicitly shown that the Jacobian matrix of the resultant set of equations is a non-singular M-matrix. Therefore, the non-negativeness of the solution (the isotropic component of the particle distribution function) is guaranteed. Dimensional splitting and box integration are applied. Discretization of the boundary conditions is also discussed. At the end of this chapter, stabilization of the linearized equation system for the small-signal analysis is presented.

In this chapter, the valley index ν is suppressed.

3.1 H-Transformation

As shown in Sect. 2.5, the part of the free-streaming operator due to the magnetic force, $B_{\mathrm{B},l,m,l',m'}$ in (2.118), does not contain partial derivatives. Therefore it can be easily treated as an operator local in both real space and energy. The scattering integral $\hat{S}_{l,m}$ is also local in real space. Since the stabilization is mainly concerned about the coupling between the spatial derivative and derivatives along other variables, the terms local in real space can be treated easily. Therefore, in the following discussion, we concentrate on the part of the free steaming-operator, which is non-local in real space. Treatment of the scattering integral and the boundary condition will be discussed in Sects. 3.3 and 3.4, respectively.

If the energy dependence of the distribution function is resolved by an expansion with polynomials as in [1], the free-streaming operator includes only the partial derivative w.r.t. position. In this case we have only one partial derivative in the free-streaming operator, and a stabilization scheme can be applied easily. However,

S.-M. Hong et al., *Deterministic Solvers for the Boltzmann Transport Equation*, Computational Microelectronics, DOI 10.1007/978-3-7091-0778-2_3,

since the energy is discretized with a grid in this work, we have two distinct partial derivatives w.r.t. position and energy. This can cause numerical problems. To overcome these difficulties, we use the H-transformation [2]. We introduce a variable transformation from $(x_\varepsilon, y_\varepsilon, z_\varepsilon, \varepsilon)$ to $(x_{\mathrm{H}}, y_{\mathrm{H}}, z_{\mathrm{H}}, H)$:

$$x_{\mathrm{H}} = x_\varepsilon, \tag{3.1}$$

$$y_{\mathrm{H}} = y_\varepsilon, \tag{3.2}$$

$$z_{\mathrm{H}} = z_\varepsilon, \tag{3.3}$$

$$H = \varepsilon \mp q\Psi(\mathbf{r}_\varepsilon), \tag{3.4}$$

where $\Psi(\mathbf{r}_\varepsilon)$ can be an arbitrary function of $\mathbf{r}_\varepsilon$ and the upper (lower) sign is for electrons (holes). The partial derivative w.r.t. ε is given by

$$\frac{\partial}{\partial \varepsilon} = \frac{\partial}{\partial H}, \tag{3.5}$$

while the del operator in the real space is given by

$$\nabla_{\mathbf{r}_\varepsilon} = \nabla_{\mathbf{r}_{\mathrm{H}}} \mp q \left(\nabla_{\mathbf{r}_\varepsilon} \Psi\right) \frac{\partial}{\partial H}. \tag{3.6}$$

Neglecting the sum over the indices l', m' and the magnetic field in the free-streaming operator (2.119), we obtain a linear operator of the form

$$\begin{aligned} L_{l,m,l',m'} f_{l',m'}(\mathbf{r}_\varepsilon, \varepsilon) = {} & \nabla_{\mathbf{r}_\varepsilon} \cdot (\mathbf{A}_{l,m,l',m'} f_{l',m'}) \\ & + \mathbf{F}_{\mathrm{E,pot}} \cdot \frac{\partial}{\partial \varepsilon} (\mathbf{A}_{l,m,l',m'} f_{l',m'}) \\ & - B_{\mathrm{E},l,m,l',m'} f_{l',m'}. \end{aligned} \tag{3.7}$$

According to the H-transformation, the free-streaming operator is transformed as follows:

$$\begin{aligned} L_{l,m,l',m'} f_{l',m'}(\mathbf{r}_{\mathrm{H}}, H) = {} & \nabla_{\mathbf{r}_{\mathrm{H}}} \cdot (\mathbf{A}_{l,m,l',m'} f_{l',m'}) \\ & + (\mathbf{F}_{\mathrm{E,pot}} \mp q(\nabla_{\mathbf{r}_\varepsilon} \Psi)) \cdot \frac{\partial}{\partial H} (\mathbf{A}_{l,m,l',m'} f_{l',m'}) \\ & - B_{\mathrm{E},l,m,l',m'} f_{l',m'}. \end{aligned} \tag{3.8}$$

Usually, H is chosen in such a way that the force in front of the derivative w.r.t. H cancels in this expression. Therefore, $\Psi(\mathbf{r}) = V(\mathbf{r}) - E^\nu(\mathbf{r})/q$ is imposed.[1] In this

[1]This definition is used throughout the book besides Sect. 3.6.

case, from the definition of $\mathbf{F}_{\text{E,pot}}$ shown in (2.10), we get

$$\mathbf{F}_{\text{E,pot}} \mp q(\nabla_{\mathbf{r}_\varepsilon} \Psi) = \mathbf{0}, \tag{3.9}$$

and $L_{l,m,l',m'}$ reduces to

$$L_{l,m,l',m'} f_{l',m'} = \nabla_{\mathbf{r}_\text{H}} \cdot (\mathbf{A}_{l,m,l',m'} f_{l',m'}) - B_{\text{E},l,m,l',m'} f_{l',m'}. \tag{3.10}$$

With this specific choice of the H function, the free-streaming operator contains only differentials w.r.t. the position ($\nabla_{\mathbf{r}_\text{H}}$). One obvious strength of the H-transformation over the discretization scheme based on the kinetic energy [3] is that the free-streaming operator can be treated correctly even in the ballistic limit [4]. Of course, this advantage comes at the cost of a potential-dependent energy grid.

Since the set of balance equations in (2.139) for a given H constitutes a system of hyperbolic conservation laws, in principle, any numerical method suitable for hyperbolic systems would be applicable [1]. When the spherical harmonics expansion is restricted to the first order [2], a specific second-order differential equation for the zeroth-order component can be found. Its discretization can be done easily because a self-adjoint differential operator is involved [5]. However, higher-order expansion is more difficult, and there have been considerable research efforts. For example, in [6] an upwind discretization is applied to the kinetic-energy-based formula. In [3, 7], the maximum entropy dissipation scheme [1] is directly applied to the kinetic-energy-based formulation. This leads to an artificial broadening of the particle distribution in the ballistic limit. In this work, the maximum entropy dissipation scheme is applied together with the H-transformation in order to avoid this problem and to improve the stability of the resultant equations [8].

3.2 Maximum Entropy Dissipation Scheme

The even part of the particle distribution function yields densities like the particle density or the energy density. In this sense the even components of the particle distribution function can be viewed as densities. Since the odd part yields the current density, the odd components can be viewed as fluxes, for which the balance equations have to be stabilized [1]. In the following, the free-streaming operator which couples $f_{l',m'}$ with an odd l' into the balance equation (2.139) for an even l is labeled "even", while the opposite one is the "odd" operator.[2]

If the entropy, which is defined as a convex function of the electron distribution function, is dissipated by the Boltzmann equation, it guarantees that the Boltzmann equation is well posed [1]. The name "maximum entropy dissipation scheme"

---

[2]The free-streaming operator due to the magnetic force does not have this property of even-odd coupling as shown in [8].

implies that the semi-discretized system is built in such a way that it preserves this property for a specific form of the entropy function. With the choice of the simplest possible entropy function, $\frac{1}{2}\exp\left(\frac{H}{k_B T_0}\right) f^2$ [1], the free streaming operator is a skew self adjoint operator under the weight function $\exp\left(\frac{H}{k_B T_0}\right)$. $k_B T_0$ is the thermal energy evaluated at the lattice temperature T_0. Therefore, in the maximum entropy dissipation scheme, the odd operator is expressed as the negative adjoint form of the corresponding even operator with the weighting factor.

When we apply this scheme to the kinetic-energy-based formula directly [3, 7], this weighting factor plays an important role, because it is a function of position. However, in the case of the total-energy-based formula, H is an independent variable. Therefore, the evaluation of the negative adjoint operator with the weighting factor becomes much simpler. Following this scheme, a different operator, which is equivalent to (3.8), is obtained

$$\begin{aligned} L_{l,m,l',m'} f_{l',m'}(\mathbf{r}_\mathrm{H}, H) = {} & \mathbf{A}_{l',m',l,m} \cdot \nabla_{\mathbf{r}_\mathrm{H}} f_{l',m'} \\ & + (\mathbf{F}_\mathrm{E,pot} \mp q(\nabla_{\mathbf{r}_\varepsilon} \Psi)) \cdot \mathbf{A}_{l',m',l,m} \frac{\partial}{\partial H} f_{l',m'} \\ & + B_{\mathrm{E},l',m',l,m} f_{l',m'}. \end{aligned} \tag{3.11}$$

Note that the index pairs (l, m) and (l', m') are interchanged in the coefficients ($\mathbf{A}$ and B_E) compared to the original form (3.8). In addition, the sign of the B_E term has changed.

Since (3.11) plays a central role in the stabilization scheme, its equivalence with the original form (3.8) should be verified. This procedure reveals that the stabilization scheme has been constructed based on an implicit use of identities shown in Chap. 2. Starting from (3.8), the equivalence of two operators (3.8) and (3.11) can be shown explicitly. The key observation is that the quantity $\nabla_{\mathbf{r}_\mathrm{H}} \cdot \mathbf{A}_{l,m,l',m'}$ can be rewritten as

$$\nabla_{\mathbf{r}_\mathrm{H}} \cdot \mathbf{A}_{l,m,l',m'} = \nabla_{\mathbf{r}_\varepsilon} \cdot \mathbf{A}_{l,m,l',m'} \pm q \nabla_{\mathbf{r}_\varepsilon} \Psi \cdot \frac{\partial}{\partial H} \mathbf{A}_{l,m,l',m'}, \tag{3.12}$$

with (3.6). Using (2.121) and (3.5),

$$\begin{aligned} \nabla_{\mathbf{r}_\mathrm{H}} \cdot \mathbf{A}_{l,m,l',m'} = {} & B_{\mathrm{E},l,m,l',m'} + B_{\mathrm{E},l',m',l,m} \\ & - (\mathbf{F}_\mathrm{E,pot} \mp q(\nabla_{\mathbf{r}_\varepsilon} \Psi)) \cdot \frac{\partial}{\partial H} \mathbf{A}_{l,m,l',m'}. \end{aligned} \tag{3.13}$$

By inserting the above relation to (3.8) and observing the symmetry of $\mathbf{A}_{l,m,l',m'}$ shown in (2.120), we obtain the expression shown in (3.11).

For the even operator, the original form of (3.8) is used, while the alternative form of (3.11) is used for the odd operator.

At equilibrium, the distribution function depends on neither the real space nor the angle variables,

$$f^{\mathrm{eq}}_{l,m}(\mathbf{r}_{\mathrm{H}}, H) = f^{\mathrm{eq}}_{0,0}(H)\delta_{l,0}\delta_{m,0}, \tag{3.14}$$

where the particular form of $f^{\mathrm{eq}}_{0,0}(H)$ depends on the statistics [9, 10]. Since $f_{l,m}$ with an odd l value identically vanishes at equilibrium, only the odd free-streaming operator is relevant. The derivatives w.r.t. H and position vanish as does the B_{E} coefficient because of (2.122) and the odd operator (3.11) is exactly zero at equilibrium. Therefore, at equilibrium, the free-streaming operator does not contribute to the balance equations at all, and the equilibrium distribution is solely determined by the scattering integral. If the scattering integral is formulated appropriately in order to satisfy the detailed balance relation, the correct equilibrium distribution is ensured. This is a desirable numerical property.

When the BTE is expanded only up to the first order (the lowest order expansion), we obtain the following simplified expressions due to (2.122)

$$L_{0,0,1,m'} f_{1,m'}(\mathbf{r}_{\mathrm{H}}, H) = \nabla_{\mathbf{r}_{\mathrm{H}}} \cdot [\mathbf{A}_{0,0,1,m'} f_{1,m'}(\mathbf{r}_{\mathrm{H}}, H)], \tag{3.15}$$

$$L_{1,m',0,0} f_{0,0}(\mathbf{r}_{\mathrm{H}}, H) = \mathbf{A}_{0,0,1,m'} \cdot \nabla_{\mathbf{r}_{\mathrm{H}}} f_{0,0}(\mathbf{r}_{\mathrm{H}}, H). \tag{3.16}$$

Let us have a look at $L_{0,0,1,m'} L_{1,m',0,0} f_{0,0}(\mathbf{r}_{\mathrm{H}}, H)$, which is obtained, when the first order equation is inserted into the zeroth order equation. Besides the inverse scattering operator for $f_{1,m'}(\mathbf{r}_{\mathrm{H}}, H)$, it corresponds to the free-streaming operator for $f_{0,0}(\mathbf{r}_{\mathrm{H}}, H)$. Using the above expressions for $L_{1,m',0,0}$ and $L_{0,0,1,m'}$ shown in (3.15) and (3.16), we get

$$L_{0,0,1,m'} L_{1,m',0,0} f_{0,0} = \sum_{d} \sum_{d'} \frac{\partial}{\partial d_{\mathrm{H}}} \left[A_{0,0,1,m',d} A_{0,0,1,m',d'} \frac{\partial}{\partial d'_{\mathrm{H}}} f_{0,0} \right], \tag{3.17}$$

where d and d' run over x, y, z. If the property $A_{0,0,1,m',d} A_{0,0,1,m',d'} = 0$ holds when $d \neq d'$,[3] it becomes

$$L_{0,0,1,m'} L_{1,m',0,0} f_{0,0} = \sum_{d} \frac{\partial}{\partial d_{\mathrm{H}}} \left[A^2_{0,0,1,m',d} \frac{\partial}{\partial d_{\mathrm{H}}} f_{0,0} \right]. \tag{3.18}$$

Note that the coefficient $A^2_{0,0,1,m',d}$ is obviously non-negative for every d and m'. This ensures that the contribution of the free-streaming operator to the partial differential equation for $f_{0,0}(\mathbf{r}_{\mathrm{H}}, H)$ satisfies the maximum/minimum principle [12]. If this equation is properly discretized, the free-streaming operator leads to a

[3] For example, this property holds for isotropic band models such as the Modena model [11], which has been extensively discussed in [8]. Moreover, the dimensional splitting used in the odd free-streaming operator effectively avoids the coupling terms between different directions.

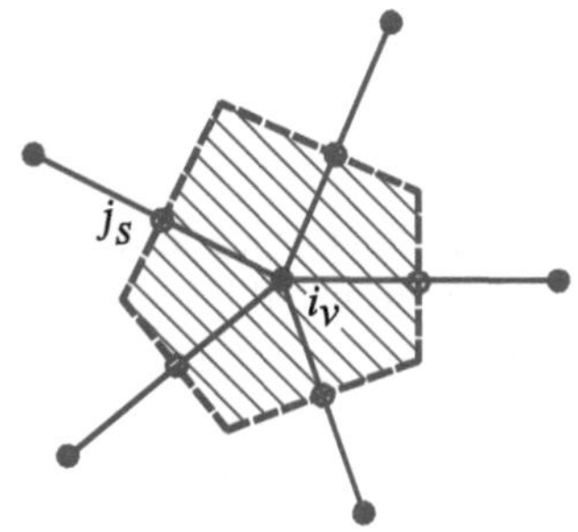

Fig. 3.1 Direct nodes (*filled circles*) and adjoint nodes (*open circles*)

contribution to the corresponding matrix, which has property M and is diagonally dominant [13].[4]

The balance equation after the stabilization, which is a partial differential equation over the real space, is discretized by the box integration method [14]. It is integrated over the control volume in the energy/real space. For box integration a special grid for the real space is introduced. Figure 3.1 shows a part of the two-dimensional real space grid. The particle distribution function $f_{l,m}$ with even l is defined on the direct grid nodes (filled circles). The box G^{i_v} for the direct grid node i_v contains all points which are closest to the grid node i_v. The position of the center direct grid node i_v is denoted as $\mathbf{r}_i$. For the particle distribution function with odd l, adjoint nodes are assigned, which are given by the nodes on the midpoint between two neighboring direct nodes. In Fig. 3.1, the adjoint nodes are denoted as opened circles and labeled as j_s. Dimensional splitting along the lines between two neighboring direct nodes (solid lines) is employed for the odd operator.

In the following, a Cartesian tensor grid for the two-dimensional real space is assumed for the sake of brevity. However, it is stressed that its extension to unstructured grids and/or the three-dimensional real space is straightforward. By an integer subscript we denote the direct grid points. The adjoint point between two adjacent direct grid points is denoted by the subscript whose value is the average of the two subscripts. For example, $x_{i+0.5}$ denotes the adjoint point between x_i and x_{i+1}.

In the case of the energy space, a uniform grid is employed. Since the uniform energy grid is shared by all valleys, intervalley scattering can be treated easily even if the valley minimum is shifted from the conduction band edge due to the position-dependent Ge content [15]. The total energy range for a given nth energy point, whose discretized H value is H_n, is denoted by (H_n^-, H_n^+). If the valley minimum is located within this range, the integral over the energy space is valid only from that value, not from H_n^-. For a given valley, the minimum value of the valid total energy for the given nth energy range is denoted by $H_n^{\min}(\mathbf{r}_{\mathrm{H}})$. Then $H_n^{\min}(\mathbf{r}_{\mathrm{H}})$ can

[4]When the absolute value of the diagonal component is larger than the absolute sum of the off-diagonal components of this matrix, the diagonal dominance strictly holds.

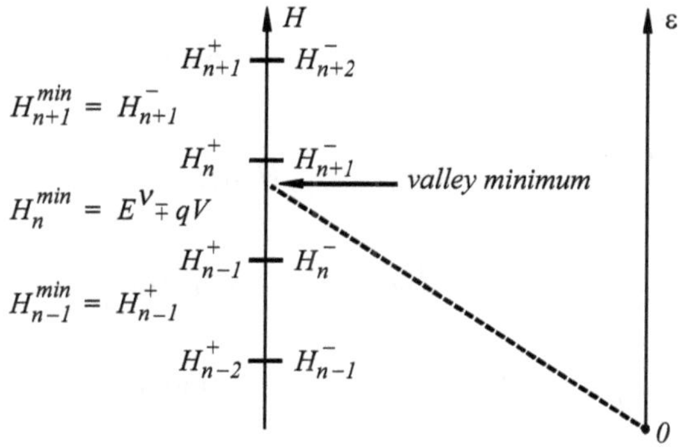

Fig. 3.2 Determination of $H_n^{\min}(\mathbf{r}_\mathrm{H})$. Three different cases for H_{n-1}, H_n, and H_{n+1} are shown, when the valley minimum is located on the range of (H_n^-, H_n^+)

be obtained by

$$H_n^{\min}(\mathbf{r}_\mathrm{H}) = \min\{H_n^+, \max\{H_n^-, \pm E^\nu(\mathbf{r}_\mathrm{H}) \mp qV(\mathbf{r}_\mathrm{H})\}\}, \tag{3.19}$$

where again the upper (lower) sign is for electrons (holes). Note that $H_n^{\min}(\mathbf{r}_\mathrm{H})$ may depend on the electrostatic potential $V(\mathbf{r}_\mathrm{H})$. The determination of $H_n^{\min}(\mathbf{r}_\mathrm{H})$ is schematically shown in Fig. 3.2.

In the following two subsections, l is even and l' odd.

3.2.1 *Odd Free-Streaming Operator*

First, the discretization of the odd free-streaming operator is shown. Since the dimensional splitting is employed for the odd operator, we may consider only one spatial derivative along the direction of the line on which the adjoint node is located, for example x. The y-directional position is fixed as y_j. Then the odd free-streaming operator defined on the line is given by

$$L_{l',m',l,m} f_{l,m}(x, y_j, H) = A_{l,m,l',m',x}(x, y_j, H) \frac{\partial}{\partial x} f_{l,m}(x, y_j, H) + B_{\mathrm{E},l,m,l',m',x}(x, y, H) f_{l,m}(x, y_j, H). \tag{3.20}$$

$B_{\mathrm{E},l,m,l',m',x}$ is interpreted as $B_{\mathrm{E},l,m,l',m'}$ calculated with only the x-directional electric force, which is consistent with the dimensional splitting, where only the force aligned with the direction of the dimensional splitting is considered,

$$B_{\mathrm{E},l,m,l',m',x} = \oint F_\mathrm{E} \mathbf{e}_x \cdot \left[\frac{\partial Y_{l,m}}{\partial \vartheta} \frac{Z}{\hbar} (\nabla_\mathbf{k} \vartheta) + \frac{\partial Y_{l,m}}{\partial \varphi} \frac{Z}{\hbar} (\nabla_\mathbf{k} \varphi) \right] Y_{l',m'} \mathrm{d}\Omega. \tag{3.21}$$

The odd operator is integrated over x (from x_i to x_{i+1}) and H (from H_n^- to H_n^+),

$$\int_{H_n^-}^{H_n^+} \int_{x_i}^{x_{i+1}} L_{l',m',l,m} f_{l,m}(x, y_j, H) \mathrm{d}x \, \mathrm{d}H$$
$$= \int_{H_n^-}^{H_n^+} A_{l,m,l',m',x}(x_{i+0.5}, y_j, H) \mathrm{d}H$$
$$\times \left[f_{l,m}(x_{i+1}, y_j, H_n) - f_{l,m}(x_i, y_j, H_n) \right]$$
$$+ \int_{H_n^-}^{H_n^+} B_{\mathrm{E},l,m,l',m',x}(x_{i+0.5}, y_j, H) \mathrm{d}H$$
$$\times \frac{f_{l,m}(x_i, y_j, H_n) + f_{l,m}(x_{i+1}, y_j, H_n)}{2} (x_{i+1} - x_i). \tag{3.22}$$

The integrals of the two quantities, $A_{l,m,l',m',x}$ and $B_{\mathrm{E},l,m,l',m',x}$, over the energy space are important coefficients. Note that the integrals are performed over H. Since the physical quantities available from the band model are functions of ε, the integrals should be converted into those over ε. From H_n^+ and $H_n^{\min}$, the upper and lower limits in terms of ε, denoted as ε_n^+ and $\varepsilon_n^{\min}$, are directly calculated using (3.4),

$$\varepsilon_n^+ = H_n^+ \pm qV, \tag{3.23}$$

$$\varepsilon_n^{\min} = H_n^{\min} \pm qV. \tag{3.24}$$

Since H_n^+ is a constant, ε_n^+ always linearly depends on the change of V, whereas $\varepsilon_n^{\min}$ may or may not depend on V due to definition (3.19). When the integrals are performed, the dependence of ε_n^+ and $\varepsilon_n^{\min}$ on V should be kept. Then, we have

$$\int_{H_n^-}^{H_n^+} A_{l,m,l',m',x}(x_{i+0.5}, y_j, H) \mathrm{d}H = \int_{\varepsilon_n^{\min}}^{\varepsilon_n^+} A_{l,m,l',m',x}(x_{i+0.5}, y_j, \varepsilon) \mathrm{d}\varepsilon, \tag{3.25}$$

$$\int_{H_n^-}^{H_n^+} B_{\mathrm{E},l,m,l',m',x}(x_{i+0.5}, y_j, H) \mathrm{d}H = \int_{\varepsilon_n^{\min}}^{\varepsilon_n^+} B_{\mathrm{E},l,m,l',m',x}(x_{i+0.5}, y_j, \varepsilon) \mathrm{d}\varepsilon. \tag{3.26}$$

Although analytical expressions for these quantities can be found in the case of the Modena model for electrons [11], as explicitly shown in [8], some band models described in Chap. 4 do not allow analytical expressions for them, due to the lack of analytical representation for the dispersion relation. In such cases, these coefficients are calculated by a means of numerical integration. A small energy step of 1 meV is used in this work. When the numerical integration is performed, the dependence of these coefficients on $V(x_i, y_j)$ and $V(x_{i+1}, y_j)$ is also calculated. This is a key characteristic which makes the fully-coupled Newton scheme and the small-signal analysis possible. Note that both, the direct dependence on $V(x_i, y_j)$ and $V(x_{i+1}, y_j)$ and the dependence through $H_n^{\min}(x_{i+0.5}, y_j)$ should be considered.

3.2.2 Even Free-Streaming Operator

Next, the discretization of the even free-streaming operator is shown. Since we employ dimensional splitting on a Cartesian grid for the odd operator, we have four adjoint nodes surrounding the given direct node.[5] For example, let us assume that the position of the direct node at the center of the box is given by x_i and y_j. The surrounding four adjoint nodes are $(x_{i\pm 0.5}, y_j)$ and $(x_i, y_{j\pm 0.5})$. Explicitly in the two-dimensional real space,

$$\begin{aligned} L_{l,m,l',m'} f_{l',m'}(x,y,H) &= \frac{\partial}{\partial x}\Big[A_{l,m,l',m',x}(x,y,H) f_{l',m'}(x,y,H)\Big] \\ &\quad - B_{l,m,l',m',x}(x,y,H) f_{l',m'}(x,y,H) \\ &\quad + \frac{\partial}{\partial y}\Big[A_{l,m,l',m',y}(x,y,H) f_{l',m'}(x,y,H)\Big] \\ &\quad - B_{l,m,l',m',y}(x,y,H) f_{l',m'}(x,y,H), \end{aligned} \tag{3.27}$$

where x and y are limited to the box (from $x_{i-0.5}$ to $x_{i+0.5}$ and from $y_{j-0.5}$ to $y_{j+0.5}$), and $B_{l,m,l',m',x}$ and $B_{l,m,l',m',y}$ are calculated with the directional components of the electric force. Defining the even operator along x-direction as follows,

$$\begin{aligned} L_{l,m,l',m',x} f_{l',m'}(x,y_j,H) &= \frac{\partial}{\partial x}\Big[A_{l,m,l',m',x}(x,y_j,H) f_{l',m'}(x,y_j,H)\Big] \\ &\quad - B_{l,m,l',m',x}(x,y_j,H) f_{l',m'}(x,y_j,H), \end{aligned} \tag{3.28}$$

and the one along y-direction in the same manner, we can treat the contributions from the specific directions separately. When the grid node is located at the interface of silicon and oxide, only the part of the box which lies inside the silicon region is considered. Without loss of generality, let us consider $L_{l,m,l',m',x} f_{l',m'}(x,y,H)$ only. The integral of the even operator over x (from $x_{i-0.5}$ to $x_{i+0.5}$) and H (from H_n^- to H_n^+) is considered.

$$\begin{aligned} &\int_{H_n^-}^{H_n^+} \int_{x_{i-0.5}}^{x_{i+0.5}} L_{l,m,l',m',x} f_{l',m'}(x,y_j,H) \mathrm{d}x\, \mathrm{d}H \\ &\quad = \int_{H_n^-}^{H_n^+} A_{l,m,l',m',x}(x_{i+0.5},y_j,H) \mathrm{d}H f_{l',m'}(x_{i+0.5},y_j,H_n) \\ &\quad\quad - \int_{H_n^-}^{H_n^+} A_{l,m,l',m',x}(x_{i-0.5},y_j,H) \mathrm{d}H f_{l',m'}(x_{i-0.5},y_j,H_n) \end{aligned}$$

[5] In the three dimensional real space with a Cartesian grid, there are six adjoint nodes surrounding the given direct node.

$$
\begin{aligned}
&- \int_{H_n^-}^{H_n^+} B_{l,m,l',m',x}(x_{i-0.5}, y_j, H)\mathrm{d}H \left[f_{l',m'}(x_{i-0.5}, y_j, H_n) \frac{x_i - x_{i-1}}{2} \right] \\
&- \int_{H_n^-}^{H_n^+} B_{l,m,l',m',x}(x_{i+0.5}, y_j, H)\mathrm{d}H \left[f_{l',m'}(x_{i+0.5}, y_j, H_n) \frac{x_{i+1} - x_i}{2} \right],
\end{aligned} \tag{3.29}
$$

where the coefficients are already defined in (3.25) and (3.26). Of course, when combined with the y-directional contributions, the control length along the y direction, $y_{j+0.5} - y_{j-0.5}$, should be multiplied.

3.3 Scattering Integral

Since the scattering integral is local in the real space, its form is not affected by the H-transformation. The in- and out-scattering terms shown in (2.131) and (2.132) can be easily formulated in the H space. From the discussion about the stabilized form of the free-streaming operator it follows that the preferred form of the unknown variable is not $g_{l,m}(\mathbf{r}, H)$ but $f_{l,m}(\mathbf{r}, H)$. Therefore, in the actual implementation $f_{l,m}(\mathbf{r}, H)$ is used together with (2.105).

Like for the free-streaming operator, the dependence of the DOS on the electrostatic potential can be calculated by

$$
\int_{H_n^-}^{H_n^+} Z_{l,m,l',m'}(\mathbf{r}_\mathrm{H}, H)\mathrm{d}H = \int_{\varepsilon_n^{\min}}^{\varepsilon_n^+} Z_{l,m,l',m'}(\mathbf{r}_\varepsilon, \varepsilon)\mathrm{d}\varepsilon. \tag{3.30}
$$

Note that the dependence of $c_{\eta,l}$ in (2.131) and (2.132) on the electrostatic potential also should be considered. In the following, we discuss two additional issues related with the scattering integral.

Recall that the contribution of the free-streaming operator to the Jacobian matrix for $f_{0,0}(\mathbf{r}, H)$ does not destroy the diagonal dominance in the case of the lowest order expansion. In this section, for the lowest order expansion, it is also shown that the contribution of the scattering integral to the Jacobian matrix for $f_{0,0}(\mathbf{r}, H)$ does not destroy the diagonal dominance.[6]

Let us consider a pair of states, whose energies are ε and $\varepsilon + \hbar\omega_\eta$. Here a positive energy $\hbar\omega_\eta$ corresponds to absorption of a phonon of the type η. Emission of this phonon involves a negative energy and is labeled with η'. Therefore, $\hbar\omega_\eta = -\hbar\omega_{\eta'}$. When the lowest order expansion is assumed, the in- and out- scattering integrals at the corresponding H are given by

[6]The Pauli principle is neglected in this analysis, because it makes the BTE nonlinear, therefore, the direct relation between the linear response and the physical solution is lost. Also velocity-randomizing scattering considered in (2.134) and (2.135) is assumed.

$$\hat{S}^{\text{in}}_{\eta',0,0}(\mathbf{r}, H) = Z_{0,0}(\mathbf{r}, H) Z_{0,0}(\mathbf{r}, H + \hbar\omega_\eta) c_{\eta',0}[\mathbf{r}, \varepsilon] f_{0,0}(\mathbf{r}, H + \hbar\omega_\eta), \quad (3.31)$$

$$\hat{S}^{\text{out}}_{\eta,0,0}(\mathbf{r}, H) = Z_{0,0}(\mathbf{r}, H) Z_{0,0}(\mathbf{r}, H + \hbar\omega_\eta) c_{\eta,0}[\mathbf{r}, \varepsilon + \hbar\omega_\eta] f_{0,0}(\mathbf{r}, H). \quad (3.32)$$

For the transition rate, the detailed balance relation should hold[7]

$$c_{\eta'}[\mathbf{r}, \varepsilon] = c_\eta[\mathbf{r}, \varepsilon + \hbar\omega_\eta] \exp\left(\frac{\hbar\omega_\eta}{k_B T_0}\right). \quad (3.33)$$

From the detailed balance relation, it is obvious that

$$\frac{\partial}{\partial f_{0,0}(\mathbf{r}, H)} \hat{S}^{\text{out}}_{\eta,0,0} < \frac{\partial}{\partial f_{0,0}(\mathbf{r}, H + \hbar\omega_\eta)} \hat{S}^{\text{in}}_{\eta',0,0}, \quad (3.34)$$

which destroys the diagonal dominance, because the absolute value of the diagonal term is smaller than that of the off-diagonal term.[8] However, by introducing a scaling factor,

$$\tilde{f}_{l,m}(\mathbf{r}, H) = \exp\left(\frac{H}{k_B T_0}\right) f_{l,m}(\mathbf{r}, H), \quad (3.35)$$

the two terms are exactly balanced

$$\frac{\partial}{\partial \tilde{f}_{0,0}(\mathbf{r}, H)} \hat{S}^{\text{out}}_{\eta,0,0} = \frac{\partial}{\partial \tilde{f}_{0,0}(\mathbf{r}, H + \hbar\omega_\eta)} \hat{S}^{\text{in}}_{\eta',0,0}. \quad (3.36)$$

Therefore, the contribution of the scattering integral to the Jacobian matrix for $\tilde{f}_{0,0}(\mathbf{r}_H, H)$ does not destroy the diagonal dominance, when the scaling law of (3.35) is applied. Note that the free-streaming operator is not affected by this scaling, because it is constructed for a given H value. Moreover, since the scaling law is related with a diagonal matrix having all positive entries, the solution of the unscaled system has the same sign as the solution of the scaled system. In other words, if the scaled system has non-negative solutions, the unscaled system also has non-negative solutions.

Another issue related with the discretization of the scattering integral is the mismatch of the energy transfer of the scattering mechanisms and the actual energy spacing used in the simulation. In a simple approach, the phonon energy is mapped onto the energy grid which simplifies the discretization of the scattering integral because in this case the energy transfer is a multiple of the energy spacing and

---

[7]In our implementation, impact ionization scattering does not satisfy this relation. However, impact ionization events are usually very rare and the degradation of stability is thus negligible.

[8]A pair of states with energies $\varepsilon - \hbar\omega_\eta$ and ε yields the opposite result. By the scaling shown in (3.35), this pair also gives exactly balanced contributions to the Jacobian matrix.

the final energies always lie on grid nodes. Using this simple approximation, the detailed balance relation can be preserved. However, due to the alignment of the phonon energies, an energy grid with a fine energy spacing is required in order to reduce the error.

In this work, instead, a scheme shown in [15] is used. Let us consider the particle transition between two control volumes in the H space represented by H_n and H_m. As before, H_n^- and H_m^- denote the lower values of the control volumes, and H_n^+ and H_m^+ the upper values. Without loss of generality, we assume that H_m is higher than H_n.

We find the maximum possible energy sub-range of (h_n^-, h_n^+) and (h_m^-, h_m^+) within the control volumes of (H_n^-, H_n^+) and (H_m^-, H_m^+), respectively, where

$$h_m^- = h_n^- + \hbar\omega_\eta, \tag{3.37}$$

$$h_m^+ = h_n^+ + \hbar\omega_\eta, \tag{3.38}$$

are fulfilled. This energy sub-range is the relevant energy range for the phonon with energy $\hbar\omega_\eta$, as much as two control volumes represented by H_n and H_m are concerned. Note that the original control volume, (H_n^-, H_n^+), can be divided into several energy sub-ranges corresponding to different H_m values. When applied to an equidistant H grid, it yields at most two energy sub-ranges. Then, the transition rate in each energy sub-range is approximated to ensure the detailed balance relation in the discretized sense. The division of the H axis into energy sub-ranges is schematically shown in Fig. 3.3.

Effectively, the phonon is divided into "small" phonons, whose energy transfers are differences between the initial energies and the final energies. The magnitude of each small phonon is proportional to the size of its relevant energy range. Because every phonon process is divided and adjusted according to the given H grids, this scheme is applicable to non-aligned and non-equidistant energy grids. Numerical property of this scheme will be discussed in Chap. 5.

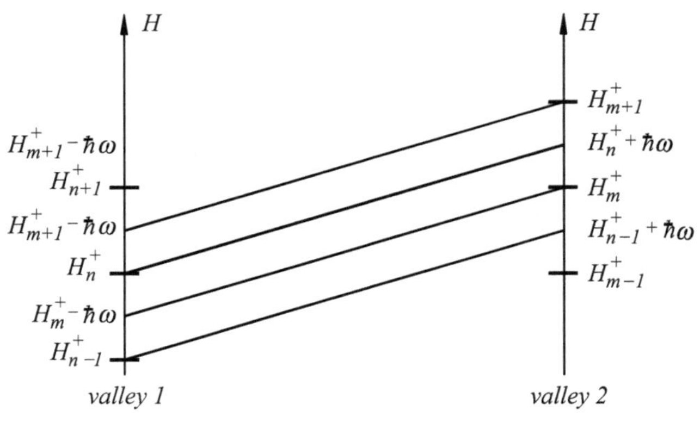

Fig. 3.3 Division of the H axis into energy sub-ranges for the energy transfer $\hbar\omega$

3.4 Boundary Condition

The projection of the boundary condition yields the interface recombination rate (2.136). It is implemented only for the boxes at the terminal nodes. Therefore, only the balance equations with even l values have to be considered.

The sum of $D_{l,m,l',m'}$ and $E_{l,m,l',m'}$ is given by

$$D_{l,m,l',m'} + E_{l,m,l',m'} = \oint \mathbf{v}Z \cdot \mathbf{n} Y_{l,m} Y_{l',m'} \mathrm{d}\Omega = \mathbf{n} \cdot \mathbf{A}_{l,m,l',m'}. \tag{3.39}$$

Since $\mathbf{v}Z \cdot \mathbf{n}$ has odd parity under inversion due to the valley-wise inversion symmetry of the underlying band structure, $D_{l,m,l',m'} + E_{l,m,l',m'}$ is nonzero only for odd l'. Moreover, for even l – which is understood – and odd l', $D_{l,m,l',m'} = E_{l,m,l',m'}$ holds. Therefore, for odd l', $D_{l,m,l',m'}$ and $E_{l,m,l',m'}$ can be calculated easily and exactly.

The difference between $D_{l,m,l',m'}$ and $E_{l,m,l',m'}$ is given by

$$D_{l,m,l',m'} - E_{l,m,l',m'} = -\oint |\mathbf{v}Z \cdot \mathbf{n}| Y_{l,m} Y_{l',m'} \mathrm{d}\Omega. \tag{3.40}$$

Since $|\mathbf{v}Z \cdot \mathbf{n}|$ has even parity under inversion, $D_{l,m,l',m'} - E_{l,m,l',m'}$ is nonzero only for even l'. In addition, for even l and even l', $D_{l,m,l',m'} = -E_{l,m,l',m'}$ holds. When the sign of $\mathbf{v}Z \cdot \mathbf{n}$ changes in a simple manner, the integral in (3.40) can be solved easily as shown in [8]. However, in the general case, numerical integration has to be used to evaluate this quantity [16].

When box integration is used, the integrated interface generation rate

$$\int_{A_t} \gamma_{s,l,m} \mathrm{d}a = \sum_{l',m'} \int_{A_t} D_{l,m,l',m'} f_{l',m'} \mathrm{d}a + \int_{A_t} E_{l,m,0,0} f_{0,0}^{\mathrm{eq}} \mathrm{d}a, \tag{3.41}$$

should be added to the RHS of the balance equation for l and m. A_t is the area of the contact. Depending on the parity of l', two cases should be distinguished. In the case of an even l' value, $D_{l,m,l',m'}$ and $E_{l,m,0,0}$ can be calculated with (3.40), and the distribution function $f_{l',m'}$ can be taken at the direct node. The area of the terminal contact is used for the surface integration.

In the case of an odd l' value, $E_{l,m,0,0}$ is not relevant. Although $D_{l,m,l',m'}$ can be calculated with (3.39), the distribution function for odd l' is defined on the surrounding adjoint nodes. In order to be consistent with the dimensional splitting scheme, when $D_{l,m,l',m'}$ is calculated on an adjoint node, $\mathbf{n}$ is adjusted to the direction of the line on which the adjoint node lies (normal to the box's surface). Accordingly, the actual area concerning an adjoint node is given by the associated surface area of the adjoint node.[9] This approach is justified, because the integral of

[9]For a Cartesian tensor grid, where $\mathbf{n}$ is aligned to $\mathbf{e}_x$ or $\mathbf{e}_y$, this procedure results in a trivial mapping.

n over the surface of the box always vanishes. Using (3.39) and combined with the free-streaming operator for that adjoint node, this approach gives the particle flux through the associated surface, which is reduced by a factor of 2.

In the previous sections, it is shown that both the free-streaming operator and the scattering integral do not destroy the diagonal dominance of the Jacobian matrix for the lowest order expansion. Finally, a test for the boundary condition is shown here. Since $E_{l,m,0,0}$ is connected with $f_{0,0}^{\text{eq}}$, which is not an unknown variable, only the $D_{l,m,l',m'}$'s are important. For the lowest order expansion, $D_{0,0,0,0}$ and $D_{0,0,1,m'}$ are relevant. Note that $D_{0,0,0,0}$ itself is always negative by its definition. Therefore, $D_{0,0,0,0} f_{0,0}$ always contributes to the positive diagonal component without contributing to the off-diagonal terms. Moreover, $D_{0,0,1,m'} f_{1,m'}$ can be considered as a modification (reduction by a factor of two) of the free-streaming operator. Therefore, $D_{0,0,1,m'} f_{1,m'}$ does not change the argument about the free-streaming operator.

3.5 Nonsingular M-Matrix

From the tests for the free-streaming operator (Sect. 3.2), the scattering integral (Sect. 3.3), and the boundary condition (Sect. 3.4), it is found that the Jacobian matrix for $f_{0,0}(\mathbf{r}, H)$, which results from the lowest order expansion, is diagonally dominant.[10] Moreover, the diagonal entries are always positive, and the off-diagonal ones negative. It means that the Jacobian matrix is an M-matrix [13], as briefly stated in [8].

For contact nodes, the strict diagonal dominance is satisfied thanks to the $D_{0,0,0,0}$ term, while the equality between the diagonal entry and the absolute sum of the off-diagonal entries holds for the bulk nodes. Since every bulk node is always connected to certain contact nodes by the free-streaming operator and/or the scattering integral, the Jacobian matrix is irreducibly diagonally dominant [14]. From Theorem 1.21. of [14], nonsingularity of the Jacobian matrix is obtained.

The fact that the Jacobian matrix is a nonsingular M-matrix ensures the non-negativeness of the solution. In other words, at least the isotropic component of the particle distribution function is always nonnegative, which is quite an attractive numerical property. Note that this desirable property cannot be drawn in the case of the kinetic-energy-based scheme [3]. Since the scaling shown in (3.35) is related with a diagonal matrix with all positive entries, the nonnegativeness of the solution is guaranteed even for the unscaled system.

When the higher-order expansion is considered, the above discussion cannot be easily extended analytically. However, from the numerical experiment, the numerical stability of the resultant set of equations is nevertheless greatly improved by this approach.

[10] The scaling shown in (3.35) is to be used.

3.6 Small-Signal Analysis

As discussed in Sect. 2.6, small-signal analysis requires that the BTE is linearized together with the Poisson equation. Therefore, the fluctuations of the electrostatic potential plays a role. In the case of the kinetic-energy-based scheme, the linearization of the BTE can be easily done, because the energy grid is independent of the electrostatic potential. However, after the H-transformation, as pointed out in [3], the energy grid is strongly dependent on the electrostatic potential, and this makes the linearization of the BTE non-trivial.

In DC simulations, in order to eliminate the partial derivative w.r.t. the energy variable, H is set equal to the total energy. However, when this approach is naively applied to small-signal analysis, it yields time-varying coefficients like the DOS. In order to avoid the numerical difficulties arising from the time-varying coefficients, following the idea in [17], we fix $\Psi(\mathbf{r})$.

When we set $\Psi(\mathbf{r}) = V_s(\mathbf{r}) - E^{\nu}(\mathbf{r})/q$, where V_s is the stationary part of the electrostatic potential, the linearization of the even free-streaming operator shown in (3.8) yields the following two terms,

$$L_{l,m,l',m'}\delta f_{l',m'}(\mathbf{r}, H, \omega) = \nabla_{\mathbf{r}} \cdot (\mathbf{A}_{l,m,l',m'}\delta f_{l',m'}) - B_{\mathrm{E},l,m,l',m'}\delta f_{l',m'}, \tag{3.42}$$

$$\delta L_{l,m,l',m'} f_{s,l',m'}(\mathbf{r}, H) = \pm q(\nabla_{\mathbf{r}}\delta V(\mathbf{r}, \omega)) \cdot \frac{\partial}{\partial H} (\mathbf{A}_{l,m,l',m'} f_{s,l',m'}), \tag{3.43}$$

where $f_{s,l',m'}$ denotes the steady-state solution. Since $\delta V(\mathbf{r}, \omega)$ cannot be neglected in the small-signal analysis, the coupling between different H-points should be treated. When integrated over H (from H_n^- to H_n^+), the coefficient of $\nabla_{\mathbf{r}}\delta V(\mathbf{r}, \omega)$ yields

$$\begin{aligned}\int_{H_n^-}^{H_n^+} &\frac{\partial}{\partial H} (\mathbf{A}_{l,m,l',m'}(\mathbf{r}, H) f_{s,l',m'}(\mathbf{r}, H))\,\mathrm{d}H \\ &= \mathbf{A}_{l,m,l',m'}(\mathbf{r}, H_n^+) f_{s,l',m'}(\mathbf{r}, H_n^+) \\ &\quad - \mathbf{A}_{l,m,l',m'}(\mathbf{r}, H_n^-) f_{s,l',m'}(\mathbf{r}, H_n^-).\end{aligned} \tag{3.44}$$

Since $\mathbf{A}_{l,m,l',m'}$ and $f_{s,l',m'}$ are defined on neither H_n^- nor H_n^+, an interpolation scheme is required. $\mathbf{A}_{l,m,l',m'}$ is directly calculated with the band model while $f_{s,l',m'}$ is linearly interpolated between two adjacent H nodes. Once the coefficient is calculated and fixed, the term is linear in the potential perturbation.

In order to be consistent with the stabilization scheme used in the DC simulation, the odd free-streaming operator should be treated separately. From numerical experiments, it is found that this separate treatment is crucial for stable small-signal analysis. Following a similar procedure, the linearization of the odd free-streaming operator shown in (3.11) yields the two terms,[11]

$$L_{l',m',l,m}\delta f_{l,m}(\mathbf{r}, H, \omega) = \mathbf{A}_{l,m,l',m'} \cdot \nabla_{\mathbf{r}}\delta f_{l,m} + B_{\mathrm{E},l,m,l',m'} f_{l,m}. \tag{3.45}$$

$$\delta L_{l',m',l,m} f_{s,l,m}(\mathbf{r}, H) = \pm q(\nabla_{\mathbf{r}}\delta V(\mathbf{r}, \omega)) \cdot \mathbf{A}_{l,m,l',m'} \frac{\partial}{\partial H} f_{l,m}. \tag{3.46}$$

Now the integration of the coefficient of $\nabla_{\mathbf{r}}\delta V(\mathbf{r}, \omega)$ yields

$$\begin{aligned}
&\int_{H_n^-}^{H_n^+} \mathbf{A}_{l,m,l',m'}(\mathbf{r}, H)\left(\frac{\partial}{\partial H} f_{s,l',m'}(\mathbf{r}, H)\right) \mathrm{d}H \\
&= \mathbf{A}_{l,m,l',m'}(\mathbf{r}, H_n^+)\left.\frac{\partial f_{s,l',m'}}{\partial H}\right|_{(\mathbf{r},H_n^+)}(H_n^+ - H_n) \\
&\quad + \mathbf{A}_{l,m,l',m'}(\mathbf{r}, H_n^-)\left.\frac{\partial f_{s,l',m'}}{\partial H}\right|_{(\mathbf{r},H_n^-)}(H_n - H_n^-) \\
&= \mathbf{A}_{l,m,l',m'}(\mathbf{r}, H_n^+)\frac{f_{s,l,m}(\mathbf{r}, H_{n+1}) - f_{s,l,m}(\mathbf{r}, H_n)}{2} \\
&\quad + \mathbf{A}_{l,m,l',m'}(\mathbf{r}, H_n^-)\frac{f_{s,l,m}(\mathbf{r}, H_n) - f_{s,l,m}(\mathbf{r}, H_{n-1})}{2}.
\end{aligned} \tag{3.47}$$

Because the scattering integral and the boundary condition are local in the real space, they can be easily treated, which is not shown here.

References

1. Ringhofer, C.: Numerical methods for the semiconductor Boltzmann equation based on spherical harmonics expansions and entropy discretizations. Transport. Theor. Stat. Phys. **31**(4–6), 431–452 (2002)
2. Gnudi, A., Ventura, D., Baccarani, G., Odeh, F.: Two-dimensional MOSFET simulation by means of a multidimensional spherical harmonics expansion of the Boltzmann transport equation. Solid State Electron. **36**(4), 575–581 (1993)
3. Jungemann, C., Pham, A.-T., Meinerzhagen, B., Ringhofer, C., Bollhöfer, M.: Stable discretization of the Boltzmann equation based on spherical harmonics, box integration, and a maximum entropy dissipation principle. J. Appl. Phys. **100**, 024502–1–13 (2006)
4. Jin, S., Fischetti, M.V., Tang, T.w.: Theoretical study of carrier transport in silicon nanowire transistors based on the multisubband Boltzmann transport equation. IEEE Trans. Electron Dev. **55**, 2886–2897 (2008)

[11]Compared with (3.11), l and l' are interchanged in order to be consistent with our parity convention.

5. Hennacy, K.A., Wu, Y.-J., Goldsman, N., Mayergoyz, I.D.: Deterministic MOSFET simulation using a generalized spherical harmonic expansion of the Boltzmann equation. Solid State Electron. **38**, 1485–1495 (1995)
6. Rahmat, K., White, J., Antoniadis, D.A.: Simulation of semiconductor devices using a Galerkin/spherical harmonic expansion approach to solving the coulped Poisson-Boltzmann system. IEEE Trans. Comput. Aided Dev. **15**(10), 1181–1196 (1996)
7. Jungemann, C.: Transport and noise calculations for nanoscale Si devices based on the Langevin Boltzmann equation expanded with spherical harmonics. J. Comput. Theor. Nanosci. **5**(6), 1152–1169 (2008)
8. Hong, S.-M., Jungemann, C.: A fully coupled scheme for a Boltzmann-Poisson equation solver based on a spherical harmonics expansion. J. Comput. Electron. **8**(3), 225–241 (2009)
9. Madelung, O.: Introduction to Solid State Theory. Springer, Berlin (1978)
10. Hong, S.-M., Jungemann, C.: Inclusion of the Pauli principle in a deterministic Boltzmann equation solver based on a spherical harmonics expansion. J. Comput. Electron. **9**, 153–159 (2010)
11. Brunetti, R., Jacoboni, C., Nava, F., Reggiani, L., Bosman, G., Zijlstra, R.J.J.: Diffusion coefficient of electrons in silicon. J. Appl. Phys. **52**, 6713–6722 (1981)
12. Larsson, S., Thomee, V.: Partial differential equations with numerical methods, texts in applied mathematics. Springer, New York (2003)
13. Berman, A., Plemmons, R.J.: Nonnegative Matrices in the Mathematical Sciences, Computer Science and Applied Mathematics. Academic, New York (1979)
14. Varga, R.S.: Matrix iterative analysis, series in automatic computation. Prentice-Hall, New Jersey (1962)
15. Hong, S.-M., Jungemann, C.: Deterministic simulation of SiGe HBTs based on the Boltzmann equation. In: Proceedings of ESSDERC, pp. 170–173 (2008)
16. Hong, S.-M., Matz, G., Jungemann, C.: A deterministic Boltzmann equation solver based on a higher-order spherical harmonics expansion with full-band effects. IEEE Trans. Electron Dev. **57**, 2390–2397 (2010)
17. Lin, C.-K., Goldsman, N., Han, Z., Mayergoyz, I., Yu, S., Stettler, M., Singh, S.: Frequency domain analysis of the distribution function by small signal solution of the Boltzmann and Poisson equations. In: Proceedings of SISPAD, pp. 39–42 (1999)

Chapter 4
Band Structure and Scattering Mechanisms

The models for the band structure and scattering are discussed in this chapter. The electronic band structure of Si or SiGe is quite complex, and analytical expressions for the dispersion relation involve rather stringent approximations, especially at high energies. On the other hand, direct inclusion of the full band structure into the SHE simulator is at least not possible for the conduction bands. Instead, a valley model of the conduction bands is preferable for SHE, and the development of such a model based on the full band structure is quite important for physically sound SHE simulations. Several band models for electrons, which can be used within the SHE solver, are introduced. In the last section of this chapter, the scattering mechanisms considered in this work are briefly presented.

4.1 Overview of the Band Structure of Silicon

The band energy of the first four conduction bands and the first three valence bands is calculated with the nonlocal empirical pseudopotential method [1–4].

Let us consider the conduction bands first. There are six equivalent energy minima in the first conduction band (Δ valleys). The whole Brillouin zone for all conduction bands is divided into three regions, according to the biggest absolute value among k_x, k_y, and k_z. They are labeled x-, y-, and z-valleys, respectively. Since there are two valleys in each region, the degeneracy factor for each region is two. Of the full band structure the following moments are considered,

$$V_0^{\nu}(\varepsilon) = \frac{1}{2}\frac{1}{(2\pi)^3}\sum_{\beta}\int_{BZ^{\nu}} \delta[\varepsilon - \varepsilon^{\beta}(\mathbf{k})]\mathrm{d}^3k, \tag{4.1}$$

and for a positive integer n,

$$V_n^{\nu}(\varepsilon) = \frac{1}{2}\frac{1}{V_0^{\nu}(\varepsilon)}\frac{1}{(2\pi)^3}\sum_{\beta}\int_{BZ^{\nu}} \delta[\varepsilon - \varepsilon^{\beta}(\mathbf{k})]|\mathbf{v}^{\beta}(\mathbf{k})|^n\mathrm{d}^3k, \tag{4.2}$$

S.-M. Hong et al., *Deterministic Solvers for the Boltzmann Transport Equation*, Computational Microelectronics, DOI 10.1007/978-3-7091-0778-2_4,

Fig. 4.1 Contribution from each conduction band of relaxed silicon to V_0^ν

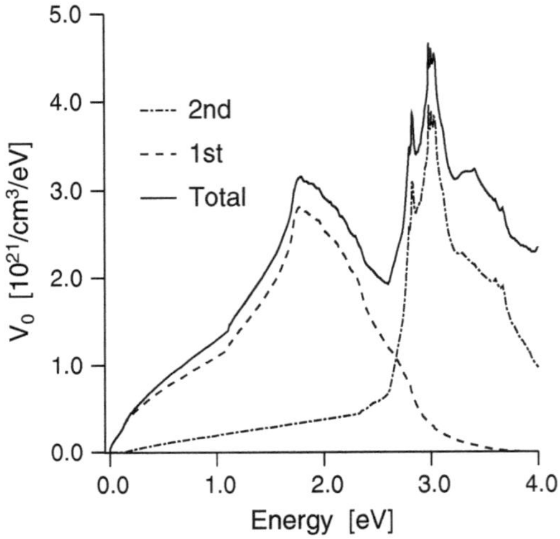

where β is the band index and the leading coefficient $1/2$ is included to consider only a single valley. The integral over the Brillouin zone is restricted to the region corresponding to the νth valley. Note that $V_0^\nu(\varepsilon)$ is the DOS of the νth valley for one spin direction, integrated over the equienergy surface. In Fig. 4.1 the contribution to V_0^ν from each conduction band up to the fourth one is shown for relaxed silicon. The second conduction band starts to contribute to V_0 at a rather low energy (0.13 eV). The peaks of the DOS observed at 1.9 and 3 eV are mainly due to states near the L points of the Brillouin zone [5]. The minimum energy at the L points is 1.1 eV for relaxed silicon. Therefore, above this energy value, the valleys (x-, y-, and z-valleys) actually become mixtures of Δ valleys and L valleys. In Fig. 4.2, the average velocity estimated by taking $\left(V_n^\nu\right)^{1/n}$ is shown. For an isotropic band structure, the average velocities estimated from different moments will be identical. On the other hand, for an anisotropic one, the average velocity estimated from V_n^ν increases when n increases, because the contribution of the higher velocity part is more emphasized. The observed difference between different estimation methods clearly represents the anisotropy of the conduction bands.

Next, let us consider the valence bands. In the case of the valence bands, no partitioning into valleys is introduced. Therefore, each valence band (the heavy hole, light hole, or split off hole band) can be treated as a single valley. They are labeled hh-, lh-, and so-valleys (or bands), respectively. In Fig. 4.3, V_0^ν is shown for each valence band of relaxed silicon. Since the heavy hole band hits the surface of the Brillouin zone at 1.27 eV (light hole band: 1.31 eV, split off band: 7.09 eV), a direct application of the SHE method based on the exact valence band structure is possible only up to this energy value [6, 7]. Above this value, an approximation is required [8].

Both, the conduction bands and the valence bands introduce some difficulties for SHE modeling. However, the reasons for the difficulties and the solutions are very different for the two cases. In the case of the conduction bands, the SHE of the full

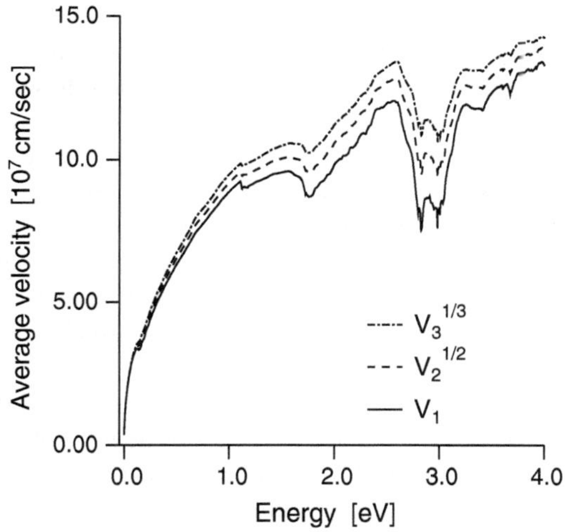

Fig. 4.2 Average velocity estimated from V_n^ν for relaxed silicon

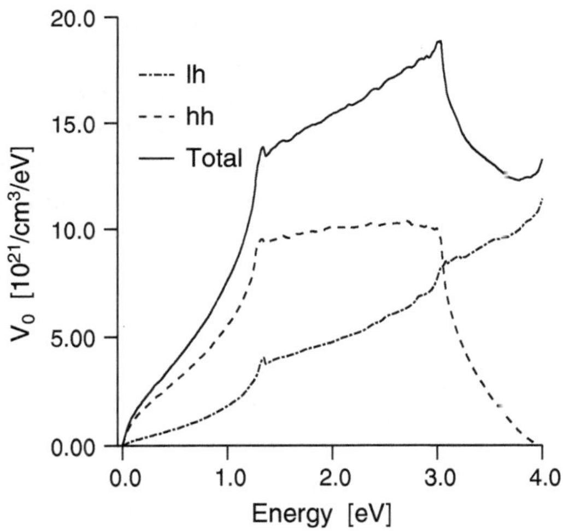

Fig. 4.3 V_0^ν for each valence band of relaxed silicon

band structure is difficult, therefore, a certain way to approximate the band structure is required. On the other hand, in the case of the valence bands, in principle, the SHE of the full band structure is straightforward. However, as pointed out in [7], it requires a high maximum order of SHE.

4.2 Band Structure Modeling for Electrons

As explained in Sect. 4.1, in contrast to the valence band [6, 8], the conduction bands cannot be directly expanded with spherical harmonics easily because the first conduction band hits the surface of the Brillouin zone at a rather low energy

(0.13 eV). In order to overcome this difficulty, several models have been proposed and applied to SHE solvers [9–13]. These models can be divided into three different categories, and will be discussed in the following.

When there exists an invertible dispersion relation, we can rigorously derive the balance equations. In the first approach, we try to construct a band model, for whose valleys the dispersion relation is invertible. The analytical band model proposed by the Modena group [9] and an anisotropic model [12] belong to this case. However, due to the intrinsic difference between the full band structure and the assumed band model, a perfect match cannot be expected even for the anisotropic model, although significant improvement is already obtained in [12] at the cost of increased numerical complexity.

Another heuristic attempt can be found in [10], where the isotropic band model is explicitly assumed without considering the underlying dispersion relation. The biggest drawback of this approach is its inherent restriction to the lowest order expansion. An extension of this approach to a higher-order expansion is proposed in [13], and it has been numerically shown that its accuracy is comparable with the more elaborated anisotropic band model [12].

The third approach is heavily relying on the generalized coordinate transform shown in Chap. 2. The description for the low energy range is further improved by considering a direct mapping from the full band structure to the resultant valley. By taking the transport coefficients at its lowest-order available after the mapping, an improved full band model can be constructed without sacrifying numerical efficiency compared to the isotropic models.

In the following subsections, the above mentioned approaches are discussed.

4.2.1 Models with Inverse Dispersion Relation

When the dispersion relation is invertible, we have a vector $\mathbf{k}^\nu(\varepsilon, \vartheta, \varphi)$ uniquely defined on the energy space. Since the angle-preserving coordinate transformation is understood in this approach, the wave vector $\mathbf{k}^\nu(\varepsilon, \vartheta, \varphi)$ is aligned with the radial direction, and can be written in terms of its magnitude, $k^\nu(\varepsilon, \vartheta, \varphi)$, as follows:

$$\mathbf{k}^\nu(\varepsilon, \vartheta, \varphi) = k^\nu(\varepsilon, \vartheta, \varphi)\mathbf{e}_\varepsilon. \tag{4.3}$$

In this case, the quantities required for calculating the transport coefficients can be readily obtained,

$$\frac{\partial \mathbf{k}^\nu}{\partial \vartheta} \times \frac{\partial \mathbf{k}^\nu}{\partial \varphi} = (k^\nu)^2 \sin\vartheta\, \mathbf{e}_\varepsilon - k^\nu \frac{\partial k^\nu}{\partial \vartheta} \sin\vartheta\, \mathbf{e}_\vartheta - k^\nu \frac{\partial k^\nu}{\partial \varphi} \mathbf{e}_\varphi, \tag{4.4}$$

$$\det(\hat{\mathbf{J}}^{-1}) = \frac{\partial \mathbf{k}^\nu}{\partial \varepsilon} \cdot \left(\frac{\partial \mathbf{k}^\nu}{\partial \vartheta} \times \frac{\partial \mathbf{k}^\nu}{\partial \varphi} \right) = (k^\nu)^2 \frac{\partial k^\nu}{\partial \varepsilon} \sin\vartheta. \tag{4.5}$$

With (2.5), (2.65), and (2.74), the DOS and the group velocity are given by

$$Z^{\nu}(\varepsilon,\vartheta,\varphi) = \frac{1}{(2\pi)^3}(k^{\nu})^2\frac{\partial k^{\nu}}{\partial \varepsilon}, \tag{4.6}$$

$$\begin{aligned}\mathbf{v}^{\nu}(\varepsilon,\vartheta,\varphi) = &\frac{1}{\hbar}\left(\frac{\partial k^{\nu}}{\partial \varepsilon}\right)^{-1}\mathbf{e}_{\varepsilon}\\ &-\frac{1}{\hbar}\frac{1}{k^{\nu}}\frac{\partial k^{\nu}}{\partial \vartheta}\left(\frac{\partial k^{\nu}}{\partial \varepsilon}\right)^{-1}\mathbf{e}_{\vartheta}-\frac{1}{\hbar}\frac{1}{k^{\nu}\sin\vartheta}\frac{\partial k^{\nu}}{\partial \varphi}\left(\frac{\partial k^{\nu}}{\partial \varepsilon}\right)^{-1}\mathbf{e}_{\varphi}.\end{aligned} \tag{4.7}$$

Note that $\nabla_{\mathbf{k}}\vartheta$ and $\nabla_{\mathbf{k}}\varphi$ can be calculated with (2.66) and (2.67), respectively.

In the analytical band model of the Modena group [9], the conduction band is described by six nonparabolic and elliptical valleys. We start directly from the valley. When a coordinate system which diagonalizes the effective mass tensor is employed, the dispersion relation is given by [14]

$$\varepsilon(1+\alpha\varepsilon) = \frac{\hbar^2}{2m_{xx}}k_x^2 + \frac{\hbar^2}{2m_{yy}}k_y^2 + \frac{\hbar^2}{2m_{zz}}k_z^2, \tag{4.8}$$

where α is the nonparabolicity factor. m_{xx}, m_{yy}, and m_{zz} are the effective masses along each direction, respectively. In order to simplify the SHE, the elliptical valleys are mapped onto spherical ones by the Herring–Vogt transform [14, 15]

$$\hat{T}^{\nu} = \begin{pmatrix} \left(\frac{m_d^{\nu}}{m_{xx}^{\nu}}\right)^{1/2} & 0 & 0 \\ 0 & \left(\frac{m_d^{\nu}}{m_{yy}^{\nu}}\right)^{1/2} & 0 \\ 0 & 0 & \left(\frac{m_d^{\nu}}{m_{zz}^{\nu}}\right)^{1/2} \end{pmatrix}, \tag{4.9}$$

where m_d is the DOS mass given by

$$m_d^{\nu} = (m_{xx}^{\nu}m_{yy}^{\nu}m_{zz}^{\nu})^{1/3}. \tag{4.10}$$

The Herring–Vogt transformed wave vector $\mathbf{k}_{\mathrm{HV}}$ is given by

$$\mathbf{k}_{\mathrm{HV}}^{\nu} = \hat{T}^{\nu}\mathbf{k}^{\nu}. \tag{4.11}$$

As an immediate result of the Herring–Vogt transformation, the dispersion relation becomes isotropic,

$$\varepsilon(1+\alpha\varepsilon) = \frac{\hbar^2}{2m_d}k_{\mathrm{HV}}^2. \tag{4.12}$$

Therefore, terms like $\partial k_{\mathrm{HV}}^{\nu}/\partial\vartheta$ and $\partial k_{\mathrm{HV}}^{\nu}/\partial\varphi$ vanish. Then the group velocity in the Herring–Vogt transformed space evaluates to

$$\mathbf{v}_{\mathrm{HV}} = v_{\mathrm{HV}}(\varepsilon)\mathbf{e}_{\varepsilon} = \frac{1}{1+2\alpha\varepsilon}\frac{\hbar k_{\mathrm{HV}}(\varepsilon)}{m_d}\mathbf{e}_{\varepsilon}. \tag{4.13}$$

Note that the group velocity in the original $\mathbf{k}$ space is

$$\mathbf{v} = \hat{T}^{\nu}\mathbf{v}_{\mathrm{HV}}. \tag{4.14}$$

Since the band structure is position-independent,[1] $\tilde{\mathbf{F}}_{\mathrm{E}}(\mathbf{r},\mathbf{k})$ does not appear. The total force is given by

$$\mathbf{F} = \mathbf{F}_{\mathrm{E,pot}} + \mathbf{F}_{\mathrm{B}} = \nabla_{\mathbf{r}}(-E^{\nu}(\mathbf{r}) + qV(\mathbf{r},t)) - q(\hat{T}^{\nu}\mathbf{v}_{\mathrm{HV}})\times\mathbf{B}. \tag{4.15}$$

The DOS, which is constant over the solid angle after the transform,[2] is given by

$$Z(\varepsilon) = \frac{k_{\mathrm{HV}}^2}{(2\pi)^3}\frac{\partial k_{\mathrm{HV}}}{\partial\varepsilon}. \tag{4.16}$$

From the above discussion, the coefficients $Z_{l,m,l',m'}$ and $\mathbf{A}_{l,m,l',m'}$ can be derived,

$$Z_{l,m,l',m'} = Z(\varepsilon)\delta_{l,l'}\delta_{m,m'}, \tag{4.17}$$

$$\begin{aligned}\mathbf{A}_{l,m,l',m'}^{\nu} &= \oint \hat{T}^{\nu}v_{\mathrm{HV}}(\varepsilon)\mathbf{e}_{\varepsilon}Z(\varepsilon)Y_{l,m}(\vartheta,\varphi)Y_{l',m'}(\vartheta,\varphi)\mathrm{d}\Omega \\ &= \hat{T}^{\nu}v_{\mathrm{HV}}(\varepsilon)Z(\varepsilon)\mathbf{a}_{l,m,l',m'}.\end{aligned} \tag{4.18}$$

Since the angle variables are kept as the same between the transformed $\mathbf{k}$ space and the energy space, $B_{\mathrm{E},l,m,l',m'}$ is given by [17]

$$B_{\mathrm{E},l,m,l',m'} = \mathbf{F}_{\mathrm{E,pot}}\cdot\hat{T}^{\nu}\mathbf{b}_{l,m,l',m'}\frac{Z(\varepsilon)}{\hbar k_{\mathrm{HV}}(\varepsilon)}. \tag{4.19}$$

In this model, we have started directly from the valley. No attempt to exactly fit the full band structure of the conduction bands is made. Nevertheless, since the transport parameters in the low energy range is properly described by this model, it can reproduce the low field transport reasonably well. Throughout part 2, the simulation results obtained with this analytic model will be compared with the other models in order to emphasize the effect of band model.

---

[1] In the case of SiGe, the position-dependent band edge of the conduction band due to the position-dependent Ge content is taken into account [16].

[2] Before the Herring–Vogt transformation is applied, the DOS is not isotropic.

The main problem of the Modena model is its inability to predict the correct distribution in the high energy range. In order to overcome this difficulty an anisotropic valley model is presented in [18]. Like the Modena model, where six valleys are considered, a six-valley model, which consists of three different pairs of valleys (x-valleys, y-valleys, and z-valleys), is adopted in order to consider the anisotropic nature of the conduction band. For each valley, it is assumed that there exists an inverse dispersion relation between the energy and the wave vector in the Herring–Vogt transformed space [15]. Its inverse dispersion relation is expanded with spherical harmonics:

$$k^{\nu}_{\mathrm{HV}}(\varepsilon, \vartheta, \varphi) = \sum_{l,m} \left\{k^{\nu}_{\mathrm{HV}}\right\}_{l,m}(\varepsilon) Y_{l,m}(\vartheta, \varphi), \tag{4.20}$$

where $\left\{k^{\nu}_{\mathrm{HV}}\right\}_{l,m}(\varepsilon)$ is the fitting parameter. Note that still $\mathbf{k}^{\nu}(\varepsilon, \vartheta, \varphi)$ is aligned with the radial direction. By positioning the origin of the coordinate frame in the minimum of the valley and orienting the polar axis along the $\langle 001 \rangle$ axis of the crystal structure we can assume a large share of the fitting coefficients to be zero due to symmetries. This leaves us with only positive m values which are multiples of four and even l values.[3]

Fitting parameters are calculated through optimization using the Levenberg-Marquardt algorithm [19, 20]. The quadratic error for the moments V_0^{ν} in (4.1) and V_n^{ν} in (4.2) for n = 1, ..., 5, evaluated for the full-band structure [21] serves as the objective function. Since the relation between the energy and the wave vector remains monotonic up to 0.13 eV, the exact full-band structure is used in this range. For higher energy values, an optimized result is used. The derivative of $k^{\nu}_{\mathrm{HV}}(\varepsilon, \vartheta, \varphi)$ with respect to energy, which is required for evaluating the group velocity, is obtained using a backward finite difference scheme. The constraint of a positive derivative with respect to energy is enforced explicitly during the optimization procedure. Once after the minimization is finished at a certain energy value, the fitted coefficients are used for the next energy value. In order to ensure the convergence of the backward scheme, a small energy interval (1 meV) is taken in the optimization.

The resultant fitting parameters for the anisotropic band structure expanded up to the fourth order are shown as functions of energy in Fig. 4.4. It can be seen that the coefficients $\{k_{\mathrm{HV}}\}_{2,0}$, $\{k_{\mathrm{HV}}\}_{4,0}$, and $\{k_{\mathrm{HV}}\}_{4,4}$ have considerable nonzero values, which represent the anisotropy of the resultant band structure even in the Herring–Vogt transformed space.

[3] When the device coordinate frame is not aligned with the coordinate system used to perform the SHE of the inverse dispersion relation, a rotation of the spherical harmonics is required. As a result of this rotation more m values can appear for which the fitting coefficients do not vanish. However, since the rotation of the spherical harmonics preserves the original l value, even after an arbitrary rotation, only even l components are required.

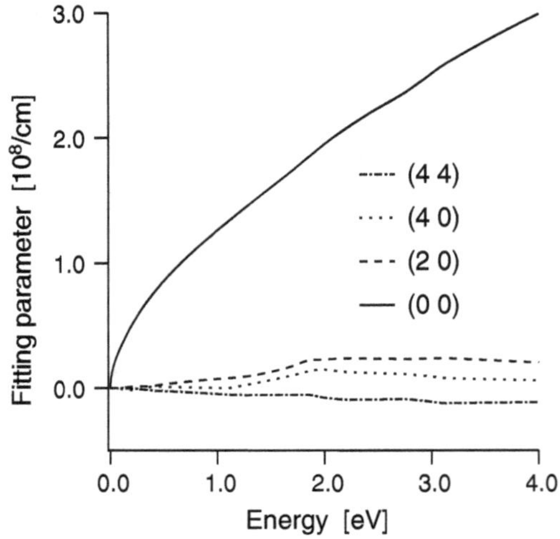

Fig. 4.4 Fitting parameters for the anisotropic model up to the fourth order

Due to the anisotropic nature, the group velocity in the transformed space is no longer aligned with the radial direction, and the expression shown in (4.7) has to be used. The group velocity in the original **k** space is given by (4.14). Moreover, since the DOS is not isotropic, the generalized energy distribution function has to be written in terms of $f_{l',m'}$ following (2.105). This feature increases the number of coupling terms in the scattering integral considerably.

This model is named the anisotropic model.

4.2.2 Model by Vecchi and Rudan and Its Extension

A model with some full band effects has been proposed by Vecchi and Rudan [10]. In this model, only a SHE of the distribution function up to the first order (the lowest order expansion) is possible.

The main idea of this model is that the information about the dispersion relation and its inversion is not explicitly required for the lowest expansion. Recall that $B_{\mathrm{E},l,m,l',m'}$ does not explicitly appear in the expressions for the free-streaming operator (3.15) and (3.16), due to (2.122), as much as the lowest order expansion is concerned. Therefore, as soon as $\mathbf{A}_{0,0,1,m'}$ is determined, the free-streaming operator for the lowest order expansion can be determined without ambiguity. This quantity is directly extracted from the full band structure, under the isotropic valley approximation.

In the original model proposed by Vecchi and Rudan [10], one isotropic valley is assigned to describe the entire conduction band. From the isotropic approximation, the velocity is always aligned with $\mathbf{e}_\varepsilon$, and the DOS is constant over the solid angle. Following the notation used in this work, (4.1) and (4.2), the coefficients $Z(\varepsilon)$ and $\mathbf{A}_{0,0,1,m'}$ are given by

$$Z(\varepsilon) = 6Y_{0,0}^2 V_0^z(\varepsilon), \tag{4.21}$$

$$\mathbf{A}_{0,0,1,m'} = 6Y_{0,0}^2 \mathbf{a}_{0,0,1,m'} \sqrt{V_2^z(\varepsilon) V_0^z(\varepsilon)}, \tag{4.22}$$

where a leading factor of 6 is introduced in order to consider all six valleys and the valley index in the above equations (currently z in (4.21) and (4.22)) can be any direction due to symmetry. Note that the square velocity $V_2(\varepsilon)$ is used for calculating the average velocity.[4]

Although the idea of building an isotropic band model with some full band effects is appealing, its inherent restriction to the lowest order SHE prevents the correct prediction of the distribution in the high-energy region [12]. Since it has been already found that the quasi-ballistic transport in nanoscale devices requires a higher order SHE [7], this approach is in its original form not suitable for simulating nanoscale devices.

A scheme which opens a way to a higher order SHE has been proposed [13], while keeping the isotropic nature of the valley model. In order to capture the anisotropy of the conduction band, three pairs of valleys are considered and the Herring–Vogt transformation is used. Up to this point, it is the similar to the Modena model and the anisotropic valley model. However, after the Herring–Vogt transformation, the isotropic valley approximation is adopted in this model. Therefore, $Z^{\nu}_{l,m,l',m'}$ and $\mathbf{A}^{\nu}_{l,m,l',m'}$ can be extracted easily with a minor modification to the original Vecchi model [13],

$$Z^{\nu}_{l,m,l',m'} = Y_{0,0}^2 V_0^{\nu}(\varepsilon)\delta_{l,l'}\delta_{m,m'}, \tag{4.23}$$

$$\mathbf{A}^{\nu}_{l,m,l',m'} = \hat{T}^{\nu} Y_{0,0}^2 \mathbf{a}_{l,m,l',m'} \sqrt{\frac{m_c^{\nu}}{m_d^{\nu}} V_2^{\nu}(\varepsilon) V_0^{\nu}(\varepsilon)}, \tag{4.24}$$

where m_c^{ν} is the conduction mass given by

$$\frac{1}{m_c^{\nu}} = \frac{1}{3}\left(\frac{1}{m_{xx}^{\nu}} + \frac{1}{m_{yy}^{\nu}} + \frac{1}{m_{zz}^{\nu}}\right). \tag{4.25}$$

Note that the factor of m_c^{ν}/m_d^{ν} is introduced in order to have the square velocity in the transformed wave vector space. However, in higher-order SHE, $B^{\nu}_{\mathrm{E},l,m,l',m}$ does not vanish in general. Neglecting $\tilde{\mathbf{F}}^{\nu}_{\mathrm{E}}$, under the isotropic approximation and with the angle-preserving transformation, $B^{\nu}_{\mathrm{E},l,m,l',m'}$ can be written as

$$B^{\nu}_{\mathrm{E},l,m,l',m'} = \hat{T}^{\nu} \mathbf{F}^{\nu}_{\mathrm{E,pot}}(\mathbf{r},t) \cdot \frac{Z}{\hbar k} \mathbf{b}_{l,m,l',m'}. \tag{4.26}$$

[4]Note that use of V_1 instead of $\sqrt{V_2}$ degrades the quality of the simulation results, as pointed out in [13], because the transport is more closely related with the V_2 moment.

Unfortunately, the meaning of $Z/\hbar k$ is not clear, because we cannot find an isotropic valley whose $Z^{\nu}_{l,m,l',m'}$ and $\mathbf{A}^{\nu}_{l,m,l',m'}$ are identical with (4.23) and (4.24), respectively. The key observation is that the underlying dispersion relation again can be removed using (2.91), because an isotropic valley is assumed. This yields

$$B^{\nu}_{\mathrm{E},l,m,l',m'} = \hat{T}^{\nu}\mathbf{F}^{\nu}_{\mathrm{E,pot}}(\mathbf{r},t)\cdot\frac{1}{2}\left(\frac{\partial}{\partial\varepsilon}vZ\right)\mathbf{b}_{l,m,l',m'}, \tag{4.27}$$

where vZ should be considered as a quantity under the isotropic approximation. To be consistent with (4.23) and (4.24), the following relation is used,

$$vZ = Y^{2}_{0,0}\sqrt{\frac{m^{\nu}_{c}}{m^{\nu}_{d}}}V^{\nu}_{2}(\varepsilon)V^{\nu}_{0}(\varepsilon). \tag{4.28}$$

Since it is consistently constructed with the stabilization scheme and the isotropic valley approximation, this model can be regarded as the consistent generalization of the original Vecchi model [10] to the multi-valley, higher-order SHE case [13].

In these approaches, the isotropic valley approximation has been heavily used in their derivation as shown above. However, the resultant valley models do not possess properly defined dispersion relation. In other words, we cannot find an isotropic valley which fits the group velocity and the DOS taken from the full band structure simultaneously. At this point, inconsistency during the derivation can be seen. However, its use can be fully justified by numerical results, as shown in Sect. 4.2.4. On the other hand, the main advantage of this model is its efficiency. When compared with the Modena model, (4.17)–(4.19), the coefficients of the model in [13] have a similar form, except for their numerical values. Therefore, the numerical complexity of the model in [13] is expected to be similar with that of the Modena model.

These models shown in [10] and [13] will be called the original Vecchi model and the extended Vecchi model, respectively, where the later one introduces significantly improved results over the former.

4.2.3 Full Band Model

As the final model in this section, we propose yet another band model. The development of this model is largely inspired by the success of the extended Vecchi model [13] shown in Sect. 4.2.2. The main motivation why the band model is further developed is twofold. First, although some band models (the anisotropic model and the extended Vecchi model) give good agreement with full band Monte Carlo results as shown in [12, 13], there still remain systematic differences due to the band description. Especially, a considerable difference appears around the energy value at which the first conduction band hits the surface of the Brillouin zone (0.13 eV).

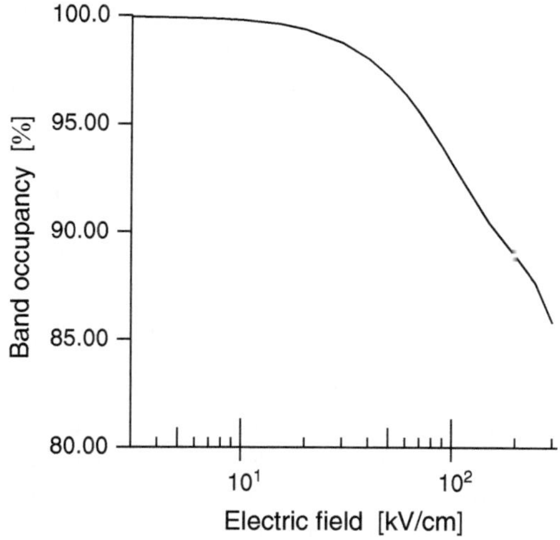

Fig. 4.5 Contribution of the first conduction band to the total electron density, obtained by the full band MC simulation

Since this is a rather low energy, it also introduces systematic differences in the simulation results, such as the drift velocity and the average electron energy. Second, although the performance of these models is quite acceptable, there is no direct linkage between the full band structure and the resultant band model. Usually, it is believed that the conduction bands cannot be directly expanded with spherical harmonics in contrast to the valence bands [12]. The model shown in this subsection can solve this difficulty, at least partially.[5]

The main idea of this model is to construct a valley directly from the full band structure, with help of the generalized coordinate transformation shown in Chap. 2. Since this model is consistently constructed from the full band structure, the accuracy of the model is expected to be improved, when compared with previous models. The problem of an increased number of coupling terms is solved by taking the transport coefficients at their lowest available order. Higher-order coefficients of the transport quantities are calculated under the isotropic valley approximation. In other words, we postpone the application of the isotropic valley approximation until the last stage of the model derivation. Since most of electrons are found in the first conduction band as shown in Fig. 4.5, only this band is treated rigorously. For the higher conduction bands, their contributions to the transport coefficients are approximated using the extended Vecchi model. In this case, V_0^ν and V_n^ν in (4.1) and (4.2) are extracted by excluding the first conduction band. The resultant

[5] Although the derivation of this model is considerably more rigorous than others, there are still missing effects. For example, the antisymmetry against the inversion operation between two valleys of a valley pair is neglected. In principle, this effect can be included at the cost of increased numerical complexity, however, it is neglected in this work. Nevertheless, the simulation result of this model is already quite acceptable.

transport coefficients are simply added to those from the first conduction band, whose construction is explained in the following.

Let us consider a transform which aligns an arbitrary vector $\mathbf{u}(\mathbf{k})$ with the radial direction,

$$\cos\vartheta = \mathbf{e}_z \cdot \frac{\mathbf{u}(\mathbf{k})}{|\mathbf{u}(\mathbf{k})|} = \frac{u_z}{u}, \tag{4.29}$$

$$\sin\vartheta\cos\varphi = \frac{u_x}{u}, \tag{4.30}$$

$$\sin\vartheta\sin\varphi = \frac{u_y}{u}, \tag{4.31}$$

where the zenith direction is aligned with the z-axis. Note that the angle-preserving transformation shown in (2.56), (2.57), and (2.58) is a special case with $\mathbf{u}(\mathbf{k}) = \mathbf{k}$. More importantly, the derivation shown in Sect. 4.2.1, where the Herring–Vogt transformation is followed by subsequent mapping to the energy space, can be considered as the special case with $\mathbf{u}(\mathbf{k}) = \hat{T}^{\nu}\mathbf{k}$. Taking the gradient, it is easily found that

$$-\sin\vartheta(\nabla_{\mathbf{k}}\vartheta) = \frac{1}{u}\nabla_{\mathbf{k}}u_z - \frac{u_z}{u^2}\nabla_{\mathbf{k}}u, \tag{4.32}$$

$$\cos\vartheta\cos\varphi(\nabla_{\mathbf{k}}\vartheta) - \sin\vartheta\sin\varphi(\nabla_{\mathbf{k}}\varphi) = \frac{1}{u}\nabla_{\mathbf{k}}u_x - \frac{u_x}{u^2}\nabla_{\mathbf{k}}u, \tag{4.33}$$

$$\cos\vartheta\sin\varphi(\nabla_{\mathbf{k}}\vartheta) + \sin\vartheta\cos\varphi(\nabla_{\mathbf{k}}\varphi) = \frac{1}{u}\nabla_{\mathbf{k}}u_y - \frac{u_y}{u^2}\nabla_{\mathbf{k}}u. \tag{4.34}$$

From these relations, $\nabla_{\mathbf{k}}\vartheta$ and $\nabla_{\mathbf{k}}\varphi$ can be determined.

Unfortunately, in the conduction bands of the full band structure for silicon $\mathbf{k}$ cannot be a good indicator for the angle variable. For example, let us consider a state near an L point in the first conduction band. Measured from the energy minimum of the first conduction band, this state yields a very sharp angular dependence. A simple solution is taking the L point as the origin of $\mathbf{k}$. However, this leads to difficulties about separating regions including the energy minima of the first conduction band and the L points. A unambiguous separation of the Brillouin zone into small volumes is not a trivial task at all. In the case of the Modena model, an equivalent transformation with $\mathbf{u}(\mathbf{k}) = \hat{T}^{\nu}\mathbf{k}$ can be found using $\mathbf{u}(\mathbf{k}) = \left(\hat{T}^{\nu}\right)^{-1}\mathbf{v}(\mathbf{k})$. Inspired from this observation, in this model, a specific case of the transformation is proposed as follows:

$$u_x(\mathbf{k}) = \tilde{T}_{xx}^{-1}(\varepsilon)v_x(\mathbf{k}), \tag{4.35}$$

$$u_y(\mathbf{k}) = \tilde{T}_{yy}^{-1}(\varepsilon)v_y(\mathbf{k}), \tag{4.36}$$

$$u_z(\mathbf{k}) = \tilde{T}_{zz}^{-1}(\varepsilon)v_z(\mathbf{k}), \tag{4.37}$$

where $\tilde{T}_{xx}^{-1}(\varepsilon)$, $\tilde{T}_{yy}^{-1}(\varepsilon)$, and $\tilde{T}_{zz}^{-1}(\varepsilon)$ are energy dependent fitting quantities. Like the Herring–Vogt transformation, the product of the three fitting quantities must be equal to one

$$\tilde{T}_{xx}^{-1}(\varepsilon)\tilde{T}_{yy}^{-1}(\varepsilon)\tilde{T}_{zz}^{-1}(\varepsilon) = 1. \tag{4.38}$$

When we consider two different fitting quantities, the longitudinal fitting quantity and the transverse one, only one independent fitting quantity remains. Then all gradients, $\nabla_{\mathbf{k}} u_d$ and $\nabla_{\mathbf{k}} u$, can be obtained by

$$\begin{aligned} \nabla_{\mathbf{k}} u_d &= \tilde{T}_{dd}^{-1}(\varepsilon) \nabla_{\mathbf{k}} v_d + v_d \nabla_{\mathbf{k}} \tilde{T}_{dd}^{-1}(\varepsilon) \\ &= \tilde{T}_{dd}^{-1}(\varepsilon) \hbar \sum_{d'} \mathbf{e}_{d'} m_{d'd}^{-1}(\mathbf{k}) + v_d \hbar \mathbf{v} \frac{\partial}{\partial \varepsilon} \tilde{T}_{dd}^{-1}(\varepsilon), \end{aligned} \tag{4.39}$$

$$\nabla_{\mathbf{k}} u = \nabla_{\mathbf{k}} \sqrt{u_x^2 + u_y^2 + u_z^2} = \frac{1}{u} \left(u_x \nabla_{\mathbf{k}} u_x + u_y \nabla_{\mathbf{k}} u_y + u_z \nabla_{\mathbf{k}} u_z \right), \tag{4.40}$$

where $m_{d'd}^{-1}(\mathbf{k})$ is an entry of the inverse mass tensor defined as [21]

$$m_{d'd}^{-1} = \frac{1}{\hbar^2} \frac{\partial^2 \varepsilon}{\partial k_{d'} \partial k_d}. \tag{4.41}$$

The calculation of the inverse mass tensor follows the method shown in [21]. Although the choice of $\tilde{T}_{dd}^{-1}(\varepsilon)$'s is largely arbitrary, we want to suppress the magnitude of coupling terms between two components for which the absolute difference of l values are larger than 1, because the isotropic valley approximation is to be applied at the last stage of the model derivation. When the zenith direction is aligned with the z-axis, these coefficients are adjusted in order to satisfy the following relation:

$$(A_{0,0,1,-1,y}^{\nu})^2 + (A_{0,0,1,0,z}^{\nu})^2 + (A_{0,0,1,1,x}^{\nu})^2 = Y_{0,0}^4 V_2^{\nu} (V_0^{\nu})^2, \tag{4.42}$$

which holds for the Vecchi models. By this constraint, the final isotropic model preserves V_2 like the Vecchi models. Figure 4.6 shows the resultant fitting parameters, $\tilde{T}_{dd}^{-1}$, satisfying the above constraint. Rapid change around 0.13 eV reflects the distortion of the dispersion relation near the surface of the Brillouin zone. Two directional fitting coefficients (longitudinal and transverse) heavily depend on the energy variable.

Recall that the L valleys start contributing to V_0 at 1.1 eV. From this energy value, the valleys considered in this model become a mixture of Δ valleys and L valleys. The principal axis of the L valleys is aligned with the $\langle 111 \rangle$ direction, instead of the $\langle 001 \rangle$ direction assumed in the transformation rule, (4.35)–(4.37). This misalignment degrades the validity of the assumed transformation. However, in order to avoid any further increase in complexity, the assumed transformation is applied without modifications.

As a result of the coordinate transformation, we have an anisotropic valley, which preserves most of the features of the original full band structure. Although all the coupling terms can be included exactly in principle, this approach has almost no practical merit. Inspired by the quite acceptable performance of the extended Vecchi model reported in [13], a simplification is performed at the final step of

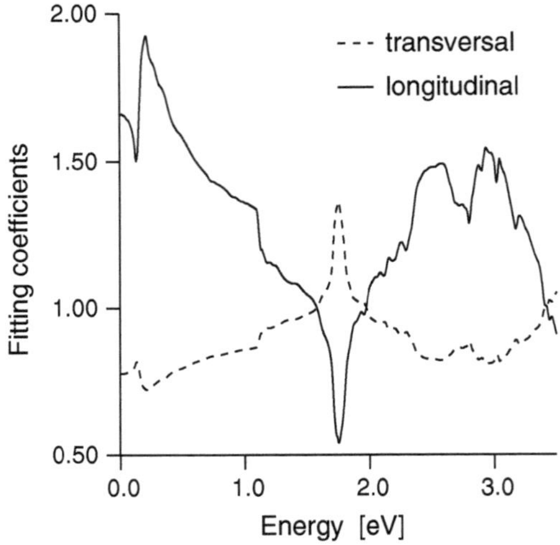

Fig. 4.6 Fitting coefficients $\widetilde{T}_{dd}^{-1}$ along different directions

the model derivation. The transport coefficients at their lowest available order are directly extracted from the full band structure after the coordinate transformation. More precisely, seven quantities – $Z^{\nu}_{0,0}$, $A^{\nu}_{0,0,1,0,z}$, $A^{\nu}_{0,0,1,1,x}$, $A^{\nu}_{0,0,1,-1,y}$, $B^{\nu}_{\mathrm{E},1,0,0,0,z}$, $B^{\nu}_{\mathrm{E},1,1,0,0,x}$, and $B^{\nu}_{\mathrm{E},1,-1,0,0,y}$ – are extracted.[6] From these quantities, applying the isotropic valley approximation to each direction, the transport coefficients are obtained as follows:

$$Z^{\nu}_{l,m,l',m'} = Y_{0,0} Z^{\nu}_{0,0} \delta_{l,l'} \delta_{m,m'}, \tag{4.43}$$

$$A^{\nu}_{l,m,l',m',d}(\varepsilon) = A^{\nu}_{0,0,1,m'',d}(\varepsilon) \frac{\mathbf{e}_d \cdot \mathbf{a}_{l,m,l',m'}}{\mathbf{e}_d \cdot \mathbf{a}_{0,0,1,m''}}, \tag{4.44}$$

$$B^{\nu}_{\mathrm{E},l,m,l',m',d}(\varepsilon) = B^{\nu}_{\mathrm{E},1,m'',0,0,d}(\varepsilon) \frac{\mathbf{e}_d \cdot \mathbf{b}_{l,m,l',m'}}{\mathbf{e}_d \cdot \mathbf{b}_{1,m'',0,0}}, \tag{4.45}$$

where m'' is 0, 1, or -1 for d being z, x, or y, respectively. Note that the denominators in the above equations do not vanish. Within the limitation of the isotropic valley, this is a consistent way to include full band effects. Of course, the validity of this final simplification should be carefully checked by the actual numerical results of the resultant band model. Note that this simplification step does not violate any of the relations between transport coefficients.

The resultant band model will be called the full band model for electrons, although in a rigorous sense it does not describe the "perfect" full band structure due to the final simplification step explained in the last paragraph.

[6]When $B^{\nu}_{\mathrm{E},1,m,0,0,d}$ coefficients are extracted, unit magnitude of the d-directional force is understood.

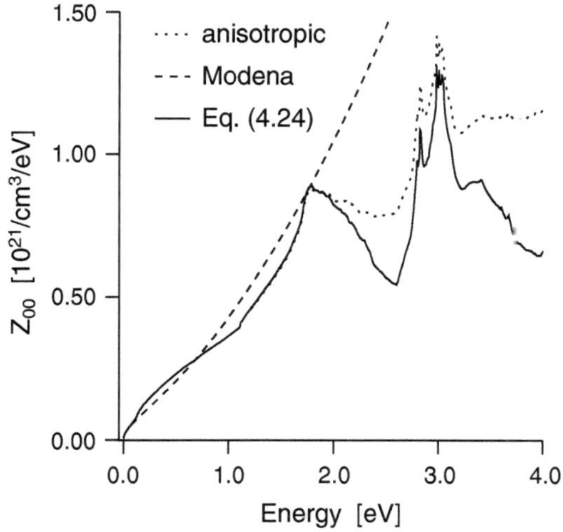

Fig. 4.7 Isotropic component of the DOS of a valley for one spin direction, $Z^{\nu}_{0,0}$, for relaxed silicon

4.2.4 Comparison

The models of the previous subsections are compared for the relaxed silicon. Since the failure of the original Vecchi model is described in [12, 13], it is not included in the comparison. As shown in the previous subsections, the models are denoted as the Modena model [9], the anisotropic model [12], the extended Vecchi model [13], and the full band model.

First, internal transport coefficients are compared. The DOS per spin, more precisely its isotropic component $Z^{\nu}_{0,0}(\varepsilon)$, is shown in Fig. 4.7. For the Modena model, up to about 1.8eV, the average DOS shows reasonable agreement. However, beyond this value, the DOS is too high. This is the main reason why this model fails to predict correctly the electron distribution function at high energies. For the anisotropic valley model shown in Sect. 4.2.1, up to about 1.8 eV the DOS of the full band structure is perfectly reproduced. Although the DOS cannot be fitted perfectly due to the constraint for the positive $\partial k^{\nu}/\partial\varepsilon$, a remarkable improvement is obvious compared to the Modena model. For the models shown in Sects. 4.2.2 and 4.2.3, the average DOS fits perfectly, due to the virtue of construction.

The transport parameters, $\mathbf{e}_z \cdot \{\mathbf{v}Z\}^z_{1,0}$ and $\mathbf{e}_x \cdot \{\mathbf{v}Z\}^z_{1,1}$, are shown in Figs. 4.8 and 4.9. Since the z-direction is the longitudinal direction for the z-valley, these quantities correspond to the longitudinal velocity and the transverse velocity multiplied by the DOS, respectively. Before the first conduction band hits the surface of the Brillouin zone (below 0.13 eV), all models agree well. Considerable differences can be found above this energy value according to the band model adopted. When compared with the anisotropic model or the extended Vecchi model, the Modena model underestimates these parameters. The anisotropic model and the extended Vecchi model show similar values, while the former has a slightly smaller magnitude for these parameters, because other higher-order coefficients such as

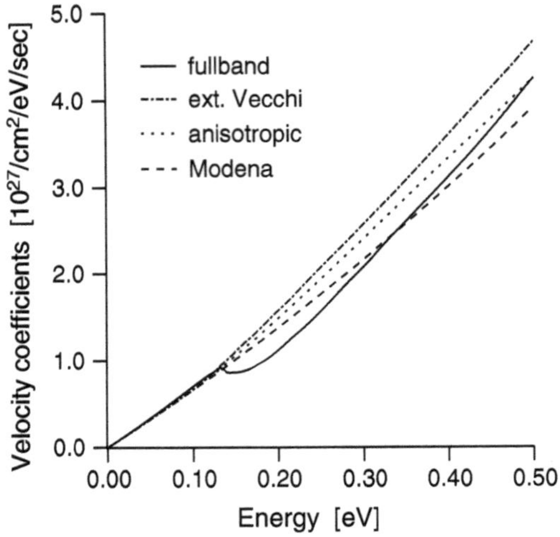

Fig. 4.8 Longitudinal velocity multiplied by the DOS, $\mathbf{e}_z \cdot \{\mathbf{v}Z\}^z_{1,0}$

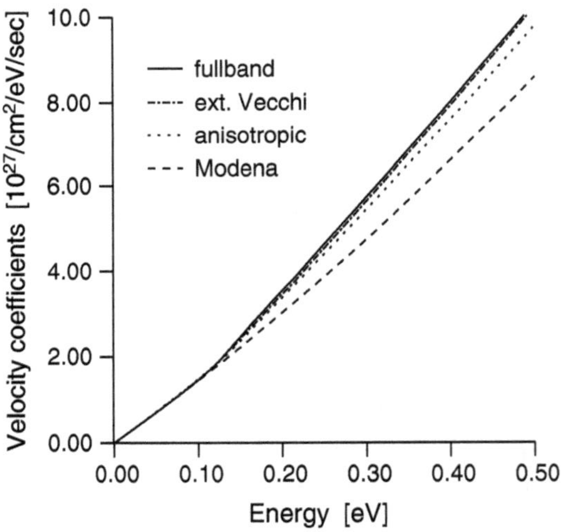

Fig. 4.9 Transverse velocity multiplied by the DOS, $\mathbf{e}_x \cdot \{\mathbf{v}Z\}^z_{1,1}$

$\mathbf{e}_z \cdot \{\mathbf{v}Z\}^z_{3,0}$ also contribute. In the case of the full band model, the transport parameter $\mathbf{e}_z \cdot \{\mathbf{v}Z\}^z_{1,0}$, which is related with the longitudinal transport, is somewhat lowered above 0.13 eV. The abrupt change of the fitting parameters around this energy value shown in Fig. 4.6 is the reason for this behavior.

In the case of the extended Vecchi model and the full band model, the same parameter of $V_2(\varepsilon)$ is used as a pivotal quantity to build the band model. Both models preserve this quantity exactly, due to the virtue of construction.[7] The differences in $\mathbf{e}_z \cdot \{\mathbf{v}Z\}^z_{1,0}$ and $\mathbf{e}_x \cdot \{\mathbf{v}Z\}^z_{1,1}$ by the band model, which are observed in Figs. 4.8

[7] Since the transverse velocity is much larger than the longitudinal one, the visible difference shown in Fig. 4.8 is counterbalanced by the small difference in Fig. 4.9.

and 4.9, reveal that the directional decomposition of the group velocity is the key modeling factor. Note that $V_2(\varepsilon)$ is isotropically distributed in the case of the original Vecchi model. It is distributed according to the diagonal components of the Herring–Vogt transformation in the extended Vecchi model. On the other hand, for the full band model for electrons, the ratio between the longitudinal and transverse velocities depends on the underlying coordinate transformation, where the fitting coefficients are contrained by (4.42).

Now, simulation results for the bulk system are shown. Results of full band Monte Carlo simulations [7], performed sufficiently long to reduce the error of the resultant drift velocity (typically below 0.2 or 0.3%), are used as reference data. A fifth-order SHE is used, which is sufficient for bulk simulations. Comparison for devices can be found in Chap. 5.

As expected, the low-field mobility shows good agreement for all band models, therefore, it is not shown here. In order to investigate high-field transport, the phonon limited drift velocity for an electric field in the ⟨001⟩ direction is shown in Fig. 4.10. The lattice temperature is 300 K. Overall, good agreement with the MC data is found for the anisotropic model, the extended Vecchi model, and the full band model. However, it is revealed that the Modena model is not very accurate for high electric fields. It erroneously predicts a much earlier roll-off of the drift velocity. At 300 K, the maximum error of the extended Vecchi model is less than 7% at 100 kV cm^{-1}. Interestingly enough, the results of the anisotropic model for the drift velocity are very similar with those of the extended Vecchi model, although elaborate anisotropic valleys are assumed. On the other hand, the full band model shows better agreement with the full band MC results. The maximum error of the drift velocity (also observed at 100 kV cm^{-1}) is less than 3%. The error of the phonon limited drift velocity compared with the accurate MC data is summarized in Fig. 4.11.

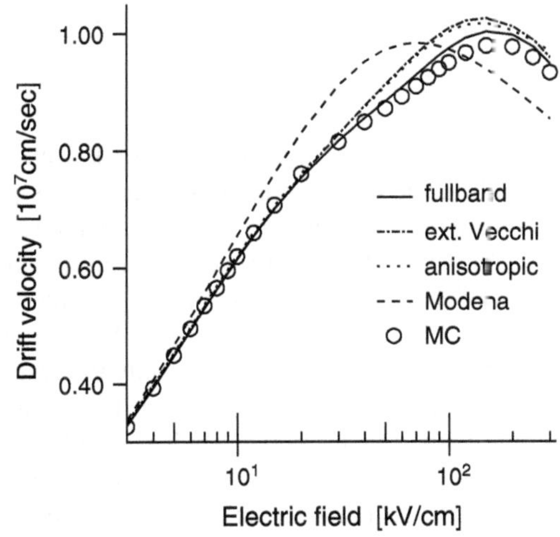

Fig. 4.10 Simulated phonon limited drift velocity of electrons in relaxed silicon for an electric field in ⟨001⟩ direction at 300 K

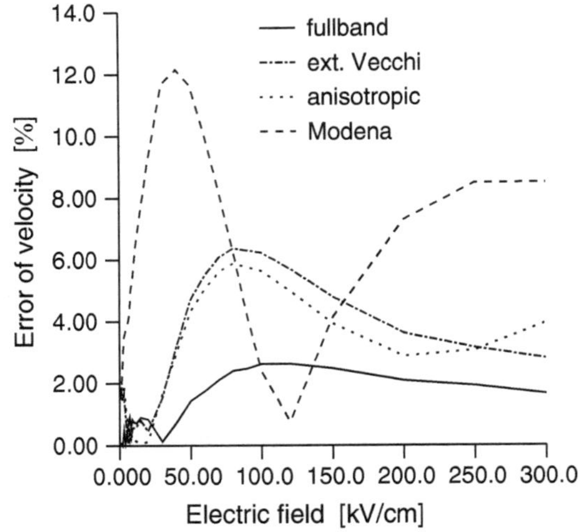

Fig. 4.11 Absolute error of the simulated phonon limited drift velocity of electrons in relaxed silicon for an electric field in ⟨001⟩ direction, compared to full band MC data

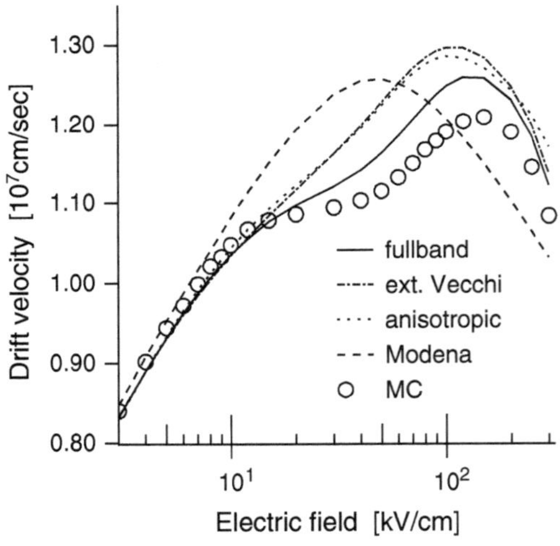

Fig. 4.12 Simulated phonon limited drift velocity of electrons in relaxed silicon for an electric field in ⟨001⟩ direction at 77 K

Although some band models work well at room temperature, their success should not be overemphasized because it is partially due to the strong scattering mechanisms in silicon at room temperature. Effects of inaccuracy in the band description are hidden by the scattering mechanisms. In order to demonstrate the effect of the scattering strength, the phonon limited drift velocity at 77 K is shown in Fig. 4.12. Since the low-field mobility is also correctly predicted by all band models even at this temperature, only the results for high electric fields are shown. The general trends are quite similar to the case of 300 K, however, the magnitude of the deviation from the FB MC results is increased. The maximum error of the drift velocity obtained by the Modena model, the anisotropic model, and the extended Vecchi model is exceeding 10%. Even for the full band model, a maximum error of about 5% is obtained. Further investigation reveals that considering the

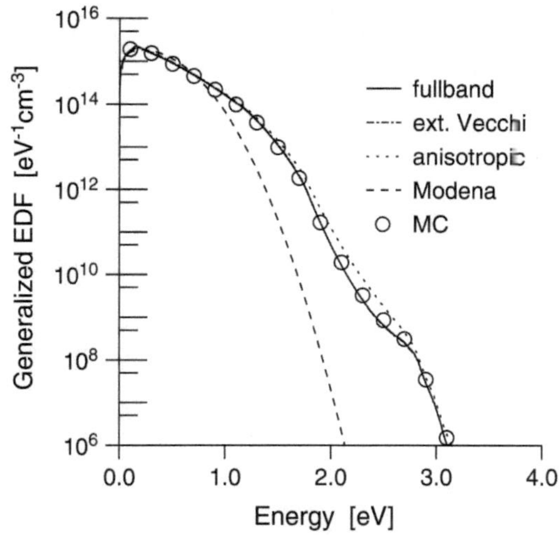

Fig. 4.13 Generalized energy distribution function of zeroth order calculated by the different models with only phonon scattering and normalized to an electron density of $10^{15}\,\mathrm{cm}^{-3}$. An electric field of $100\,\mathrm{kV\,cm}^{-1}$ is applied in $\langle 001\rangle$ direction

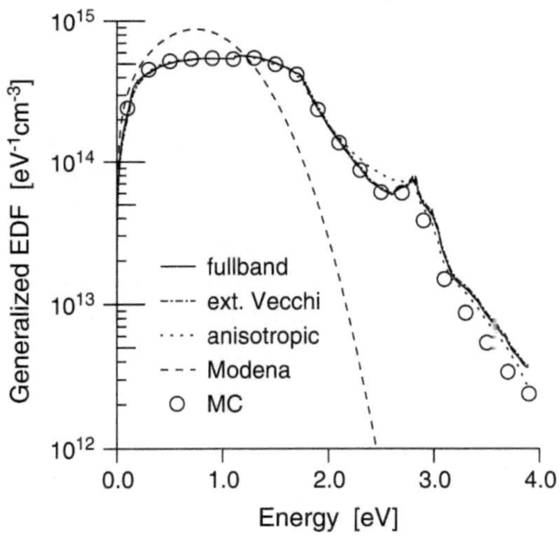

Fig. 4.14 Generalized energy distribution function of zeroth order calculated by the different models. The applied electric field is $300\,\mathrm{kV\,cm}^{-1}$. Otherwise, the conditions are the same as in Fig. 4.13

antisymmetry against the inversion operation between two valleys of a valley pair is the key factor to obtain even better agreement with the full band MC results. However, since the number of unknown variables doubles if the antisymmetric part of the band structure is taken into account, this approach is of less practical interest.

The generalized energy distribution defined in (2.105), more precisely its isotropic component $g^{\nu}_{0,0}(\varepsilon)$ summed over all six valleys, is shown in Figs. 4.13 and 4.14. The applied electric field is 100 and $300\,\mathrm{kV\,cm}^{-1}$, respectively. The importance of taking into account the FB effects can be clearly seen since the Modena model predicts a much "colder" distribution compared to other models. All other models show excellent agreement with the FB MC results for the entire range of energy [12, 13]. In these figures plotted on a semi-logarithmic scale, the

Fig. 4.15 Simulated impact ionization coefficient over the inverse electric field

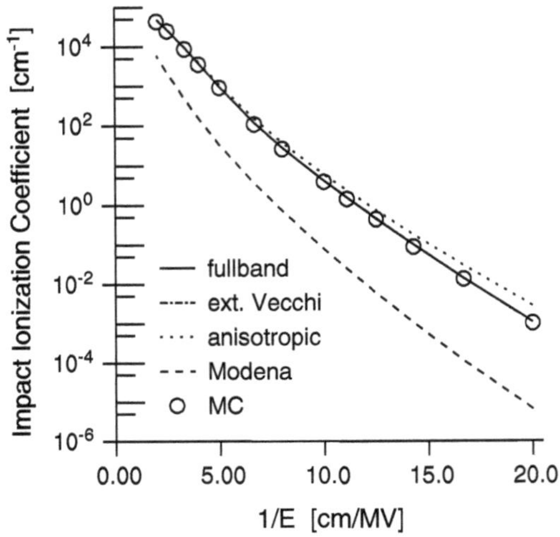

difference of the simulated distributions of the extended Vecchi model and the full band model is invisible.

Since the distribution function is closely reproduced by all band models except for the Modena model, an averaged quantity such as the impact ionization coefficient is also expected to agree well with the FB MC results. The impact ionization coefficient simulated by the different band models is shown in Fig. 4.15. Impact ionization is modeled with an exclusively energy-dependent scattering rate [21], as shown in Sect. 4.3. The Modena model with the same scattering rate for impact ionization generates a far too low impact ionization coefficient. The anisotropic model, the extended Vecchi model, and the full band model, show excellent agreement over a wide range of the applied electric field. Again, the difference of the impact ionization coefficients between the extended Vecchi model and the full band model is negligible.

Although the impact ionization coefficient, which is determined in the high energy range, can be accurately calculated by three models (the anisotropic model, the extended Vecchi model, and the full band model), the behavior in the low energy range shows some systematic differences due to the different band models. As pointed out before, the different band models have different longitudinal transport parameters ($\mathbf{e}_z \cdot \{\mathbf{v}Z\}^z_{1,0}$) as shown in Fig. 4.8. Since the full band model has lower values for this parameter around the surface of the Brillouin zone, a slightly "colder" distribution is obtained with this model. In Fig. 4.16, the generalized energy distribution is shown for two different values of the applied electric field, 100 and 150 kV cm^{-1}. Note that the maximum error of the drift velocity is observed at about 100 kV cm^{-1}. A linear scale is used in order to see the difference more clearly. The results of the Modena model are not shown, because it is very different from the others. The electron distribution is more closely reproduced by the full band model. Especially, the slope of the distribution below 0.13 eV is increased to

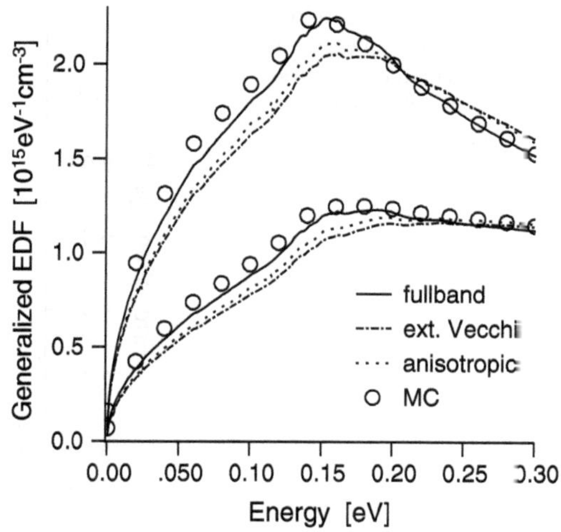

Fig. 4.16 Generalized energy distribution function of zeroth order calculated by the different models. Two different values of the applied electric field, 100 kV cm^{-1} (*upper four curves*) and 150 kV cm^{-1} (*lower four curves*), are considered. Otherwise, the conditions are the same as Fig. 4.13

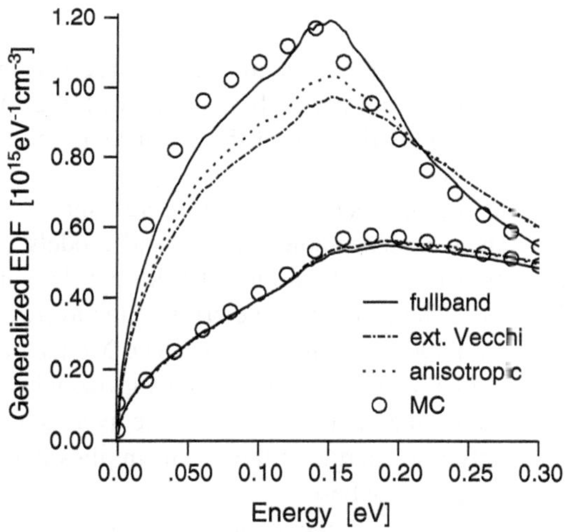

Fig. 4.17 Generalized energy distribution function of zeroth order calculated by the different models. The contribution from the z-valleys (*upper four curves*) and the x- or y-valleys (*lower four curves*) is separately shown. The applied electric field is 100 kV cm^{-1}. Otherwise, the conditions are the same as Fig. 4.13

yield a better agreement with the FB MC results.[8] Note that below 0.13 eV the transport parameters shown in Figs. 4.8 and 4.9 are almost identical among the different models. Therefore, the improved description of the distribution by the full band model should be attributed to the better consideration of the zone boundary effects. Such an argument is further confirmed by the generalized energy distribution summed over only a pair of valleys, as shown in Fig. 4.17. For all models shown in the figure, the distribution of the x- or y-valleys is accurately calculated, because the transverse transport coefficient, $\mathbf{e}_x \cdot \{\mathbf{v}Z\}^z_{1,1}$, is modeled accurately. However,

[8]The slope of the distribution in the low energy range, calculated by the full band model, is still slightly smaller than that by the FB MC simulation. It seems that the antisymmetry of the valley is the main cause of the remaining discrepancy.

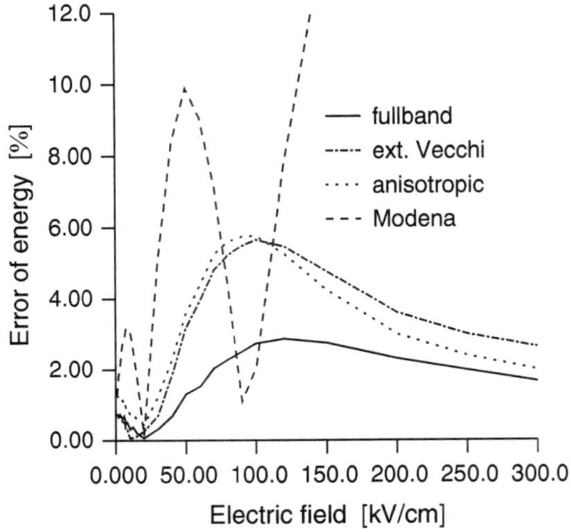

Fig. 4.18 Absolute error of the average electron energy in relaxed silicon for an electric field in ⟨001⟩ direction, compared with MC data

considerable differences are clearly observed in the distribution of the z-valleys. Although it is not perfectly matched with the FB MC results, the full band model predicts the distribution of the z-valleys in the low energy range much better than other two models.

The average energy of electrons also shows a similar behavior. Although overall agreement is acceptable for all band models except for the Modena model, for certain electric fields (around $100\,\mathrm{kV\,cm^{-1}}$), some discrepancy occur as shown in Fig. 4.18. The peak absolute error of the average energy is similar than that of the drift velocity shown in Fig. 4.11. The Modena model is not very accurate, and its error w.r.t. the FB MC results rapidly grows when the applied electric field increases. Since both, the drift velocity and the average energy are quantities determined by relatively low energies,[9] this result implies that the full band model describes the low energy range best.

As a result of the comparison, it is clear that the anisotropic model, the extended Vecchi model, and the full band model give good agreement with the MC simulation, where the last one is better than the other two. The Modena model is not very successful for describing the high-field transport, and completely fails to predict impact ionization. In the point of view of numerical efficiency, certainly the anisotropic model has a drawback due to the increased number of coupling terms. On the other hand, the extended Vecchi model and the full band model have the same numerical properties as the Modena model. Summarizing these results, it is concluded that the full band model is the best choice among the four models considered in this subsection both regarding the physical accuracy and the numerical

[9] For example, at $100\,\mathrm{kV\,cm^{-1}}$, the peak of the electron current, which is distributed over the energy, occurs at 0.3 eV.

efficiency, while the extended Vecchi model also performs quite well. These two models are extensively employed in the simulations shown in Chap. 5.

4.3 Scattering Mechanisms

The scattering processes are characterized by their respective transition rates which are calculated with Fermi's golden rule [22]

$$S^{\beta',\beta}(\mathbf{r},\mathbf{k}'|\mathbf{k}) = \frac{2\pi}{\hbar}|M^{\beta',\beta}(\mathbf{k}'|\mathbf{k})|^2\delta[\varepsilon^{\beta'}(\mathbf{k}') - \varepsilon^{\beta}(\mathbf{k}) - \varepsilon^{\mathrm{trans}}(\mathbf{k}'|\mathbf{k})], \tag{4.46}$$

where M is the matrix element of the interaction and $\varepsilon^{\mathrm{trans}}$ the corresponding energy transfer.

Scattering by phonons, alloy fluctuations, ionized impurities, and surface roughness is included. Constant matrix elements for phonon scattering are assumed. The scattering mechanisms and related parameters used in the FB Monte Carlo simulator [21] are used without modification.

In the following, for completeness, the transition rates of the scattering mechanisms are briefly shown.

4.3.1 *Phonon Scattering*

The transition rates for Si are explicitly shown here. The transition rate for the interaction with optical and intervalley acoustic phonons reads

$$S^{\beta',\beta}_{\mathrm{opt},\eta}(\mathbf{r},\mathbf{k}'|\mathbf{k}) = \frac{\pi(D_t K_\eta)^2}{\Omega_s \rho^{\mathrm{Si}}\omega^{\mathrm{Si}}_\eta}\left(n(\hbar\omega^{\mathrm{Si}}_\eta) + \frac{1}{2} \pm \frac{1}{2}\right)$$
$$\delta[\varepsilon^{\beta'}(\mathbf{k}') - \varepsilon^{\beta}(\mathbf{k}) \pm \hbar\omega^{\mathrm{Si}}_\eta] r(\eta, \nu(\mathbf{k}'), \nu(\mathbf{k})), \tag{4.47}$$

where ρ is the mass density, $D_t K_\eta$ the coupling constant, ω_η the phonon frequency, r the selection rule of the phonon mode η, n the occupation number for phonons given by the Bose-Einstein distribution [22]. The transition rate for scattering with intravalley acoustic phonon is given in the equipartition, elastic approximation by [14]

$$S^{\beta',\beta}_{ac}(\mathbf{r},\mathbf{k}'|\mathbf{k}) = \frac{2\pi k_{\mathrm{B}} T_0 (\Xi^{\mathrm{Si}})^2}{\Omega_s \hbar \rho^{\mathrm{Si}} (u^{\mathrm{Si}}_l)^2}\delta[\varepsilon^{\beta'}(\mathbf{k}') - \varepsilon^{\beta}(\mathbf{k})]\delta_{\nu'(\mathbf{k}'),\nu(\mathbf{k})}, \tag{4.48}$$

where Ξ is the deformation potential and u_l the longitudinal sound velocity.

Table 4.1 Material parameters for Si and Ge [23,24]

	Si	Ge
ρ	$2.33\,\mathrm{g\,cm^{-3}}$	$5.32\,\mathrm{g\,cm^{-3}}$
u_l	$9.05 \cdot 10^5\,\mathrm{cm\,s^{-1}}$	$5.40 \cdot 10^5\,\mathrm{cm\,s^{-1}}$
ϵ_{si}	$11.7 \cdot \epsilon_0$	$16.0 \cdot \epsilon_0$
Ξ_e	$8.90\,\mathrm{eV}$	$8.79\,\mathrm{eV}$

Table 4.2 Modes, coupling constants, energies, and selection rules of inelastic phonon scattering [9,23–25]

η	Mode	$D_t K$ ($10^8\,\mathrm{eV\,cm^{-1}}$) Si	$\hbar\omega$ (meV) Si	$D_t K$ ($10^8\,\mathrm{eV\,cm^{-1}}$) Ge	$\hbar\omega$ (meV) Ge	r
1	TA	0.470	12.1	0.479	5.60	g
2	LA	0.740	18.5	0.772	8.60	g
3	LO	10.23	62.0	9.280	37.0	g
4	TA	0.280	19.0	0.283	9.90	f
5	LA	1.860	47.4	1.940	28.0	f
6	TO	1.860	58.6	1.690	32.5	f

The transition rates for SiGe are obtained by linear interpolation of the silicon and germanium cases [21], and are not repeated here. Tables 4.1 and 4.2 summarize the parameters used for phonon scattering. Note that for the Modena model, its own phonon parameters shown in [9] have been used instead.

When the transition rates are actually implemented in the SHE solver, which is based on the valley description, $S^{\beta',\beta}$ should be changed into $S^{\nu',\nu}$. Since the selection rules for the transition rates (4.47) and (4.48) are already written in terms of valleys instead of bands, their transform can be done in a straightforward way.

4.3.2 Alloy Scattering

The transition rate of alloy scattering reads

$$S_{\mathrm{alloy}}^{\beta',\beta}(\mathbf{r},\mathbf{k}'|\mathbf{k}) = \frac{\pi a_\perp a_\parallel^2 U^2}{4\Omega_s \hbar}(1 - x_{\mathrm{Ge}})x_{\mathrm{Ge}}\delta[\varepsilon^{\beta'}(\mathbf{k}') - \varepsilon^{\beta}(\mathbf{k})], \tag{4.49}$$

where $a_\perp$ is the lattice constant in growth direction and $a_\parallel$ parallel to the SiGe layer. The alloy scattering potential U is 0.75 eV for electrons [24].

Since there is no selection rule for this scattering, the resultant $S^{\nu',\nu}$ does not include a selection rule.

4.3.3 Impurity Scattering

Scattering by ionized impurities is based on the model by Brooks and Herring [14,26],

$$S_{\text{imp}}^{\beta',\beta}(\mathbf{r},\mathbf{k}|\mathbf{k}') = \frac{2\pi q^4(N_\text{A}+N_\text{D})}{\Omega_s\hbar\kappa(\mathbf{r})(\beta^{\text{sc}\,2}(\mathbf{r})-|\mathbf{k}'-\mathbf{k}|^2)} \delta[\varepsilon^{\beta'}(\mathbf{k}')-\varepsilon^{\beta}(\mathbf{k})]\delta_{\beta',\beta}\delta_{\nu'(\mathbf{k}'),\nu(\mathbf{k})}, \tag{4.50}$$

where N_A and N_D are the acceptor and donor concentration, respectively, and the dielectric constant $\kappa(\mathbf{r})$ is obtained by linear interpolation. The inverse screening length β^{sc} is evaluated assuming local equilibrium, where the temperature coincides with the lattice temperature T_0. When the Pauli principle is included, the local equilibrium distribution is calculated based on Fermi-Dirac statistics [22], otherwise Maxwell–Boltzmann statistics.

Impurity scattering is approximated by an isotropic-elastic process [21, 27]. Using an approximation for a nonparabolic spherical band, the expression depending on $|\mathbf{k}'-\mathbf{k}|$ can be written as [21]

$$\frac{1}{\beta^{\text{sc}}-|\mathbf{k}'-\mathbf{k}|^2} = \ln\left(\frac{a+2}{a}\right) - \frac{2}{a+2}, \tag{4.51}$$

where $a = (\beta^{\text{sc}})^2/(2|\mathbf{k}|^2)$. Note that $|\mathbf{k}|$ for evaluating a should be calculated consistently with the approximated band. Furthermore, comparison of simulations with and without the approximation of an isotropic-elastic process yields no significant difference both in the linear and nonlinear transport regimes [27,28].

Since the Brooks-Herring model for impurity scattering does not correctly describe the mobility at high doping concentration, an empirical correction of the model is used [21,29]. Figure 4.19 shows different fit factors used for different statistics in order to reproduce the Caughey-Thomas expression for the mobility [30]. Therefore, the mobility limited by impurity scattering obtained from different statistics are the same.

4.3.4 Impact Ionization

An exclusively energy dependent scattering rate is used to model impact ionization. The resultant scattering rate reads for electrons in relaxed silicon [25,31]

$$S^{\text{II}}(\varepsilon) = \begin{cases} 1.49\cdot10^{11}\frac{1}{s}\left(\frac{\varepsilon}{\text{eV}}-1.128\right)^3: & 1.128\,\text{eV}<\varepsilon<1.750\,\text{eV} \\ 1.13\cdot10^{12}\frac{1}{s}\left(\frac{\varepsilon}{\text{eV}}-1.572\right)^2: & 1.750\,\text{eV}<\varepsilon \end{cases}. \tag{4.52}$$

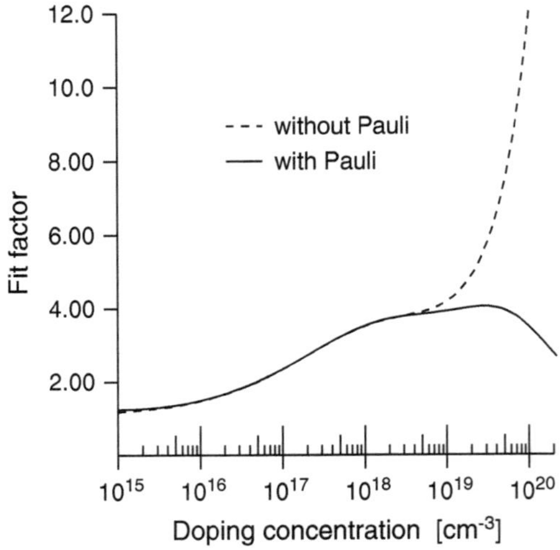

Fig. 4.19 Heuristic fit factor for impurity scattering with or without the Pauli principle

The transition rate is obtained by considering the total DOS at the final energy. It is assumed that all three particles (primary electron, secondary electron, and hole) have the same energy after scattering [32].

4.3.5 Surface Roughness Scattering

The full quantum mechanical treatment of the inversion-layer transport can be found in Part 3, where the transport in a quasi-2D hole gas is the main topic. In the 3D **k**-space of the SHE solver, a rigorous treatment on this effect cannot be expected. Here, instead, a simple approach to fit an experimentally verified mobility model is used.

In order to account for inversion-layer transport, an additional scattering mechanism is included in the simulation [33]. This scattering mechanism is assumed to be an isotropic-elastic scattering process. The dependence of the transition rate on the energy is also neglected. In this work, the parameters for this scattering process is matched to the Lombardi mobility model [34]. Note that other surface mobility models can be also incorporated easily.

References

1. Rieger, M.M., Vogl, P.: Electronic-band parameters in strained $Si_{1-x}Ge_x$ alloys on $Si_{1-y}Ge_y$ substrates. Phys. Rev. B **48**, 14276–14287 (1993)
2. Rieger, M.M., Vogl, P.: Electronic-band parameters in strained $Si_{1-x}Ge_x$ alloys on $Si_{1-y}Ge_y$ substrates. Phys. Rev. B **50**, 8138 (1994)
3. Cohen, M.L., Chelikowsky, J.R.: Electronic Structure and Optical Properties of Semiconductors, 2nd edn. Springer, New York (1989)

4. Ungersboeck, E., Dhar, S., Karlowatz, G., Sverdlov, V., Kosina, H., Selberherr, S.: The effect of general strain on the band structure and electron mobility of silicon. IEEE Trans. Electron Dev. **54**(9), 2183–2190 (2007)
5. Abramo, A., Venturi, F., Sangiorgi, E., Higman, F., Riccò, B.: A numerical method to compute isotropic band models from anisotropic semiconductor band structures. IEEE Trans. Comp. Aided Des. **12**, 1327–1336 (1993)
6. Kosina, H., Harrer, M., Vogl, P., Selberherr, S.: A Monte Carlo transport model based on spherical harmonics expansion of the valence bands. In: Proceedings of SISDEP, pp. 396–399 (1995)
7. Jungemann, C., Pham, A.-T., Meinerzhagen, B., Ringhofer, C., Bollhöfer, M.: Stable discretization of the Boltzmann equation based on spherical harmonics, box integration, and a maximum entropy dissipation principle. J. Appl. Phys. **100**, 024502–1–13 (2006)
8. Pham, A.T., Jungemann, C., Meinerzhagen, B.: A full-band spherical harmonics expansion of the valence bands up to high energies. In: Proceedings of SISPAD, pp. 361–364 (2006)
9. Brunetti, R., Jacoboni, C., Nava, F., Reggiani, L., Bosman, G., Zijlstra, R.J.J.: Diffusion coefficient of electrons in silicon. J. Appl. Phys. **52**, 6713–6722 (1981)
10. Vecchi, M.C., Rudan, M.: Modeling electron and hole transport with full-band structure effects by means of the spherical-harmonics expansion of the BTE. IEEE Trans. Electron Dev. **45**(1), 230–238 (1998)
11. Matz, G., Hong, S.-M., Jungemann, C.: A spherical harmonics expansion of the conduction bands including the full-band effects. In: Proceedings of SISPAD, pp. 167–170 (2010)
12. Hong, S.-M., Matz, G., Jungemann, C.: A deterministic Boltzmann equation solver based on a higher-order spherical harmonics expansion with full-band effects. IEEE Trans. Electron Dev. **57**, 2390–2397 (2010)
13. Jin, S., Hong, S.-M., Jungemann, C.: An efficient approach to include full band effects in deterministic Boltzmann equation solver based on high-order spherical harmonics expansion. IEEE Trans. Electron Dev. (2011), vol. 58, pp. 1287–1294
14. Jacoboni, C., Lugli, P.: The Monte Carlo Method for Semiconductor Device Simulation. Springer, New York (1989)
15. Herring, C., Vogt, E.: Transport and deformation-potential theory for many-valley semiconductors with anisotropic scattering. Phys. Rev. **101**(3), 944–962 (1956)
16. Hong, S.-M., Jungemann, C.: Deterministic simulation of SiGe HBTs based on the Boltzmann equation. In: Proceedings of ESSDERC, pp. 170–173 (2008)
17. Hong, S.-M., Jungemann, C.: A fully coupled scheme for a Boltzmann-Poisson equation solver based on a spherical harmonics expansion. J. Comput. Electron. **8**(3), 225–241 (2009)
18. Hong, S.-M., Jungemann, C.: Inclusion of the Pauli principle in a deterministic Boltzmann equation solver for semiconductor devices. In: Proceedings of SISPAD, pp. 135–138 (2010)
19. Levenberg, K.: A method for the solution of certain problems in least squares. Quart. Appl. Math. **2**, 164–168 (1944)
20. Marquardt, D.: An algorithm for least-squares estimation of nonlinear parameters. SIAM J. Appl. Math. **11**, 431–441 (1963)
21. Jungemann, C., Meinerzhagen, B.: Hierarchical Device Simulation: The Monte-Carlo Perspective, Computational Microelectronics. Springer, New York (2003)
22. Madelung, O.: Introduction to Solid State Theory. Springer, Berlin (1978)
23. Jacoboni, C., Reggiani, L.: The Monte Carlo method for the solution of charge transport in semiconductors with application to covalent materials. Rev. Mod. Phys. **55**, 645–705 (1983)
24. Bufler, F.M.: Full-band Monte Carlo simulation of electrons and holes in strained Si and SiGe. Dissertation, Universität Bremen, Bremen, 1997. H. Utz Verlag Wissenschaft, München (1998)
25. Jungemann, C., Keith, S., Bufler, F.M., Meinerzhagen, B.: Effects of band structure and phonon models on hot electron transport in silicon. Electr. Eng. **79**, 99–101 (1996)
26. Brooks, H.: Scattering by ionized impurities in semiconductors. Phys. Rev. **83**, 879 (1951)
27. Kosina, H.: A method to reduce small-angle scattering in Monte Carlo device analysis. IEEE Trans. Electron Dev. **46**(6), 1196–1200 (1999)

28. Graf, P., Bufler, F.M., Meinerzhagen, B., Jungemann, C.: A comprehensive SiGe Monte Carlo model for transient 2D simulations of HBTs. In: IEEE Tech. Dig. IEDM, pp. 881–884 (1997)
29. Hong, S.-M., Jungemann, C.: Inclusion of the Pauli principle in a deterministic Boltzmann equation solver based on a spherical harmonics expansion. J. Comput. Electron. **9**, 153–159 (2010)
30. Caughey, D.M., Thomas, R.E.: Carrier mobilities in silicon empirically related to doping and field. Proceedings of IEEE **55**, 2192–2193 (1967)
31. Thoma, R., Peifer, H.J., Engl, W.L., Quade, W., Brunetti, R., Jacoboni, C.: An improved impact-ionization model for high-energy electron transport in Si with Monte Carlo simulation. J. Appl. Phys. **69**, 2300–2311 (1991)
32. Kunikiyo, T., Takenaka, M., Kamakura, Y., Yamaji, M., Mizuno, H., Morifuji, M., Taniguchi, K., Hamaguchi, C.: A Monte Carlo simulation of anisotropic electron transport in silicon including full band structure and anisotropic impact-ionization model. J. Appl. Phys. **75**, 297–312 (1994)
33. Liang, W., Goldsman, N., Mayergoyz, I., Oldiges, P.J.: 2-D MOSFET modeling including surface effects and impact ionization by self-consistent solution of the Boltzmann, Poisson, and hole-continuity equations. IEEE Trans. Electron Dev. **44**(2), 257–267 (1997)
34. Lombardi, C., Manzini, S., Saporito, A., Vanzi, M.: A physically based mobility model for numerical simulation of nonplanar devices. IEEE Trans. Comp. Aided Des. **7**, 1164–1170 (1988)

Chapter 5
Results

In this chapter the SHE solver is applied to devices. First, a unipolar 1D n^+nn^+ structure is simulated and the impact of the different band models is investigated by comparison with full band MC results. Also noise simulations and magneto transport are demonstrated for this kind of device. Second, MOSFET devices are simulated. The impact of the Pauli principle is investigated for a double-gate MOSFET. A partially-depleted SOI MOSFET is selected to demonstrate the numerical robustness of the SHE solver. Finally, SiGe HBTs are investigated. Comparison of the SHE results with full band MC data is followed by characterization of the DC and RF performance of a 2D SiGe HBT.

5.1 n^+nn^+ Structure

An n^+nn^+ structure with a 40 nm long lowly doped region ($2\times10^{17}\,\mathrm{cm}^{-3}$) between two highly doped contact regions ($5\times10^{19}\,\mathrm{cm}^{-3}$) is investigated with the different models. The doping profile is shown in Fig. 5.1. The real space is discretized on a grid of 241 nodes with a constant spacing of 0.5 nm. Band gap narrowing and the Pauli principle are not considered for this artificial structure. Current flow is in the $\langle 100\rangle$ direction, and all results are for unstrained Si at room temperature in this section.

The conduction band edge, which is obtained by self-consistent simulation of the BTE and Poisson equation, is shown in Fig. 5.2. Note that the Fermi level at equilibrium is used as the origin of the total energy throughout Part 2. The peak value of the energy barrier at the n^+n junction is strongly reduced with increasing bias.

In Fig. 5.3, the isotropic component of the electron distribution function for the x-valleys, $f^x_{0,0}(\mathbf{r},H)Y_{0,0}$, is shown for a bias of 0.5 V. Note that the total energy instead of the kinetic energy is used as the energy variable in this figure. Several orders of magnitude of the electron distribution function can be simulated without any numerical difficulties due to the stabilization scheme. The value of

S.-M. Hong et al., *Deterministic Solvers for the Boltzmann Transport Equation*,
Computational Microelectronics, DOI 10.1007/978-3-7091-0778-2_5,

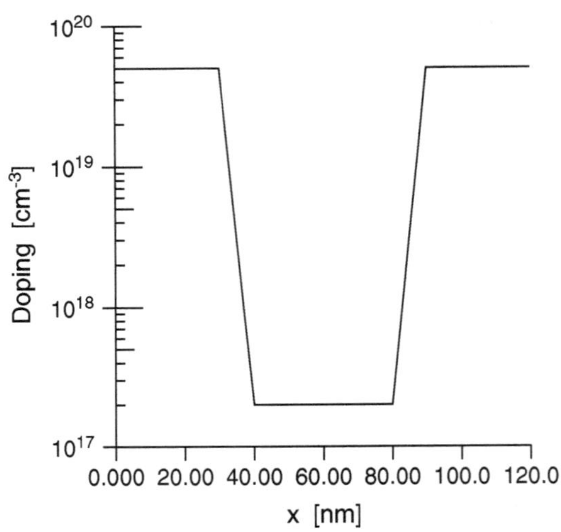

Fig. 5.1 Doping profile of an n^+nn^+ structure

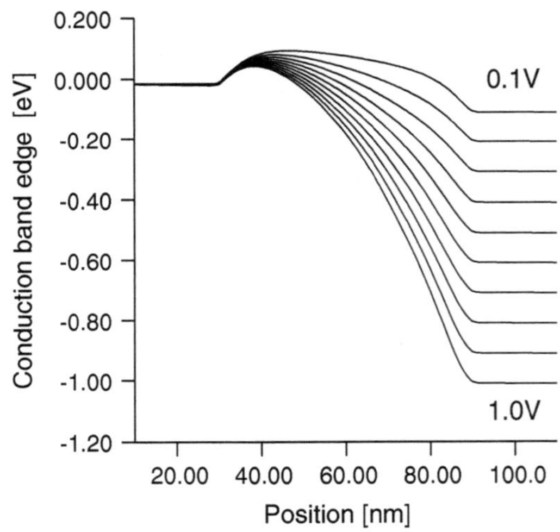

Fig. 5.2 Conduction band edge of the n^+nn^+ structure biased with voltages between 0.1 and 1.0 V

the distribution function at the bottom energy near the two contacts is slightly higher than unity, because the Pauli principle is not included in this simulation. The distribution near 0.12 μm is a mixture of an equilibrium Maxwell–Boltzmann distribution and a hot-electron tail. Also energy dissipation by the dominant LO phonon mode, which results in ripples in the distribution function, can be seen.

5.1.1 Convergence to the Real Solution

In this example, typically a 7th-order SHE and a uniform energy grid spacing of 5.1667 meV, which is about a twelfth of the Si LO phonon energy, are used. Before starting detailed discussions of the simulation results, the convergence of the SHE

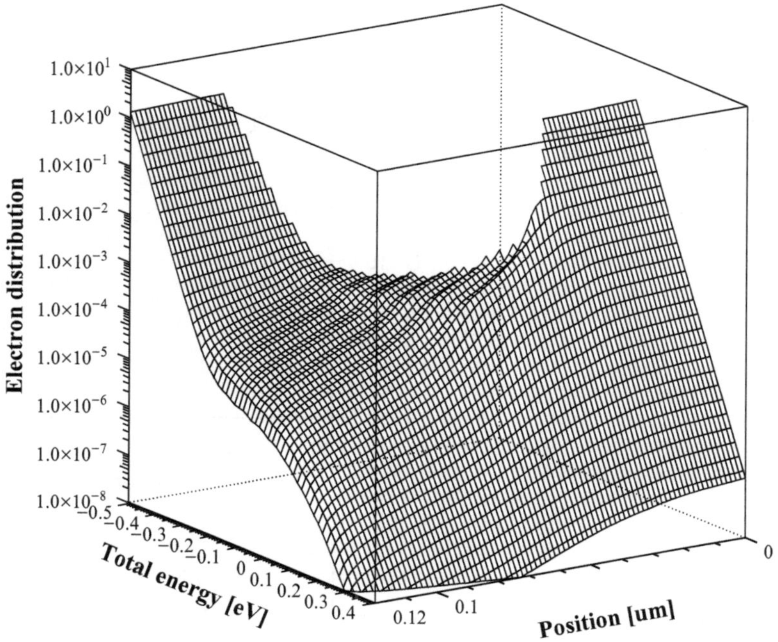

Fig. 5.3 Electron distribution function for the n^+nn^+ structure biased at 0.5 V. The quantity for the x-valleys is shown. For clarity, the resolution of both the real space and the energy space is intentionally reduced by a factor of 3

solution w.r.t. the maximum SHE order and the energy spacing is checked. The full band model for electrons introduced in Sect. 4.2.3 is used for this test.

Figure 5.4 shows the error of the terminal current for different numbers of spherical harmonics w.r.t. a simulation with a 17th-order SHE. The lowest order SHE fails to predict the terminal current accurately, because its error exceeds 30%. The error decreases exponentially with the maximum SHE order, and the error for the terminal current is well below 1% for a 7th order SHE. In Fig. 5.5, the average electron velocity for different maximum SHE order is shown. The lowest order SHE highly overestimates the velocity profile.

In Fig. 5.6 the error of the terminal current for different values of the energy spacing relative to a simulation with 1 meV spacing is shown. A 7th order SHE is used. In order to preserve the inelastic nature of phonon scattering, the energy spacing is increased only up to a sixth of the Si LO phonon energy (10.3333 meV). Even for this coarse energy spacing, the error of the terminal current is only about 0.3%. The error of the electron velocity shown in Fig. 5.7 clearly demonstrates that sufficiently converged solutions for internal quantities can be obtained with a relatively coarse energy spacing.

Similar results are obtained for different bias conditions. From the above results, it is concluded that a well converged solution can be obtained with a 7th order SHE and an energy spacing of 5.1667 meV. These values are used in the simulations for the following subsection.

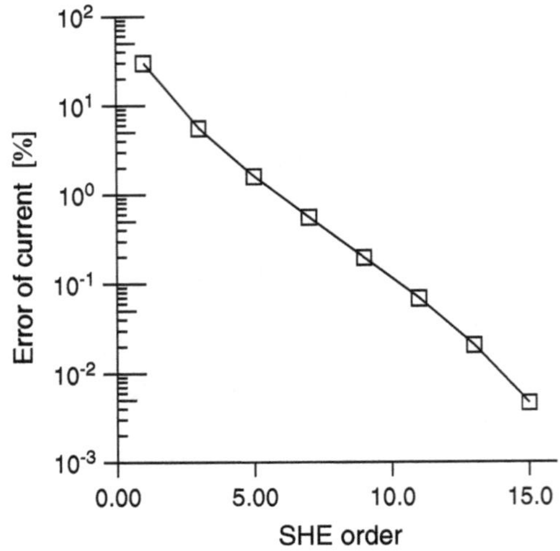

Fig. 5.4 Error of the terminal current for different maximum SHE order relative to a 17th-order SHE. The applied voltage is 1 V and the energy spacing is 5.1667 meV

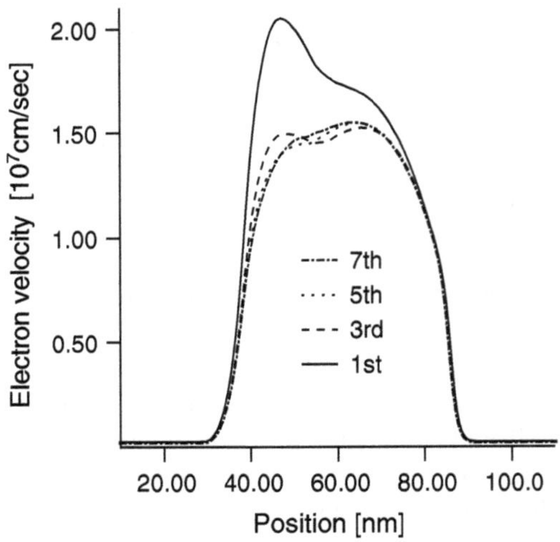

Fig. 5.5 Electron velocity for different maximum SHE order. The applied voltage is 1 V

5.1.2 Impact of the Band Model

The impact of the different band models has been investigated for the bulk system in Sect. 4.2.4. It has been found that the band model has a significant impact on important bulk properties such as the drift velocity, the electron energy, and the impact ionization coefficient. A similar comparison for the different band models is made for the n^+nn^+ structure. The results of the Modena model, the extended Vecchi model, and the full band model are compared with those of full band MC

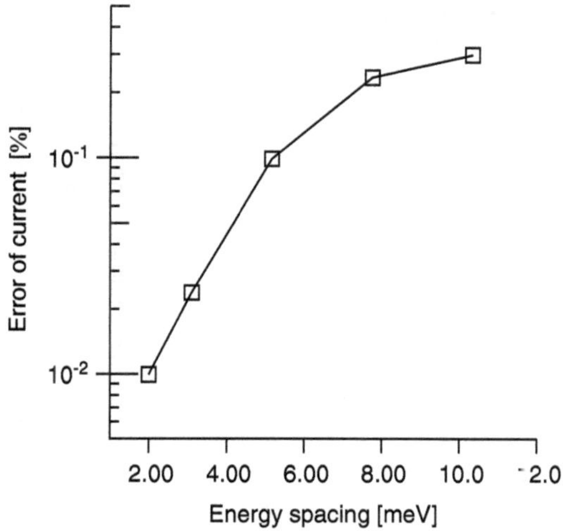

Fig. 5.6 Error of the terminal current for different values of the energy spacing relative to that for 1 meV spacing. The applied voltage is 1 V and the maximum SHE order 7

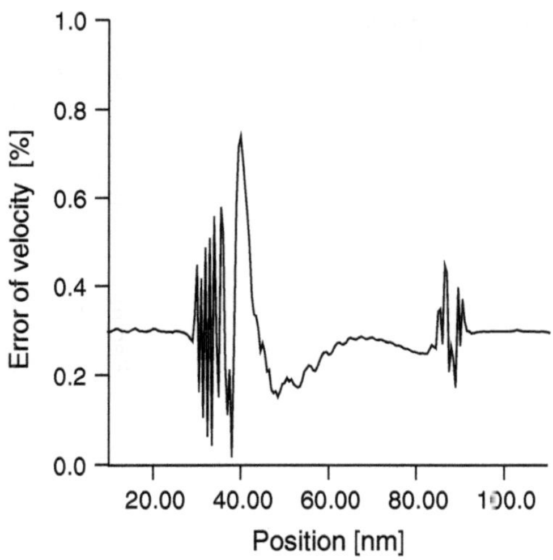

Fig. 5.7 Error of the electron velocity for 10.3333 meV spacing, relative to that for 1 meV spacing. The applied voltage is 1 V and the maximum SHE order 7

simulations. For this purpose, a set of accurate full band MC simulations with a short potential update time (0.25 fs) and a sufficiently long simulation time have been performed. The anisotropic band model is not included in this comparison, because the numerical results of the anisotropic band model are quite similar to those of the extended Vecchi model. Also the original Vecchi model is not considered as in the bulk case. Its failure for nanoscale devices has been demonstrated before [1, 2].

The calculated terminal current as a function of the applied voltage is shown in Fig. 5.8. Results of all three SHE models considered in this subsection follow well the general shape of the full band MC data. For example, even for the Modena model the error of the terminal current at 1 V is slightly below 10%. For the other

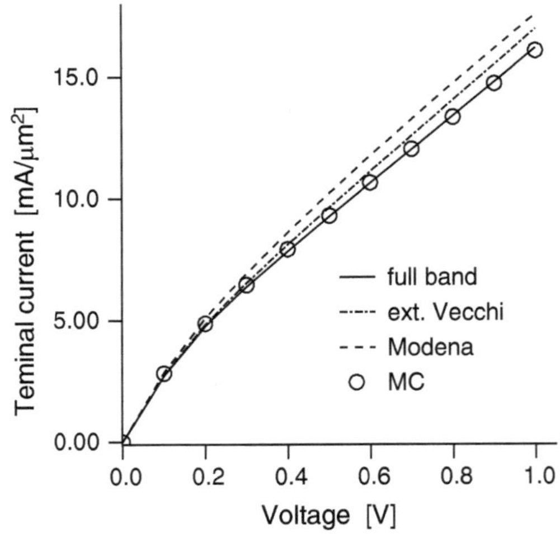

Fig. 5.8 Terminal current vs. the applied voltage

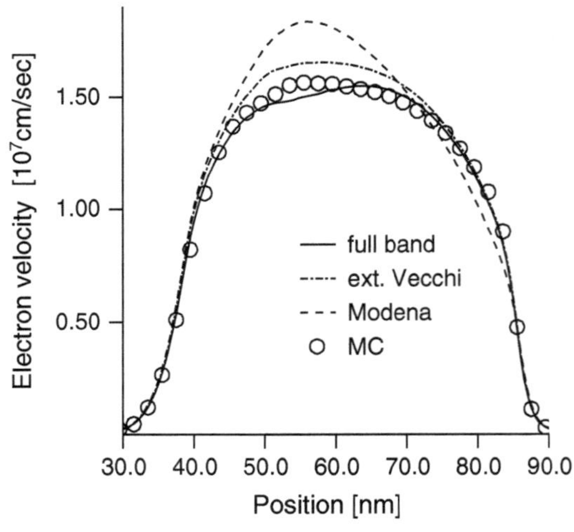

Fig. 5.9 Electron velocity for the different band models. The applied bias is 1 V

two models, an even better agreement with the full band MC data is obtained. At 1 V, the error for the extended Vecchi model and the full band model is about 5% and less than 1%, respectively. Considering the error range of the simulations, the full band model yields almost perfect agreement with the full band MC simulation.

The electron velocity evaluated with the different band models is shown in Fig. 5.9. When compared with the full band MC result, the Modena model predicts a somewhat higher peak velocity, larger than $1.8 \times 10^7\,\mathrm{cm\,s^{-1}}$. The observed higher peak velocity in the Modena model is compensated by a lower electron density resulting in a moderate agreement with the full band MC current. The extended Vecchi model predicts a slightly higher electron velocity throughout the lowly doped region, consistently with the slightly higher terminal current (about 5%). On the other hand, the full band model yields a velocity profile quite similar to the full band MC simulation, although the position of the peak velocity is slightly shifted.

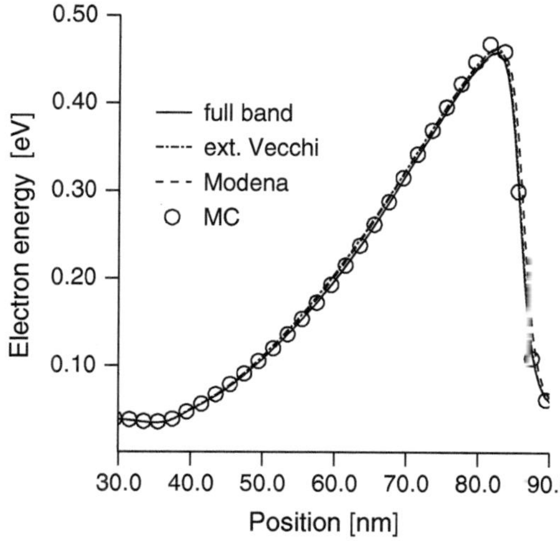

Fig. 5.10 Electron average energy for different band models. The applied bias is 1 V

Especially, the velocity profile around 40 nm, where the energy barrier has its peak value, is very accurately reproduced. This is the main reason why the full band model gives almost perfect agreement with the full band MC simulation, at least for the terminal current. Note that no hot-electron effect is directly involved in this region around 40 nm. A better description for the low energy range, which is obtained by a direct mapping from the full band structure to the band model, is the key ingredient of this improvement.

The electron average energy shown in Fig. 5.10 shows much less dependence on the band model. Although small differences due to the band model (about a few %) do exist for this quantity, it seems that with all three SHE models considered in this comparison an acceptable agreement with the full band MC simulation is obtained.[1]

It has been found that the full band model for electrons gives excellent agreement with the full band MC simulation for various quantities of interest, such as the terminal current (Fig. 5.8), the drift velocity (Fig. 5.9), and the average energy (Fig. 5.10). However, it is again noted that this model is not exactly identical with the full band structure, because of the inherent approximations. The most significant one among these approximations seems to be the fact that three pairs of valleys are considered instead of the individual six valleys of the first conduction band. It does not allow for the separate treatment of the two valleys in a valley pair. As a result, the occupancy of the valleys whose principal axis are aligned with the field direction cannot be calculated exactly. In order to demonstrate this point, the contribution from the x-valleys to the local electron density, evaluated with the different band models, is shown in Fig. 5.11. Note that in the case of the full band MC simulation, the contributions from two different valleys (x^+ and x^-) are simply

[1]In [2], it is claimed that even the original Vecchi model, which is not suitable for estimating the terminal current of nanoscale devices, can be used in order to estimate the average electron energy.

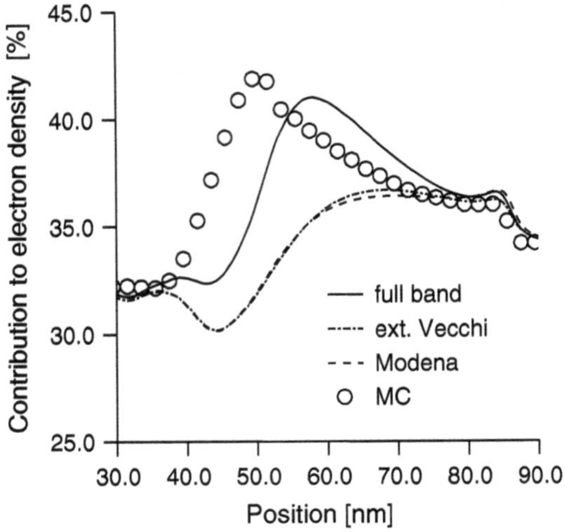

Fig. 5.11 Contribution from the x-valleys to the local electron density evaluated with different band models. The applied bias is 1 V

added in order to yield the contribution for the valley pair shown in the figure. Those two valleys behave quite differently (One is highly occupied while the other is much less occupied.), and the averaged valley assumed in the SHE models cannot capture this. For example, in the case of the full band model, although the increase of the x-valley contribution observed in the full band MC result is reproduced, the position of its peak value is slightly shifted to the r.h.s. like the velocity profile (Fig. 5.9). One final remark in this subsection is still remaining. Interestingly enough, the Modena model and the extended Vecchi model, which give different results for the terminal current and the velocity profile, predict very similar contributions from the x-valleys to the electron density. On the other hand, the extended Vecchi model and the full band model, which give more similar results for other quantities, predict rather different profiles. This strongly suggests the importance of the ratio between the directional velocities, which is discussed in Sect. 4.2.4.[2]

Given these limitations, the full band model for electrons is quite a good choice for solving the BTE for semiconductor devices.

5.1.3 Magneto Transport

The direction of the magnetic field is assumed to be perpendicular to the two-dimensional simulation plane. For the sake of simplicity the Modena model is adopted in the simulation of magneto transport.

[2]This ratio for the full band model is not exactly matched with that of the full band structure, because of the constraint (4.42). Nevertheless, a certain improvement over the energy-independent ratio used in the Modena model and the extended Vecchi model is obvious.

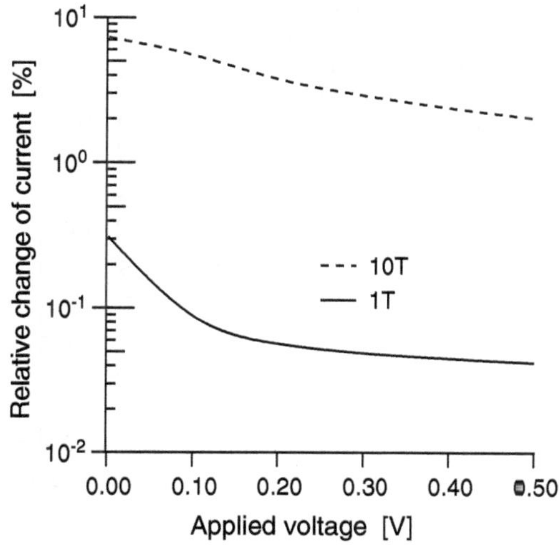

Fig. 5.12 Absolute values of the relative change of the terminal current at different applied voltages relative to 0 T. The Modena model is used

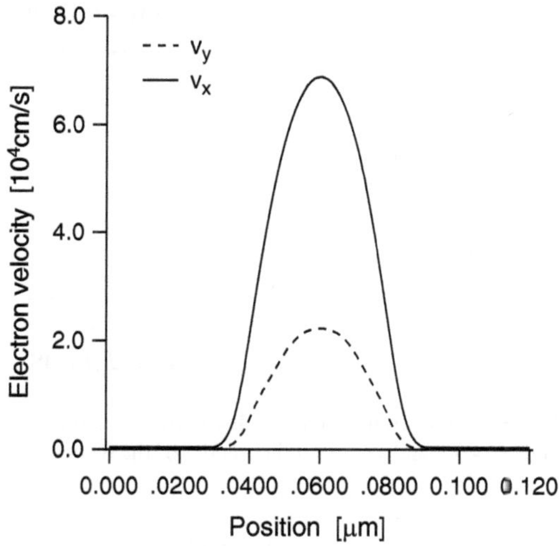

Fig. 5.13 Electron velocities along the x- and y-directions at an applied voltage of 1 mV and $B = 10$ T. The maximum SHE order is 7

In Fig. 5.12, absolute values of the relative change of the terminal current at different applied voltages relative to 0 T are shown for two magnetic fields, 1 and 10 T. The change due to the magnetic field is small even for a strong magnetic field. For example, the biggest change in the figure is about 7% at 10 T. Therefore, it is very CPU time consuming to simulate magneto transport with the MC approach in contrast to SHE. The effect of the magnetic field on the terminal current is stronger at low applied voltage than at higher applied voltage.

In Fig. 5.13, the electron velocities along the x- and y-directions at an applied bias of 1 mV and $B = 10$ T are shown. The ratio of the y- and x-components has its peak value in the lowly doped region. In Fig. 5.14, the electron velocities along x- and y-directions for an applied voltage of 0.5 V and $B = 10$ T are shown. Hot

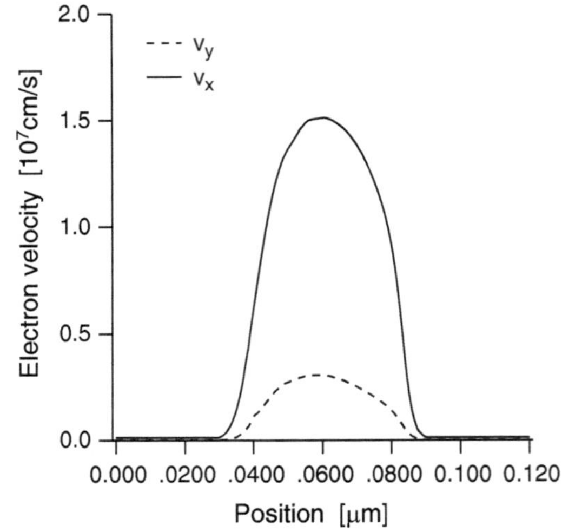

Fig. 5.14 Electron velocities along the x- and y-directions at an applied voltage of 0.5 V and $B = 10\,\text{T}$. The maximum SHE order is 9

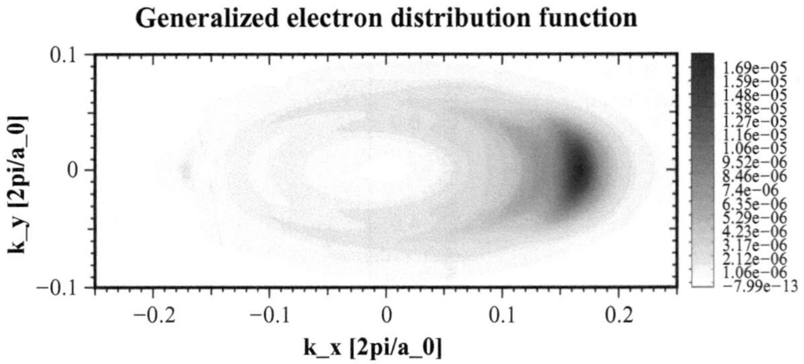

Fig. 5.15 Generalized energy distribution function in the x-valley for an applied voltage of 0.5 V and $B = 0\,\text{T}$ and $k_z = 0$ in the middle of the device

electron effects reduce the peak value of the velocity ratio compared to the result for 1 mV. This can be understood by comparing the relative magnitude of the free-streaming operator due to the magnetic force, $B_{\mathrm{B},l,m,l',m'}$ in (2.118), with those of the free streaming operator and the scattering integral. The $B_{\mathrm{B},l,m,l',m'}$ coefficient shows a much milder variation with energy than the $A_{l,m,l',m'}$ coefficient, because the $\nabla_{\mathbf{k}}\vartheta$ and $\nabla_{\mathbf{k}}\varphi$ coefficients in (2.118) tend to decrease with increasing energy.

Since the Modena model is adopted, the interpretation of the simulation results in the **k**-space is easily possible. In Fig. 5.15 the generalized energy distribution function is shown in the **k**-space for the x-valley at $V = 0.5\,\text{V}$ and $B = 0\,\text{T}$ in the middle of the device. The distribution is symmetric with respect to k_y and a ballistic peak appears at about $k_x = 0.16 \cdot 2\pi/a_0$, where a_0 is the lattice constant of the

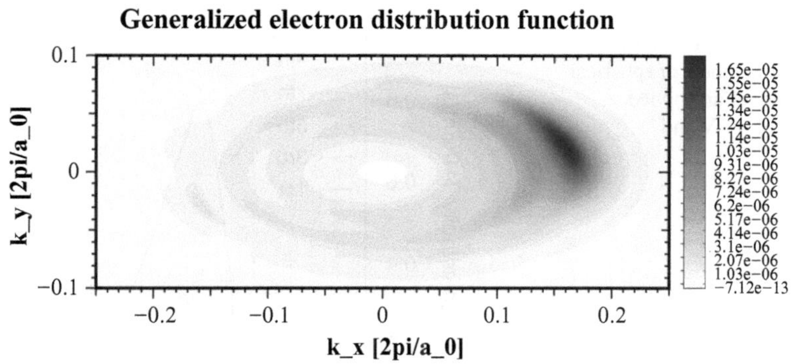

Fig. 5.16 Generalized energy distribution function for the same conditions as in Fig. 5.15 besides a magnetic field of 10 T

Fig. 5.17 Electron velocities along the x-direction for different numbers of spherical harmonics at an applied voltage of 0.5 V and $B = 10\,\mathrm{T}$

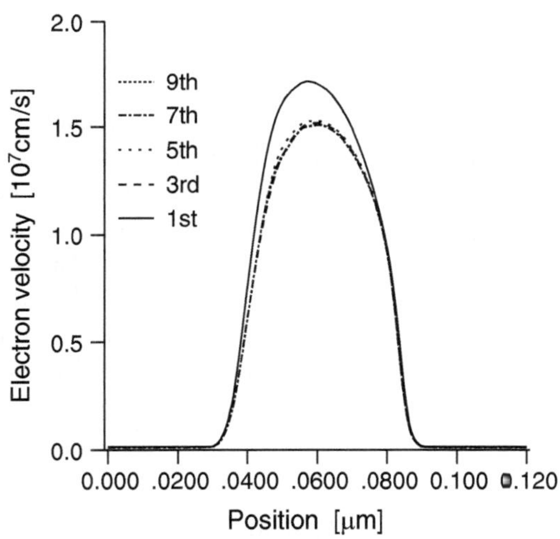

relaxed silicon.[3] In Fig. 5.16, the same quantity at $B = 10\,\mathrm{T}$ is shown. The distortion toward the positive k_y direction due to the magnetic force is clearly seen.

One important question regarding the SHE is, whether the maximum SHE order is sufficiently large or not. In Fig. 5.17, the electron velocities along the x-direction for different numbers of spherical harmonics are shown. The applied bias is 0.5 V and the magnetic field is 10 T. When compared with the previous result without a magnetic field (Fig. 5.5), the convergence of the longitudinal velocity w.r.t. the maximum order of SHE shows a similar behavior. In Fig. 5.18, the electron velocities along the y-direction for the same conditions are shown. It is obvious that

[3]In the full band structure the origin of the x-valley is locate4d at $0.85 \cdot 2\pi/a_0$ along the $\langle 100 \rangle$ direction, and this relative value of **k** lies outside of the first Brillouin zone. Recall that the Modena model is used.

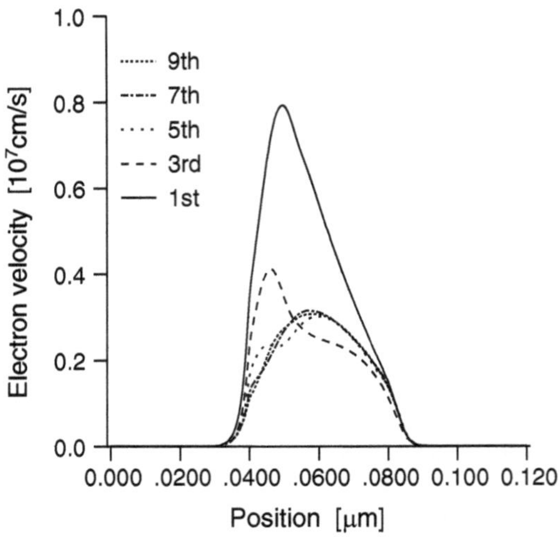

Fig. 5.18 Electron velocities along the y-direction for different numbers of spherical harmonics at an applied voltage of 0.5 V and $B = 10\,\mathrm{T}$

the velocity along the y-direction requires more spherical harmonics than that along the x-direction. This can be understood by realizing the different coupling behavior of $B_{\mathrm{B},l,m,l',m'}$ compared to other coefficients such as $\mathbf{A}_{l,m,l',m'}$ and $B_{\mathrm{E},l,m,l',m'}$. Note that $B_{\mathrm{B},l,m,l',m'}$ only couples the electron distribution $f_{l',m'}$ with odd(/even) l' to the balance equation for odd(/even) l. Therefore, in the case of isotropic bands, $B_{\mathrm{B},l,m,l',m'}$ has nonzero values for l and l', whose difference is ± 2. Since $\mathbf{A}_{l,m,l',m'}$ and $B_{\mathrm{E},l,m,l',m'}$ do not vanish when the difference between l and l' is ± 1 [1], the bandwidth of the resultant coupling coefficients due to $B_{\mathrm{B},l,m,l',m'}$ (in terms of its l dependence) is larger than those due to $\mathbf{A}_{l,m,l',m'}$ and $B_{\mathrm{E},l,m,l',m'}$.

5.1.4 Noise Simulations

Noise simulations are performed for the n^+nn^+ structure directly in the frequency domain under stationary bias conditions. The full band model is used.

Since noise simulations are based on a perturbation approach, the correct evaluation of the small-signal response is of critical importance. Recall that we have introduced a specialized discretization scheme for the small-signal analysis [3], as shown in Sect. 3.6, in order to avoid numerical difficulties arising from the time-varying transport coefficients. Therefore, first the correctness of that discretization scheme should be checked. At zero frequency, the small-signal response obtained from a solution of the linearized BTE should be in good agreement with the one by finite differences. Figure 5.19 shows the small-signal response of the electron density due to a unit voltage excitation at the l.h.s. contact. In order to make a comparison with the result obtained by the finite difference method, the excitation frequency is set to be zero. In the case of the finite difference method, an excitation voltage corresponding to one step of the energy spacing (5.1667 meV) is applied to

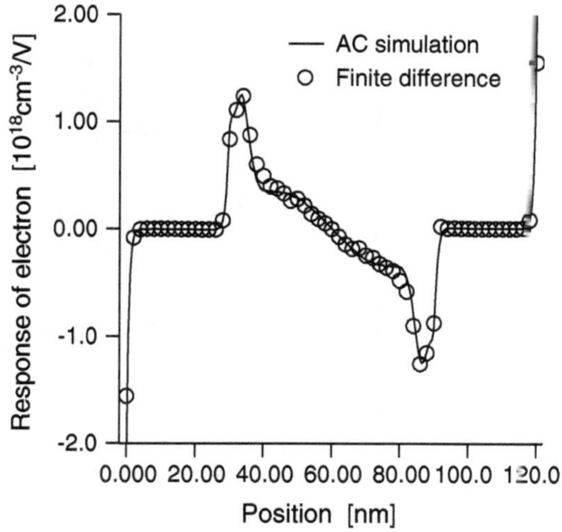

Fig. 5.19 Small-signal response of the electron density at equilibrium. Values evaluated by the finite difference method and AC simulation are shown

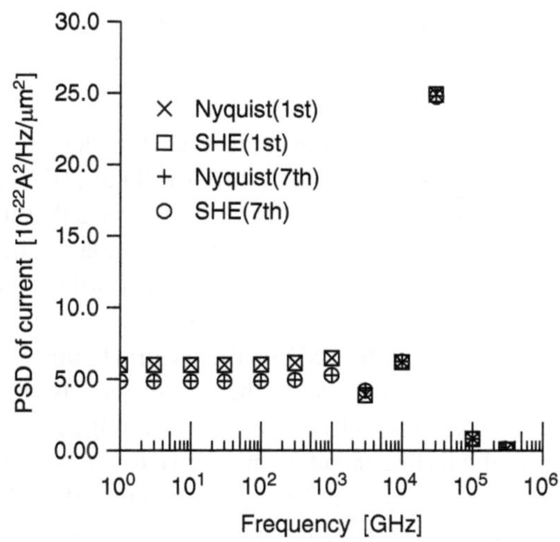

Fig. 5.20 PSD of the terminal current for the 1D 40 nm n^+nn^+ structure at equilibrium. The value evaluated by the Nyquist theorem is also shown for comparison

the l.h.s. contact, and the resultant difference of the electron densities is normalized to the unit voltage excitation. The good agreement of the two results shown in the figure reveals that the small-signal Jacobian matrix is constructed consistently with the DC simulation scheme.

Next, it is checked whether the Nyquist theorem for the terminal current noise holds at equilibrium for all frequencies. At equilibrium, the power spectral density of the lth and mth terminal current fluctuations, $S_{I_l I_m}(f)$, is directly related with the real part of the admittance $Y_{l,m}(f)$ [4, 5]:

$$S_{I_l I_m}(f) = 4k_{\mathrm{B}}T\Re\{Y_{l,m}(f)\}\,. \tag{5.1}$$

In Fig. 5.20, the PSD of the terminal current for the n^+nn^+ structure at equilibrium is shown. The Nyquist theorem is satisfied within a small numerical error. For

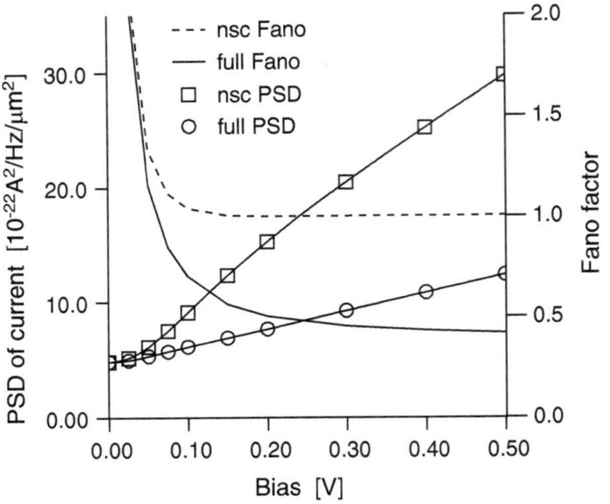

Fig. 5.21 PSD of terminal current fluctuations and the corresponding Fano factor based on self-consistent solutions (*full*) and nonself-consistent solutions (*nsc*) for 0 Hz

example, at 1 GHz, the calculated PSD deviates from the Nyquist theorem by only 0.16%. The spike in the current PSD at 30 THz is due to a plasma resonance [6]. In addition, results are shown for the case that the SHE is truncated at the 1st order and a large error occurs below the plasma frequency. Note that the Nyquist theorem is nevertheless satisfied for the lowest order SHE, as long as the admittance is also obtained by the lowest order SHE.

In Fig. 5.21 the PSD of the terminal current fluctuations is shown as a function of the applied voltage. In the nonself-consistent case the fluctuations are calculated for a "frozen" electric field. As soon as the bias is so large that the electrons injected from the RHS contact cannot move to the left terminal, the current is determined by the electrons moving from the left to the RHS. They are injected independently over the potential barrier, and the nonself-consistently calculated noise becomes full shot noise. This results in a Fano factor (PSD of noise divided by full shot noise), which is one. In the self-consistent case the noise is suppressed, because the potential barrier is determined by the number of electrons in the lowly doped region and electron injection over the barrier becomes correlated due to electrostatic interaction [7]. With respect to electronic noise self-consistent simulations are the only meaningful ones, but in MC simulations also frozen field simulations are used, because they seem to be more CPU efficient. This result shows that this must not necessarily be the case. The increase in stochastic noise might degrade the performance of the MC simulation more than what is gained from the frozen field approach [8].

It is also possible to calculate the spatial origin of the terminal current noise, which is shown in Fig. 5.22 for a bias of 0.5 V. In the nonself-consistent case the noise stems from the LHS of the potential barrier, which is located at about 40 nm for a bias of 0.5 V, in accordance with shot noise theory. In the self-consistent case

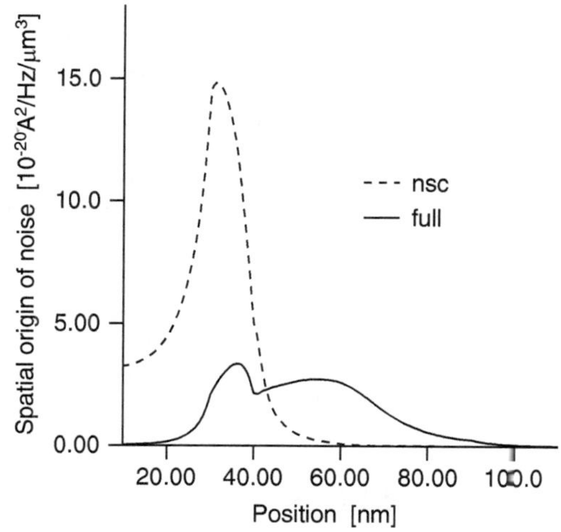

Fig. 5.22 Spatial origin of the terminal current fluctuations based on self-consistent solutions (*full*) and nonself-consistent solutions (*nsc*) for 0 Hz and a bias of 0.5 V

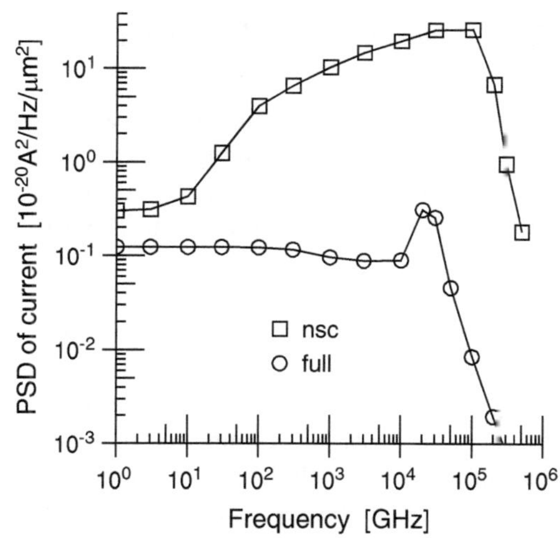

Fig. 5.23 PSD of the terminal current fluctuations based on self-consistent solutions (*full*) and nonself-consistent solutions (*nsc*) at a bias of 0.5 V

this is no longer the case and also the electrons in the lowly doped region contribute, because their number controls the barrier height and therewith the electron injection. It is found that the hot electrons on the RHS of the lowly doped region produce less noise than the cold or warm electrons near the injection point.

Not only the spatial distribution of the terminal current noise is different in the case of nonself-consistent simulations, but also the frequency behavior, as shown in Fig. 5.23. In the nonself-consistent case the noise is determined by the slow diffusion from the left terminal to the barrier leading to a strong frequency dependency already in the lower GHz range. The self-consistent result, on the other hand, shows the usual behavior and at very high frequencies (about 30 THz due to the high doping concentration in the contacts) a plasma resonance occurs [6].

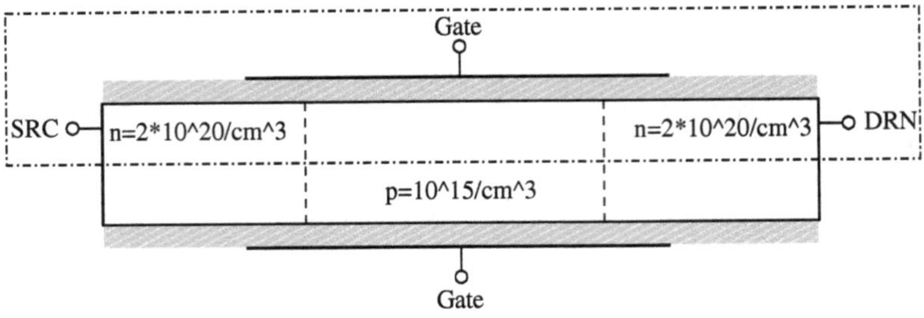

Fig. 5.24 Structure of the simulated double-gate NMOSFET. The length of the metal gate is 70 nm. The oxide thickness is 1.3 nm and the metallurgical channel length 50 nm. The body of the double-gate MOSFET is 20 nm thick

5.2 nMOSFETs

5.2.1 Double-Gate MOSFET

A two-dimensional double-gate MOSFET is simulated. The impact of the Pauli principle is investigated. In Fig. 5.24 the structure of the simulated device is shown. The length of the metal gate is 70 nm. The oxide thickness is 1.3 nm and the metallurgical channel length 50 nm. The body of the double-gate MOSFET is 20 nm thick.

A 3rd order SHE and energy spacing of 10.3333 meV are used together with the full band model for electrons. Use of other band models like the Modena model [9] or the original Vecchi model [10] can be found in [11, 12]. In this work, the quantization of the carrier motion perpendicular to the interface is not taken into account. In order to account for inversion-layer transport, additional scattering mechanisms are included in the simulation [13], and the parameters for those scattering processes are matched to the Lombardi mobility model [14]. The same parameters are used for both simulations with and without the Pauli principle.

In Fig. 5.25 the typical convergence behavior of the simulation is shown for different bias points. When the potential correction is smaller than a predefined value (10^{-10} V in this example) convergence is achieved. The initial solution is obtained by solving the BTE for a fixed electrostatic potential. Once an initial solution of the BTE is obtained, a fully-coupled scheme is used for the BTE and Poisson equation. When the maximum value of the potential correction at a certain Newton–Raphson iteration exceeds the energy spacing, additional fixed potential simulations are performed. Without the Pauli principle, only one iteration is required because the linear BTE is solved. Even with the Pauli principle, usually a few iterations are sufficient to achieve a converged solution of the nonlinear BTE, as shown in Fig. 5.26. When the temporal solution of the Newton–Raphson iteration is close enough to the converged solution, such fixed potential iterations naturally

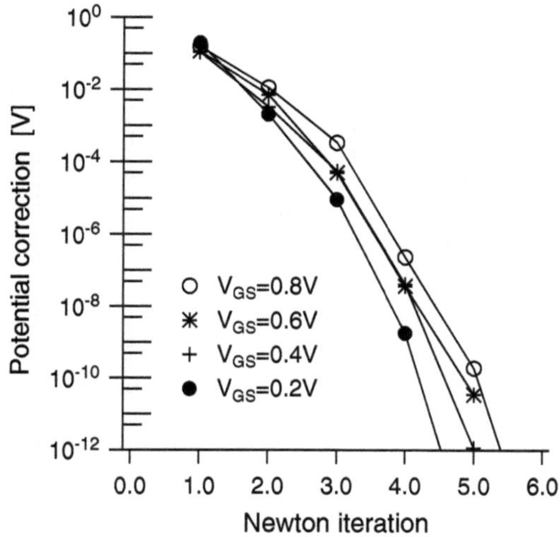

Fig. 5.25 Maximum value of the potential correction at each Newton–Raphson iteration for different V_{GS} values. The Pauli principle is included. $V_{DS} = 0.9$ V

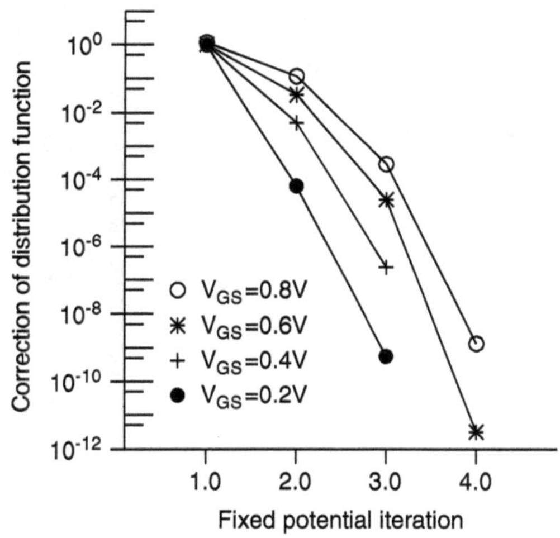

Fig. 5.26 Maximum value of the correction of the electron distribution function at each fixed potential iteration after the first Newton–Raphson iteration for different V_{GS} values. The Pauli principle is included. $V_{DS} = 0.9$ V

disappear. In every simulation the last Newton–Raphson step shows a huge decrease of the norm of the update vector.

In Fig. 5.27 the electron distribution function with or without the Pauli principle is shown. In the drain contact, where the doping density is as high as $2\times10^{20}\,\mathrm{cm}^{-3}$, the impact of degeneracy is clearly visible in the electron distribution function. The electron distribution function in the drain contact also demonstrates that the injection boundary condition discussed in Sect. 3.4 [15] can handle such highly non-equilibrium distributions without any difficulties. The number of hot electrons is larger in the simulation with the Pauli principle.

In Fig. 5.28, the conduction band edge with or without the Pauli principle is shown. The lateral position of the middle of the channel is denoted as zero. In the source and drain regions, the difference of the conduction band edge due to

Fig. 5.27 Electron distribution function at two different positions, the drain contact and the position where the peak electron velocity is observed (20 nm). $V_{GS} = 0.9$ V and $V_{DS} = 0.9$ V. The vertical position is fixed to the Si/SiO_2 interface

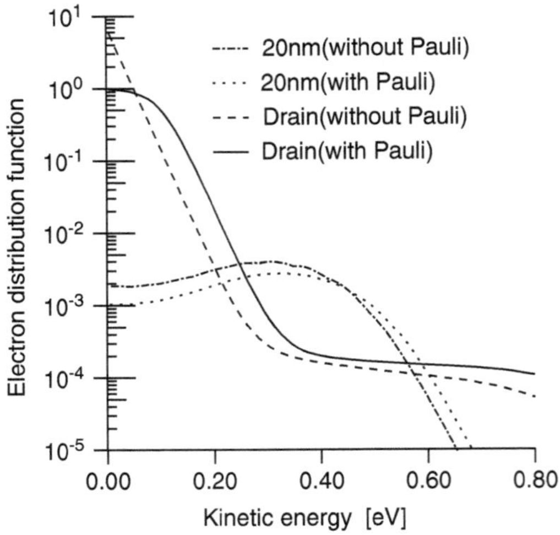

Fig. 5.28 Conduction band edge. $V_{GS} = 0.0$ V and $V_{DS} = 0.9$ V. The vertical position is fixed to the Si/SiO_2 interface

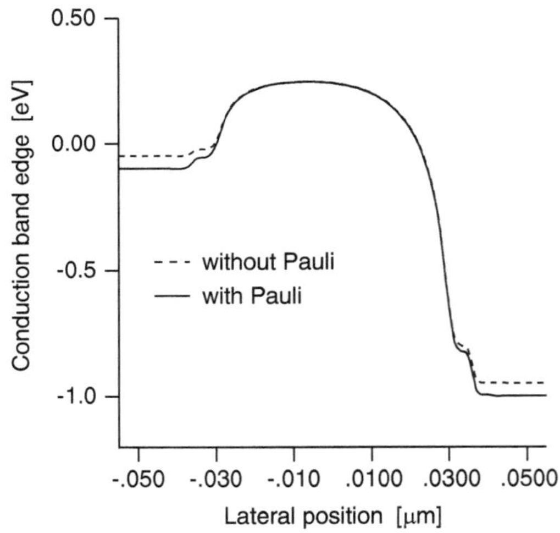

the different statistics is clearly visible. In these regions, the quasi-Fermi level for electrons is located about 0.1 eV above the conduction band edge when the Pauli principle is considered. Without the Pauli principle, the difference is reduced to about 50 meV.

In Fig. 5.29 the electron sheet density of the double-gate MOSFET is shown along the lateral position. For comparison, it is also displayed without the Pauli principle. In Fig. 5.30, the input characteristics calculated with or without the Pauli principle is shown. From these results and the output characteristics shown in Fig. 5.31, it is found that the drain current is largely insensitive to the Pauli principle. Note that from similar results for the drain current has been also obtained by multisubband MC simulations [16].

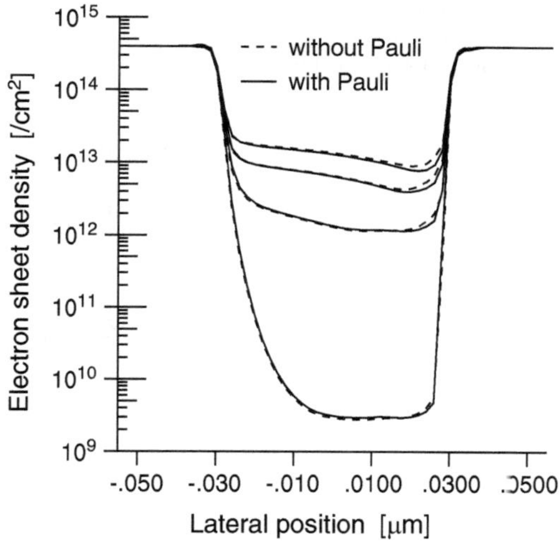

Fig. 5.29 Electron sheet density along the vertical position for different V_{GS} values, 0.0, 0.3, 0.6, and 0.9 V. $V_{DS} = 0.9$ V

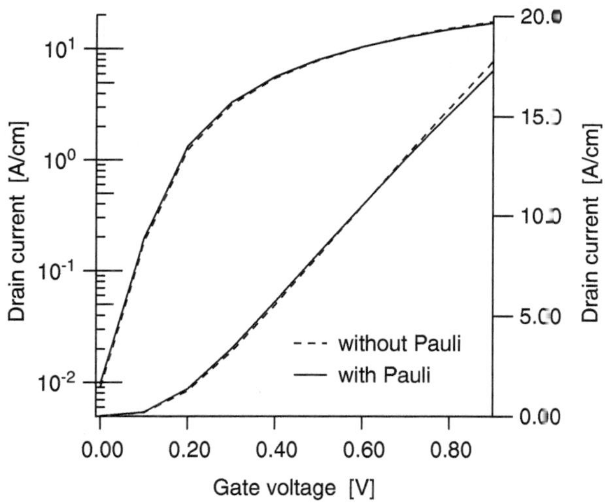

Fig. 5.30 Input characteristics for $V_{DS} = 0.9$ V

Although the drain current shows good agreement with and without the Pauli principle, other internal quantities can show larger differences. For example, the drain current shows a difference about 2.6% at $V_{GS} = 0.9$ V and $V_{DS} = 0.9$ V, however, more than 10% difference in the peak electron velocity is observed (Fig. 5.32). By inclusion of the Pauli principle, the average distance of the inversion electrons from the interface increases. With this deeper distribution, inversion electrons near the channel-drain junction experience a higher longitudinal electric field, resulting in a higher peak velocity.

Next, the RF performance of the double-gate MOSFET is investigated by simulations without the Pauli principle. Noise simulation results with the Pauli principle can be found in [12]. As shown in Sect. 3.6, for the small-signal Jacobian matrix, the linearized BTE is built upon the steady-state total energy [3]. As a

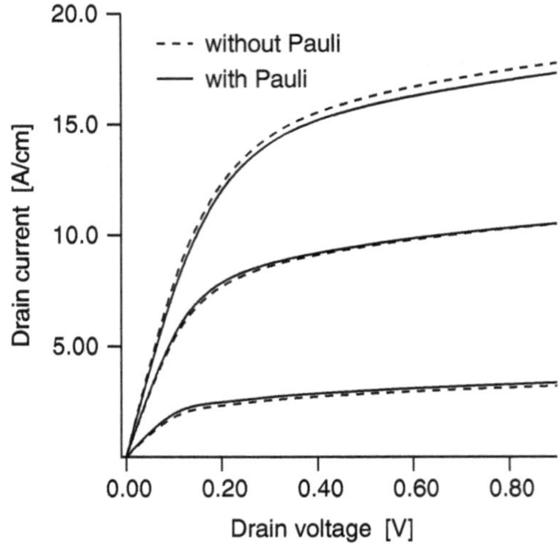

Fig. 5.31 Output characteristics for different V_{GS} values, 0.3, 0.6, and 0.9 V

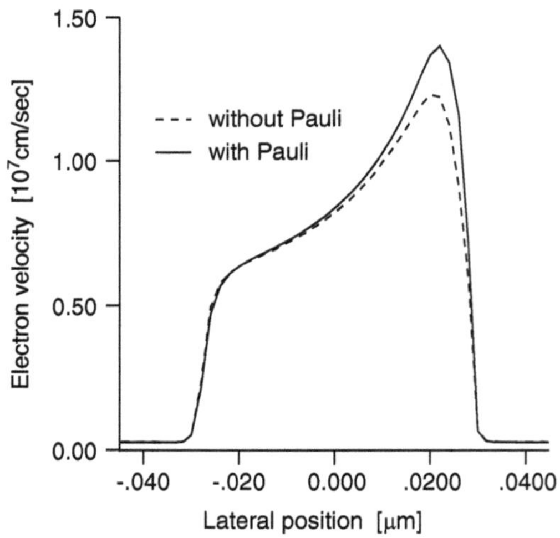

Fig. 5.32 Longitudinal electron velocity averaged perpendicular to the channel direction at $V_{GS} = 0.9$ V and $V_{DS} = 0.9$ V

result, the fluctuating electric potential introduces additional coupling between two discretized nodes with different total energy. The Green's function for the terminal noise current can be obtained by a solution of the corresponding adjoint system of equations [17]. Noise simulation is based on the method described in Sect. 2.6 [18, 19].

Since small-signal analysis is available, Y parameters can be calculated in a straightforward manner, and hence the small-signal current gain. Figure 5.33 shows the small-signal current gain as a function of frequency. In Fig. 5.34 the cutoff frequency of the double-gate MOSFET extrapolated from the small-signal current gain at 10 GHz is shown.

Figure 5.35 shows the spectral intensities of the terminal current noise as a function of V_{GS} at 10 GHz. The absolute value is shown for the cross correlation

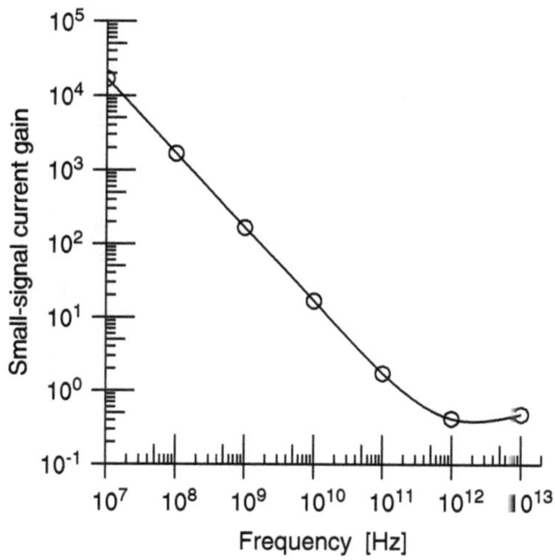

Fig. 5.33 Small-signal current gain at $V_{GS} = 0.4$ V and $V_{DS} = 0.9$ V

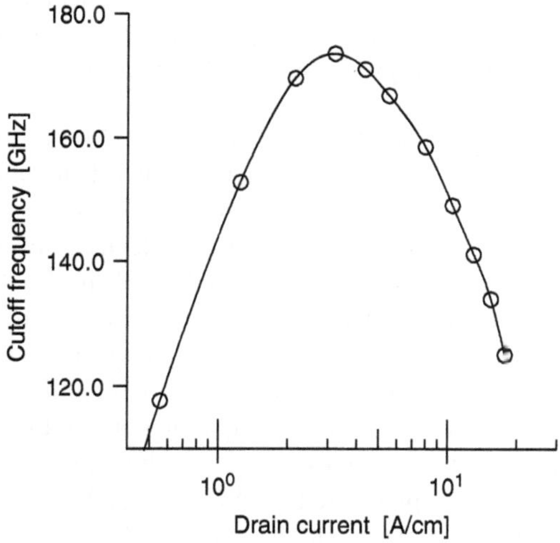

Fig. 5.34 Cutoff frequency at $V_{DS} = 0.9$ V

between the gate and drain current noise, $S_{I_G I_D}$. The spectral intensities of the drain and gate current noise, $S_{I_G I_G}$ and $S_{I_G I_D}$, which are orders of magnitude smaller than $S_{I_D I_D}$, can be evaluated without numerical problems.

5.2.2 Partially-Depleted SOI MOSFET

Numerical simulation of a partially-depleted SOI MOSFET is known to be a very challenging task [20–22]. Floating-body effects are determined by a delicate balance between various generation and recombination mechanisms (SRH generation/recombination, impact ionization, etc.) and artificially enhanced hot electron diffusion near the drain end of the MOSFET channel in the case of the HD model.

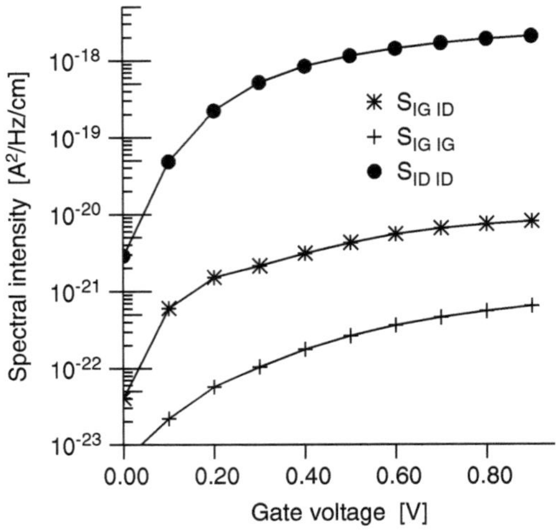

Fig. 5.35 Spectral intensities of the drain current noise as a function of V_{GS} at $V_{DS} = 0.9$ V and 10 GHz. Absolute value is shown for $S_{I_G I_D}$

In addition to the last subsection impact ionization is included in the simulations. On the other hand, SRH generation/recombination is not considered. A 3rd order SHE and energy spacing of 10.3333 meV are used together with the full band model for electrons.

The outline of the SOI MOSFET [23] is shown in Fig. 5.36. The simulated SOI NMOSFET has a body thickness of 180 nm, a top oxide of 8.5 nm, a bottom oxide of 400 nm, and a gate length of 500 nm. It is discretized with a grid of 63×81 nodes with a non-uniform spacing. The device is of the partially-depleted type and has a homogeneous body doping of 2×10^{17} cm^{-3}.

The output characteristics of the device are shown in Fig. 5.37. When the drain voltage is high enough to activate impact ionization, the secondary holes generated by impact ionization are injected into the partially-depleted body region. The additional holes lead to a rapid increase of the body potential. As a result, the drain current is increased sharply, and this rapid rise is known as the kink effect for the partially-depleted SOI MOSFETs [22]. In the figure, the kink effect is clearly visible from $V_{DS} = 0.6$ V. On the other hand, when impact ionization is neglected, such an effect does not appear at all. In Fig. 5.38, the corresponding source hole current generated by impact ionization is shown. The amount of the secondary holes rapidly increases when the drain voltage increases.

In Fig. 5.39 the electron concentration is shown for $V_{GS} = 1.0$ V and $V_{DS} = 1.0$ V, above the kink. It is seen that the vertical diffusion of electrons near the drain end of the MOSFET channel is not strong in contrast to hydrodynamic simulations [20,21].

In Fig. 5.40 the electron dynamic temperature [24] and the impact ionization generation rate are shown along the Si/SiO$_2$ interface for $V_{GS} = 1.0$ V and $V_{DS} = 1.0$ V. Both quantities are rather unrelated and impact ionization cannot be described by a local temperature model typically used in hydrodynamic simulations [25]. The peak of the impact ionization generation rate appears closer to the drain contact than that of the dynamic temperature. Thus, more holes generated by impact ionization flow out through the drain terminal than a local temperature model would predict.

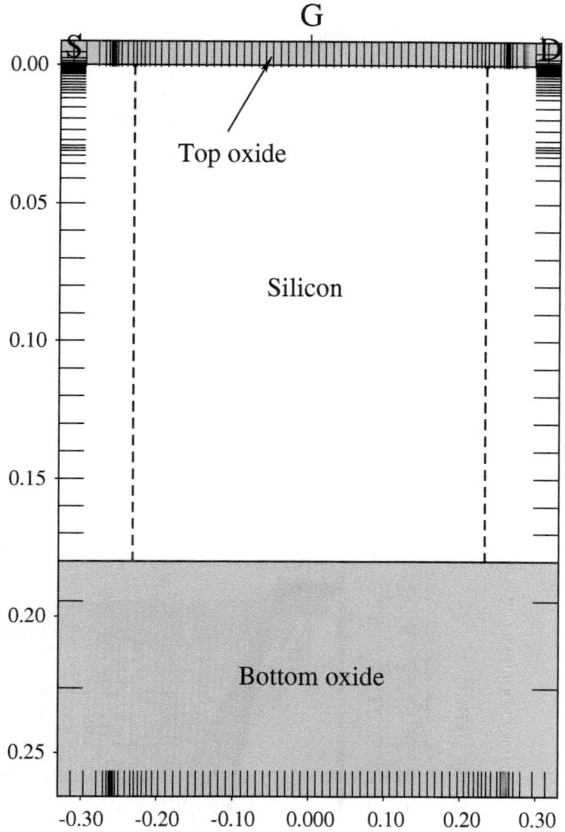

Fig. 5.36 2D structure of the SOI MOSFET. Only part of the bottom oxide whose thickness is 400 nm is shown. The *ticks* indicate the real space grid

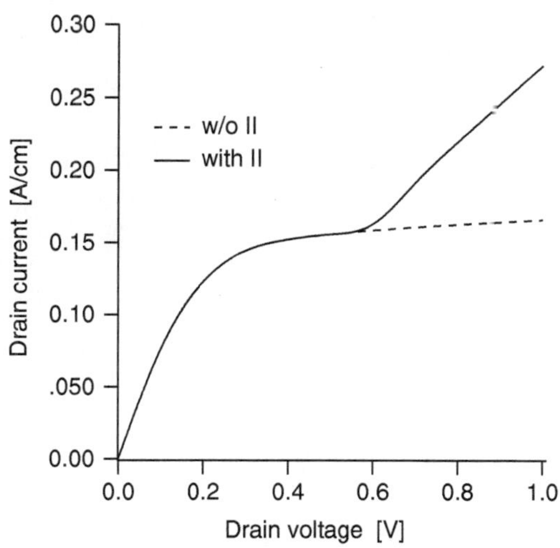

Fig. 5.37 Output characteristics of the SOI MOSFET for $V_{GS} = 1.0\,V$ with and without impact ionization

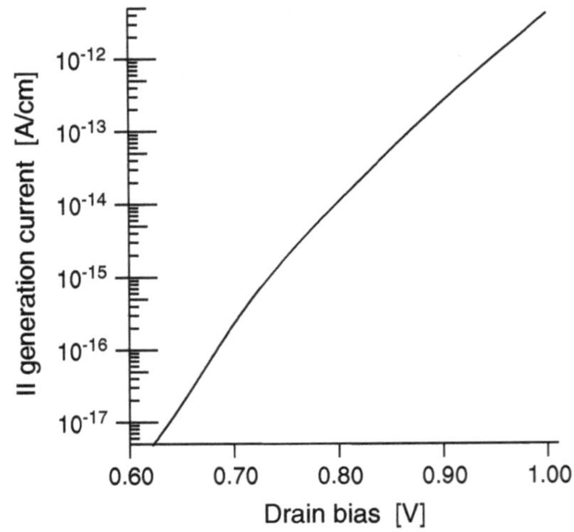

Fig. 5.38 Source hole current generated by impact ionization in the SOI MOSFET for $V_{GS} = 1.0\,\mathrm{V}$

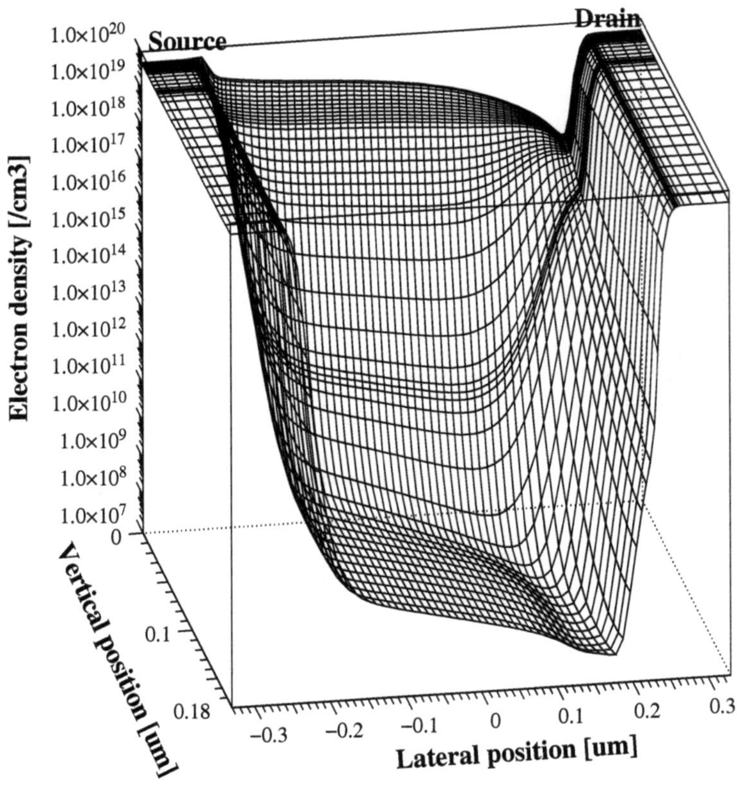

Fig. 5.39 Electron concentration in the SOI MOSFET for $V_{GS} = 1.0\,\mathrm{V}$ and $V_{DS} = 1.0\,\mathrm{V}$

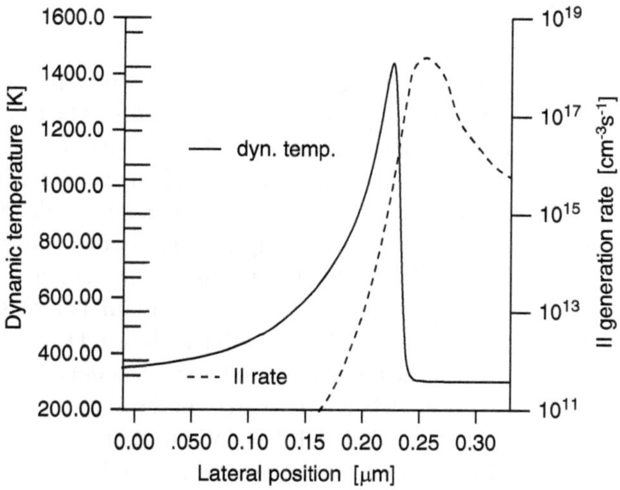

Fig. 5.40 Electron dynamic temperature and impact ionization generation rate along the Si/SiO_2 interface for $V_{GS} = 1.0\,V$ and $V_{DS} = 1.0\,V$

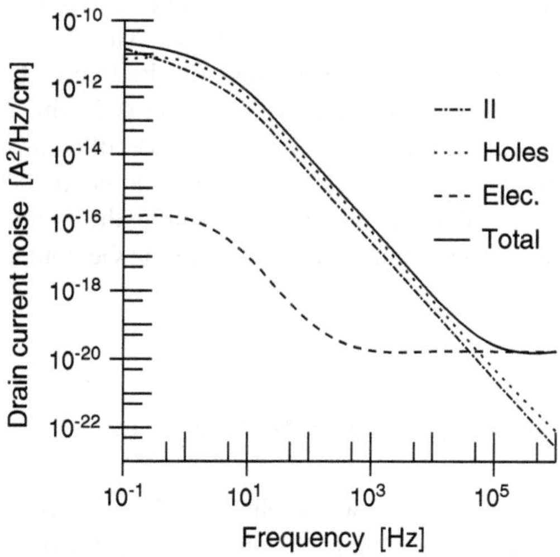

Fig. 5.41 Spectral intensity of the drain current fluctuations for $V_{GS} = 1.0\,V$ and $V_{DS} = 1.0\,V$

The low frequency noise for $V_{GS} = 1.0\,V$ and $V_{DS} = 1.0\,V$ is shown in Fig 5.41. The contributions of the different noise sources to the drain current noise are also shown. Above the kink, the noise is caused by impact ionization and hole scattering. Impact ionization and hole scattering both yield shot-like noise which is strongly amplified by the floating body effect [26]. At higher frequencies this feedback effect is short-circuited by capacitive coupling and the decrease in noise with frequency is similar to the effect of a low-pass filter of the 1st order.

5.3 SiGe HBTs

In this section the deterministic SHE solver is applied to the simulation of SiGe HBTs, which are pseudomorphically grown on unstrained Si. The base region contains a position-dependent Ge content resulting in compressive biaxial strain which modifies the band edges. The position-dependent band edges induce an additional driving force for the electrons entering the base and block the holes from spilling into the emitter making a higher base doping possible, which improves the RF performance of the HBTs [27–30]. In the first part, the accuracy of the SHE simulation result is checked by a detailed comparison with full band MC data. In the second part, full characterization of a 2D SiGe HBT is shown. Detailed comparison of the SHE result with drift-diffusion and hydrodynamic models and discussion of the accuracy of the classical models are presented. In the last part, using the SHE solver, the vertical Ge profile of a SiGe HBT is optimized in order to improve both the cutoff frequency and the breakdown voltage, which are key figures of merit of SiGe HBTs.

Apparent bandgap narrowing due to heavy doping is included [31]. SRH generation/recombination is not included.

In this section an extremely scaled SiGe HBT [32] is used to test the SHE solver. In Fig. 5.42, the outline of this SiGe HBT is shown. The device is symmetric w.r.t. the line $y = 0$ and only the right half is shown in the figure. The emitter width of the entire structure is 50 nm. The length of the spacer is 25 nm. The collector region is selectively implanted in order to reduce the base-collector capacitance. In the case of the base and collector regions, the region underneath the emitter window is denoted as intrinsic. The doping and Ge profiles along the symmetry line are shown in Fig. 5.43. The base is less than 10 nm thick and the Ge content has the form of a box profile.[4]

5.3.1 *Numerical Properties of Solutions*

Using the 1D-device approximation, the SiGe HBT whose doping and Ge profiles are shown in Fig. 5.43 is simulated with the different band models. The shift of the out-of-plane valleys relative to the in-plane valleys, which originates from the biaxial strain for a nonvanishing Ge content, has been taken into account [33, 34]. The missing base contact is simulated by setting the quasi-Fermi potential of the holes in the center of the base to the value of the base voltage [35, 36].

Using self-consistent solutions, the cutoff frequency is calculated in the quasi-stationary approximation in accordance with the definition given in [37]. Under fixed V_{CE}, the finite differences of quantities – the integrated electron density and

[4] In the last part of this section, the Ge content is optimized in order to obtain a better performance.

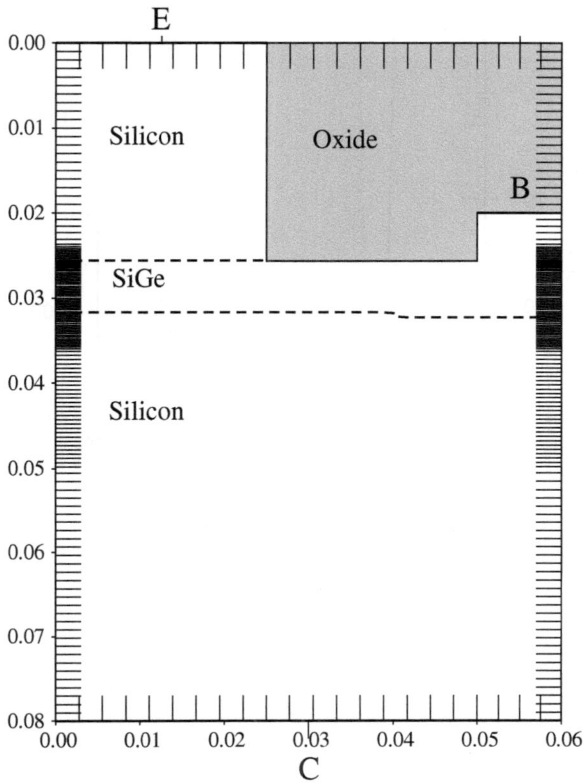

Fig. 5.42 2D structure of the SiGe HBT. The *ticks* indicate the real space grid

the terminal current – are taken between two V_{BE} values separated by 10 mV. In Fig. 5.44, the cutoff frequency of the SiGe HBT obtained with the different band models is shown. The second peak around the collector current of 7.5 mA μm^{-2} is related with the profile of the conduction band edge, and it is further investigated in Sect. 5.3.3. The Modena model predicts a by 5.6% higher peak cutoff frequency for this device. Since the extended Vecchi model and the fullband model show very close agreement, they are almost indistinguishable in the figure. Since the same maximum SHE order (7th order) is used for all band models, the differences between those models originate from the differences in underlying band models themselves. Note that the calculation of the cutoff frequency is particularly time intensive with the MC simulator, because the relative stochastic error of the output is strongly amplified by the finite differences [38]. Therefore, a comparison with MC results is not tried.

The dependence of the simulation results on the maximum order of SHE and the energy spacing is tested. In Fig. 5.45, the cutoff frequency evaluated with different maximum SHE order is shown for the full band model. As expected, the lowest order expansion yields a much higher cutoff frequency than higher order

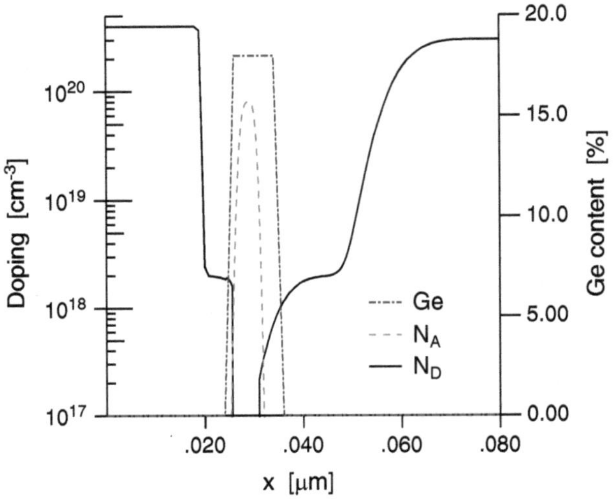

Fig. 5.43 1D doping and Ge profiles of the SiGe HBT for $y = 0$

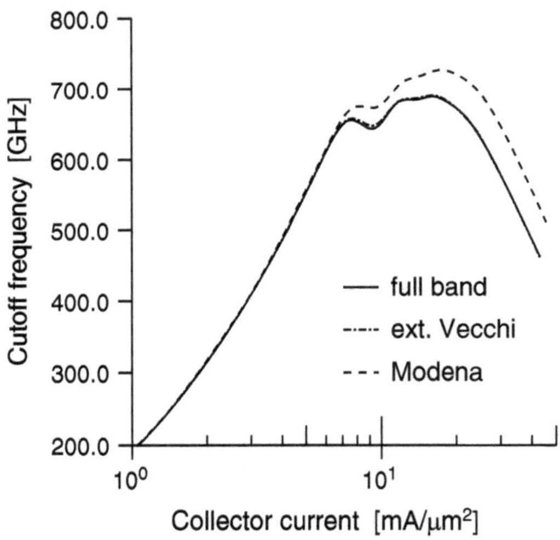

Fig. 5.44 Cutoff frequency of the 1D SiGe HBT for different band models at $V_{\mathrm{CE}} = 1.0\,\mathrm{V}$. The 7th order expansion is used

expansions. Compared with the peak cutoff frequency obtained with the 9th order expansion, the lowest order expansion shows more than 24% overestimation of the peak cutoff frequency. On the other hand, in the case of the 5th and the 7th order expansion, the overestimation of the peak cutoff frequency is about 2 and 0.5%, respectively, which is quite acceptable for practical purposes. Therefore, at least for the calculation of the cutoff frequency, the 5th order expansion can be a reasonable choice. In Fig. 5.46, the error of the collector current and the cutoff

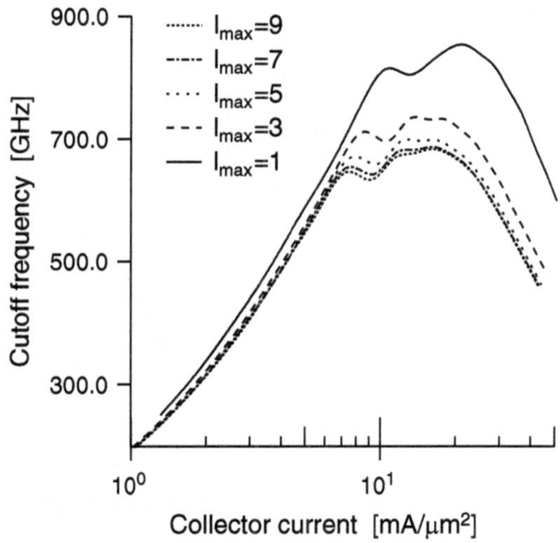

Fig. 5.45 Cutoff frequency of the SiGe HBT evaluated with different maximum SHE order at $V_{CE} = 1.0$ V for the full band model

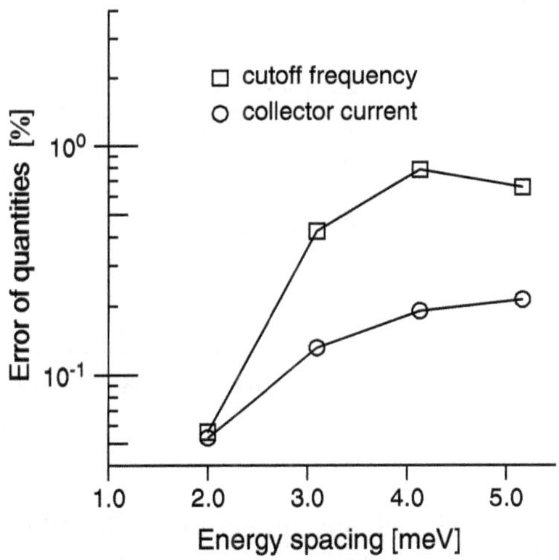

Fig. 5.46 Error of the collector current and the cutoff frequency for different values of the energy spacing, compared with the result for 1 meV spacing. The 5th order SHE is used and $V_{CE} = 1.0$ V. For the collector current, $V_{BE} = 0.91$ V. For the evaluation of the cutoff frequency, the finite difference between $V_{BE} = 0.91$ V and $V_{BE} = 0.92$ V is taken. The full band model for electrons is used

frequency for different values of the energy spacing is compared with the result for 1 meV spacing. In order to preserve the inelastic nature of phonon scattering including the Ge phonons, the energy spacing is increased only up to a twelfth of the Si LO phonon energy (5.1667 meV). It is obvious that the energy spacing of 5.1667 meV, which is the typical simulation parameter in this section, already gives quite good results within an error well below 1%.

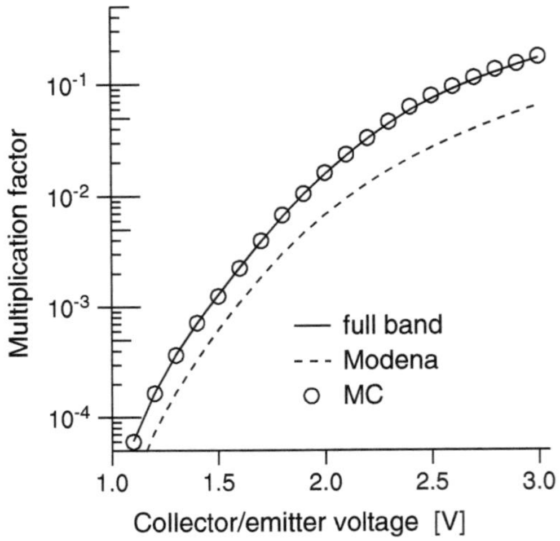

Fig. 5.47 Multiplication factor of the SiGe HBT at $V_{BE} = 0.7$ V. The frozen field is the same for all simulations

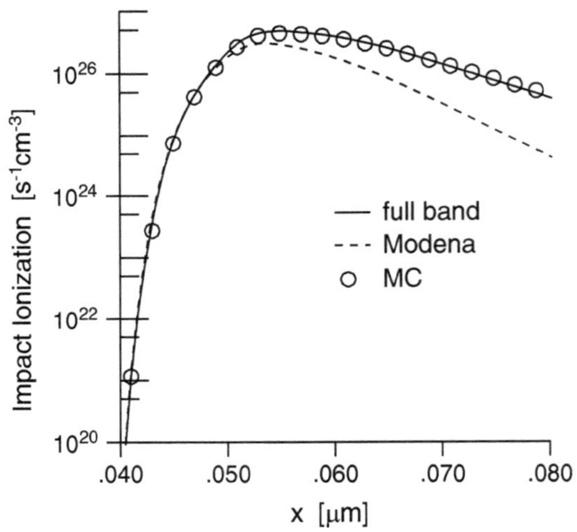

Fig. 5.48 Generation rate due to impact ionization at $V_{BE} = 0.7$ V and $V_{CE} = 1.5$ V. The frozen field is the same for all simulations. The collector current is normalized to 1 mA μm^{-2}

Breakdown voltages are key figures of merit of SiGe HBTs, and their accurate calculation is quite important for device simulation. The multiplication factor for a frozen field simulation is shown in Fig. 5.47. Note that this result is obtained without any modification of the scattering mechanisms and scattering parameters originally used in the full band MC simulation. When we employ the Modena model, on the other hand, a reasonable agreement of the multiplication factor cannot be expected. Even introducing a global scaling coefficient for the impact ionization rate cannot be a solution because the dependence on the collector/emitter voltage is somewhat different. Excellent agreement between the results of full band MC and the SHE solver with the full band model is obtained.

Not only the multiplication factor but also the local generation rate due to impact ionization shows excellent agreement, as shown in Fig. 5.48. This makes it possible to replace the very CPU intensive MC simulations for breakdown by much faster SHE simulations.

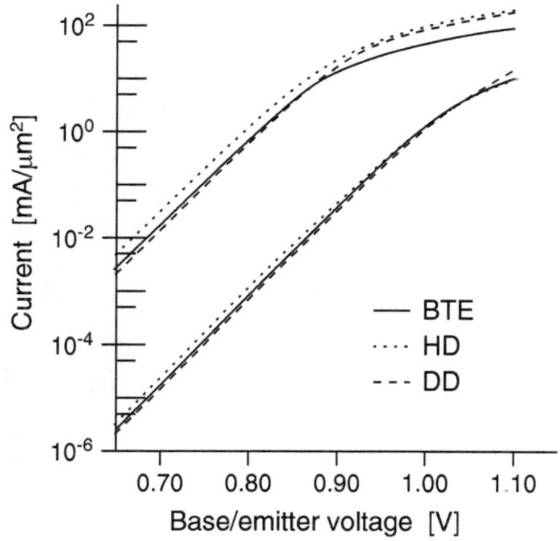

Fig. 5.49 Gummel plot. $V_{CB} = 0.1$ V. The collector current (*the three higher curves*) and the base current (*the three lower curves*) are scaled with the emitter window width

5.3.2 *Failure of Momentum-Based Models*

In this subsection, the simulation results of the SHE solver are compared with those of the classical TCAD tools (drift-diffusion (DD) and hydrodynamic (HD) models). Two-dimensional simulations of the structure shown in Fig. 5.42 are performed. A 5th order SHE and energy spacing of 5.1667 meV are used together with the full band model for electrons. The DD and HD simulations are performed with Galene III, which is based on a table model, where all transport and noise parameters are generated by consistent MC simulation [38, 39]. Impact ionization is described by local models (DD: local field model [40], HD: local temperature model [25]) with impact ionization parameters taken from [41].

The Gummel plot is shown in Fig. 5.49. For base/emitter voltages of more than 0.9 V the collector current is overestimated by a factor of about two by the DD and HD models. Below this value the result of the SHE solver is in between the DD and HD results. The current gain at $V_{BE} = 0.7$ V and $V_{CB} = 0.1$ V is about 1,000. The SHE solver is able to simulate the collector and base current without any problems over a range of several orders of magnitude, where the CPU time is much less sensitive to the bias condition than in the case of a MC simulator. Given the spatial grid and the maximum SHE order, the main factor which determines the simulation time is the span of the total energy included in the simulation. For example, in the case of the SiGe HBT a larger energy span is required in the case of lower V_{BE} values or higher V_{CE} values.

The electron density in the two-dimensional simulation plane is shown in Fig. 5.50. The density is smooth and no artificial oscillations occur due to the successful stabilization of the BTE. Note that very small densities near the base-collector junction underneath the base contact can be simulated without any difficulties. Since the electron current through the base contact is several orders-of-magnitude smaller than the electron current through the collector contact, an

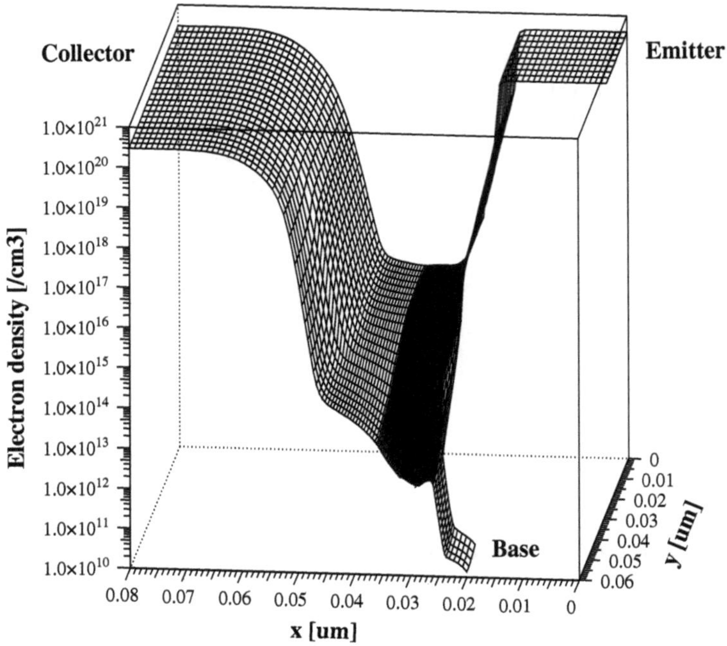

Fig. 5.50 Electron density in the two-dimensional simulation plane at $V_{\mathrm{BE}} = 0.7\,\mathrm{V}$ and $V_{\mathrm{CB}} = 0.1\,\mathrm{V}$

accurate evaluation of the electron density near the base-collector junction by MC simulations is rather CPU time consuming. On the other hand, although the electron density can be easily obtained from the classical TCAD tools, the results from those models can be inaccurate due to model deficiency. The electron current through the base contact, which is shown in Fig. 5.51, shows quite big differences between different models. The HD model predicts too strong lateral diffusion of electrons in the base region. Therefore, the electron current through the base contact is highly overestimated by a factor of almost 100.[5] On the other hand, for base/emitter voltages less than 0.9 V, the DD model underestimates the electron current through the base contact by a factor of about 2.

The cutoff frequency for the DC bias conditions with constant $V_{\mathrm{CB}} = 0.1\,\mathrm{V}$ is shown in Fig. 5.52. For this purpose, AC simulations are performed. The small-signal current gain at 100 GHz is used for extrapolation of the cutoff frequency. The values of the peak cutoff frequency are 507, 1,110, and 643 GHz for the DD, HD, and SHE models, respectively. The corresponding collector currents for which the

[5]This huge overestimation can be partially reduced by introducing a heuristic fitting of the heat flux coefficient [42]. However, even with this heuristic fitting of a transport coefficient, a quantitative agreement cannot be expected. Moreover, it has a detrimental side effect on the noise simulations [38].

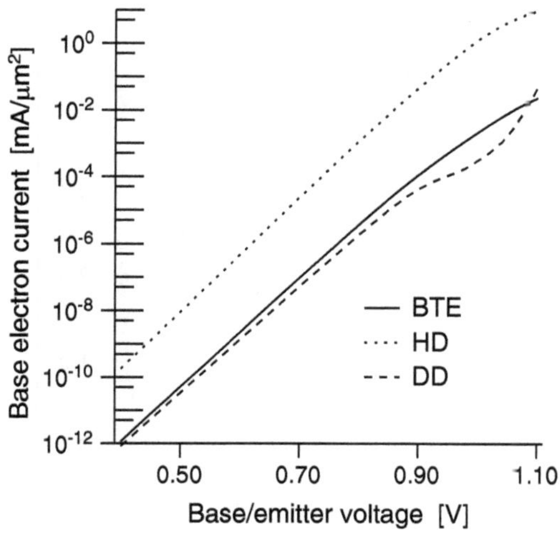

Fig. 5.51 Electron current at the base contact. $V_{CB} = 0.1\,V$

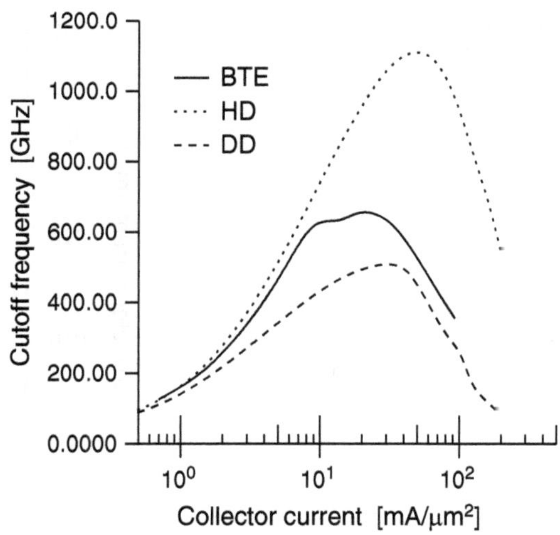

Fig. 5.52 Cutoff frequency extracted from the small-signal current gain at 100 GHz. $V_{CB} = 0.1\,V$

peak cutoff frequency occurs are 28.5, 47.9, 19.0 mA μm^{-2} for the DD, HD, and SHE model, respectively. When compared with the peak cutoff frequency obtained by 1D simulation (with the 5th order SHE), the peak cutoff frequency is reduced by about 50 GHz, as a result of including the extrinsic parts. Note that the change in the slope of the cutoff frequency versus collector current at about 10 mA μm^{-2}, which results in the second peak, is not found in the DD and HD results. This is related with the occupancy of the different valleys, which cannot be captured by the classical TCAD models. This is further investigated in Sect. 5.3.3. Although the HD model is supposed to be more accurate than the DD, its error is much larger. This is a result often seen in extremely scaled devices.

In order to predict the cutoff frequency and the corresponding collector current density correctly, accurate evaluation of the electron velocity is required. The

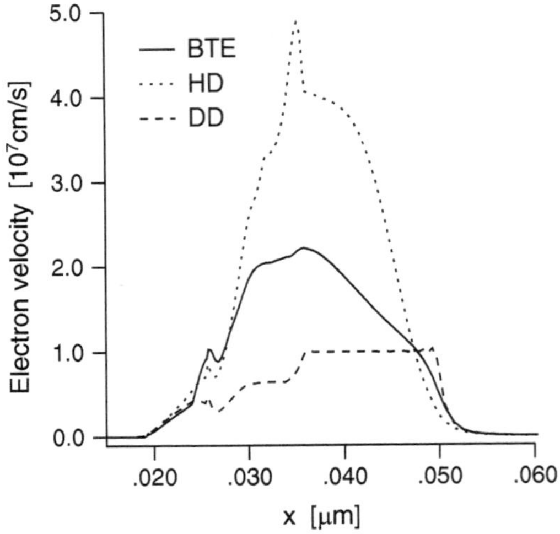

Fig. 5.53 Velocity profile for $y = 0$ at $I_C = 19.0\,\text{mA}\,\mu\text{m}^{-2}$ and $V_{CB} = 0.1\,\text{V}$

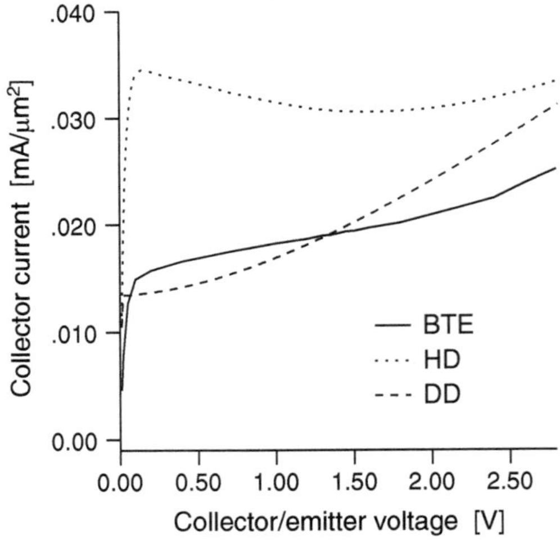

Fig. 5.54 Output characteristics. $V_{BE} = 0.7\,\text{V}$

electron velocity along the symmetry line ($y = 0$) is shown in Fig. 5.53. The collector current is the same for all three models. The maximum electron velocity of the HD model is nearly five times the saturation velocity of silicon and two times larger than the SHE result. The high electron velocity results in a lower transit time. Also the roll-off behavior of the cutoff frequency is affected by the simulation model. In the HD model a higher peak cutoff frequency occurs at a higher collector current. In the DD model the velocity cannot exceed the saturation velocity and the cutoff frequency is underestimated.

The output characteristics are shown in Fig. 5.54. The base/emitter voltage is kept as 0.7 V. The DD and HD results deviate significantly from the SHE result. The HD model yields a negative Early voltage. The negative slope of the output characteristics is related with the overestimated heat flux of the HD model. The DD model suffers from a far too strong impact of impact ionization.

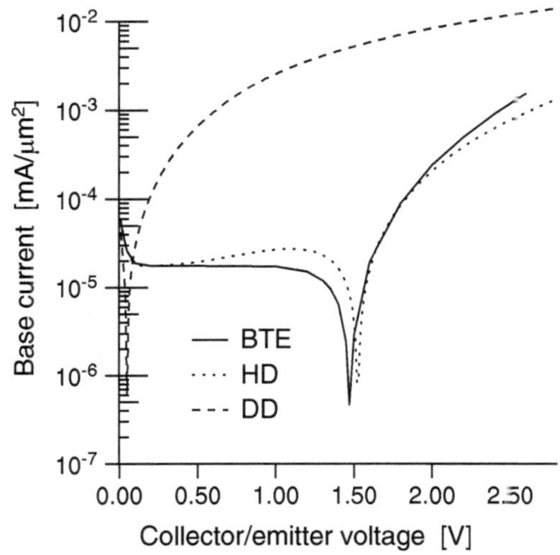

Fig. 5.55 Absolute value of the base current. $V_{BE} = 0.7\,V$

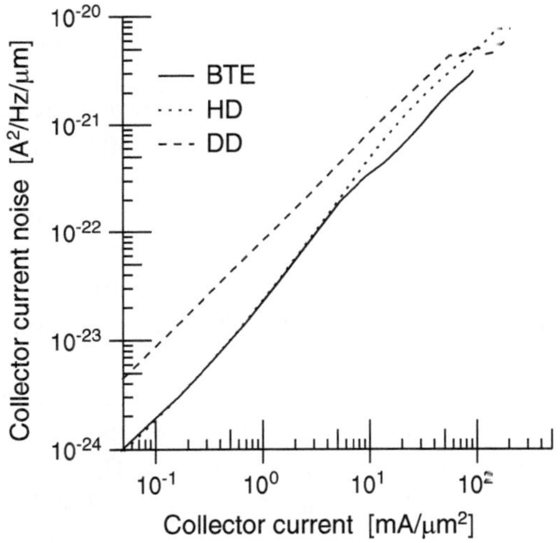

Fig. 5.56 Spectral intensity of the collector/collector current fluctuations at 100 GHz. $V_{CB} = 0.1\,V$

In Fig. 5.55 the absolute value of the corresponding base current is shown. The open base breakdown voltage evaluated from the collector/emitter bias at zero base current ($V_{BE} = 0.7\,V$) is 0.05, 1.52, and 1.47 V for the DD, HD, and SHE model, respectively. The reasonable agreement of the HD and SHE models is accidental, because the failure of the HD model can be observed for different devices. The local models for impact ionization of the DD and HD do not have a threshold and yield wrong results under strongly inhomogeneous transport conditions [43].

The spectral intensity of the collector/collector current fluctuations is shown in Fig. 5.56. For low collector current densities it follows the shot noise expression ($2qI_C$). The DD model overestimates the noise mainly due to an error in the collector electron current noise, for which it artificially produces super shot noise [44].

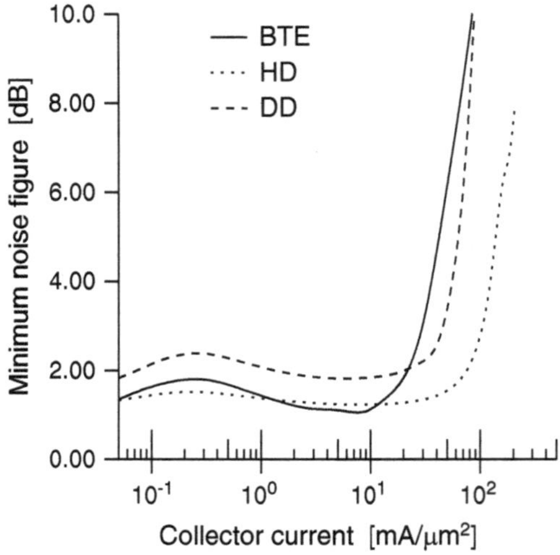

Fig. 5.57 Minimum noise figure at 100 GHz. $V_{CB} = 0.1$ V

On the other hand, the HD model follows the shot noise expression correctly for low collector current densities.

The minimum noise figure at 100 GHz is shown in Fig. 5.57. The lowest value is 1.82, 1.23, and 1.09 dB for the DD, HD, and SHE model, respectively. As explained above, the DD model overestimates the noise mainly due to an error in the collector electron current noise, for which it artificially produces super shot noise [44]. The HD model does not show this error and agrees well with the SHE model up to a current of about 10 mA μm^{-2}. For high collector current densities the minimum noise figure is affected by the simulation model, similarly with the roll-off behavior of the cutoff frequency.

As a result of the comparison, it is clear that the classical TCAD tools cannot accurately describe the transport and noise properties of the extremely scaled device considered in this section. The DD and HD models fail w.r.t. almost all aspects such as input characteristics (Fig. 5.49), output characteristics (Fig. 5.54), cutoff frequency (Fig. 5.52), and impact ionization (Fig. 5.55). Even when we must inevitably use the classical TCAD tools due to their numerical efficiency, the limitations of these models should be kept in mind.

5.3.3 *Performance Improvement*

In Fig. 5.58 the transit time distribution is shown for three different base/emitter voltages obtained by SHE. The derivative is evaluated with the finite difference technique and a change in the base/emitter voltage of 10 mV is used. The change in the slope of the cutoff frequency versus collector current in Fig. 5.52 is due to charge storage in the emitter/base space charge region, which can be seen in Fig. 5.58. Note that this behavior cannot be observed with the DD and HD models.

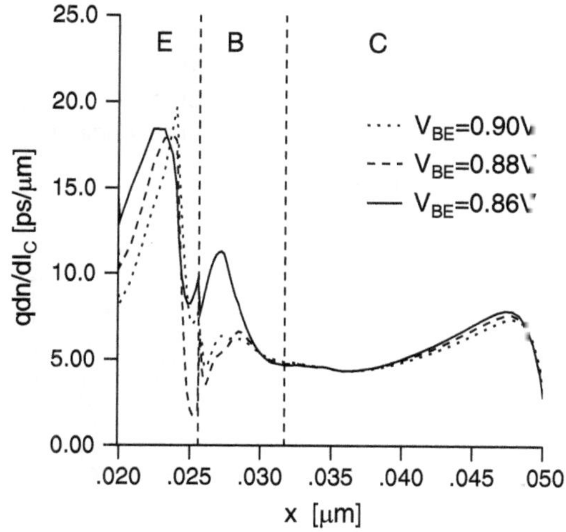

Fig. 5.58 Transit time distribution for three different base/emitter voltages. *Vertical lines* represent the metallurgical junctions

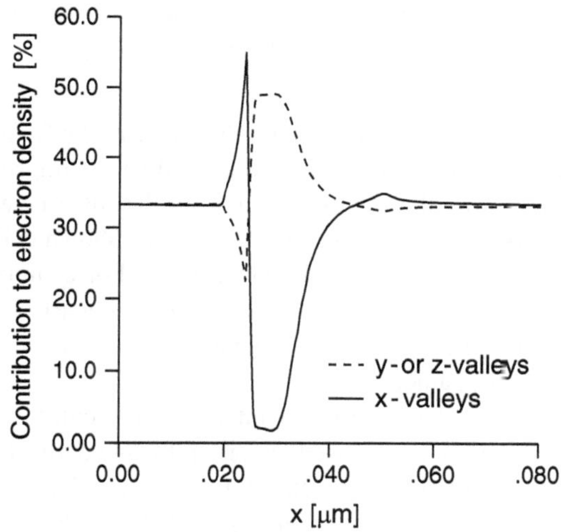

Fig. 5.59 Contribution of each valley pair to the local electron density at $V_{BE} = 0.92$ V and $V_{CE} = 1.0$ V

In Fig. 5.59 the contribution of each valley pair to the local electron density is shown. The bias condition near the peak cutoff frequency is selected. Note that the valley minimum energy of the x-valleys in the SiGe material is shifted from the band edge due to the biaxial strain. Therefore, the x-valleys are almost empty in the base region. On the other hand, the contribution of the x-valleys to the electron density is very high in the emitter/base space charge region, because the energy barrier seen by the electrons in the x-valleys is much higher than the one in the other valleys. For example, in the case of 18% Ge content considered in this section, the difference between the valley minimum of the x-valleys and the other valleys can be as large as 113 meV. Therefore, a sudden change of the Ge content introduces a steep energy barrier for electrons in the x-valleys, which results in the carrier jam in the emitter/base space charge region.

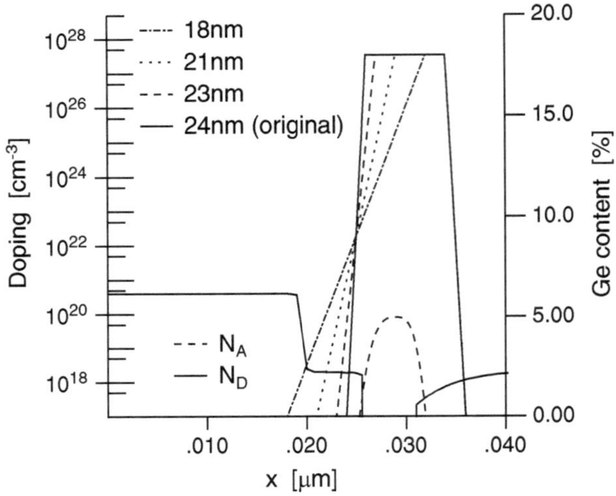

Fig. 5.60 Ge profiles with different starting points. The total Ge content integrated over the vertical direction is kept as 180 nm·%

It is expected that a smooth change of the Ge profile can reduce the severity of the carrier jam. In order to use this effect for improving the RF performance, without increasing the total Ge content in the SiGe layer (180 nm·%), the slope of the Ge profile is varied. Tested Ge profiles are shown in Fig. 5.60. Profiles are denoted using the starting point of the Ge content. For example, in the case of the one denoted "21 nm" in the figure, the Ge content linearly increases with a slope of 2.25%/nm until it reaches at 18%. Except for the Ge profile, all other simulation conditions are kept the same.

Since the quantity of interest is the cutoff frequency, the efficient 1D approximation is again used in order to avoid the huge computational burden of the 2D simulation. The cutoff frequency calculated for the different Ge profiles at $V_{CE} = 1.0$ V is shown in Fig. 5.61. When the slope of the Ge profile is decreased, the current level where the peak cutoff frequency appears is increased. In terms of the peak cutoff frequency, the profile whose Ge content starts from 21 nm is found to be optimal. Without much increase of the current level, the peak cutoff frequency is increased by more than 170 GHz. This significant improvement of the performance is achieved only by using an optimal Ge profile. Since the total Ge content in the SiGe layer is fixed in this example, there still remains a design option to increase the total Ge content as a performance booster, which are not investigated in this work.

The contribution of the x-valleys to the electron density is compared between the original Ge profile (starting from 24 nm) and the optimal one (starting from 21 nm) in Fig. 5.62. For fair comparison, the bias condition near the peak cutoff frequency corresponding to each profile is found. As expected, it is found that the huge pile-up of electrons observed in the original Ge profile is somewhat suppressed in the optimal profile.

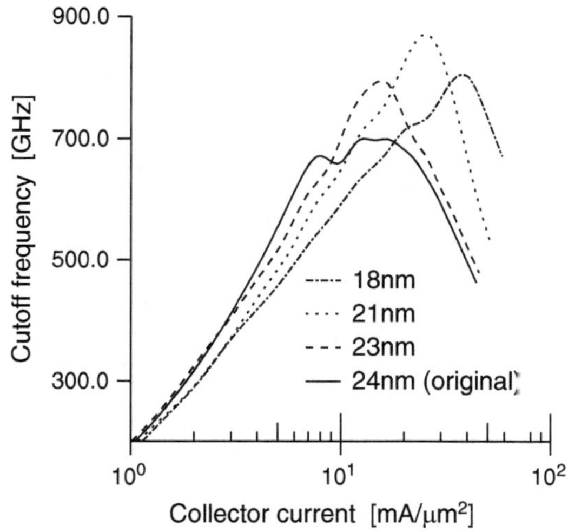

Fig. 5.61 Cutoff frequency of the 1D SiGe HBT for different Ge profiles at $V_{CE} = 1.0$ V. The 5th order expansion is used

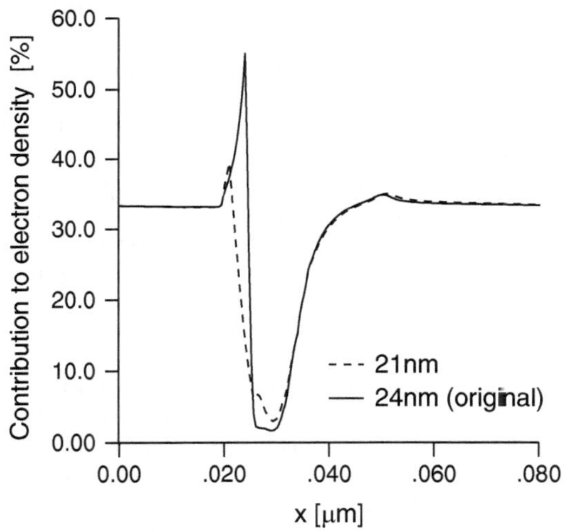

Fig. 5.62 Contribution of the x-valleys to the electron density for two different Ge profiles, 24 and 21 nm. For the profile starting from 24 nm, $V_{BE} = 0.92$ V. For the one starting from 21 nm, $V_{BE} = 0.95$ V. $V_{CE} = 1.0$ V

References

1. Hong, S.-M., Matz, G., Jungemann, C.: A deterministic Boltzmann equation solver based on a higher-order spherical harmonics expansion with full-band effects. IEEE Trans. Electron Dev. **57**, 2390–2397 (2010)
2. Jin, S., Hong, S.-M., Jungemann, C.: An efficient approach to include full band effects in deterministic Boltzmann equation solver based on high-order spherical harmonics expansion. IEEE Trans. Electron Dev. **58**(5), 1287–1294 (2011)
3. Lin, C.-K., Goldsman, N., Han, Z., Mayergoyz, I., Yu, S., Stettler, M., Singh, S.: Frequency domain analysis of the distribution function by small signal solution of the Boltzmann and Poisson equations. In: Proceedings of SISPAD, pp. 39–42 (1999)
4. Nyquist, H.: Thermal agitation of electric charge in conductors. Phys. Rev. **32**(1), 110–113 (1928)

5. Kogan, Sh.: Electronic Noise and Fluctuations in Solids. Cambridge University Press, Cambridge (1996)
6. Varani, L., Reggiani, L., Kuhn, T., González, T., Pardo, D.: Microscopic simulation of electronic noise in semiconductor materials and devices. IEEE Trans. Electron Dev. **41**(11), 1916–1925 (1994)
7. North, D.O.: Fluctuations in space-charge-limited currents at moderately high frequencies, Part II – Diodes and negative-grid triodes. RCA Rev. **4**, 441–472 (1940)
8. Jungemann, C., Meinerzhagen, B.: In-advance CPU time analysis for stationary Monte Carlo device simulations. IEICE Trans. Electron. **E86-C**(3), 314–319 (2003)
9. Brunetti, R., Jacoboni, C., Nava, F., Reggiani, L., Bosman, G., Zijlstra, R.J.J.: Diffusion coefficient of electrons in silicon. J. Appl. Phys. **52**, 6713–6722 (1981)
10. Vecchi, M.C., Rudan, M.: Modeling electron and hole transport with full-band structure effects by means of the spherical-harmonics expansion of the BTE. IEEE Trans. Electron Dev. **45**(1), 230–238 (1998)
11. Hong, S.-M., Jungemann, C.: Inclusion of the Pauli principle in a deterministic Boltzmann equation solver for semiconductor devices. In: Proceedings of SISPAD, pp. 135–138 (2010)
12. Hong, S.-M., Jungemann, C.: Inclusion of the Pauli principle in a deterministic Boltzmann equation solver based on a spherical harmonics expansion. J. Comput. Electron. **9**, 153–159 (2010)
13. Liang, W., Goldsman, N., Mayergoyz, I., Oldiges, P.J.: 2- D MOSFET modeling including surface effects and impact ionization by self-consistent solution of the Boltzmann, Poisson, and hole-continuity equations. IEEE Trans. Electron Dev. **44**(2), 257–267 (1997)
14. Lombardi, C., Manzini, S., Saporito, A., Vanzi, M.: A physically based mobility model for numerical simulation of nonplanar devices. IEEE Trans. Comp. Aided Des. **7**, 1164–1170 (1988)
15. Hong, S.-M., Jungemann, C.: A fully coupled scheme for a Boltzmann-Poisson equation solver based on a spherical harmonics expansion. J. Comput. Electron. **8**(3), 225–241 (2009)
16. Lucci, L., Palestri, P., Esseni, D., Bergagnini, L., Selmi, L.: Multisubbnad Monte Carlo study of transport, quantization, and electron-gas degeneration in ultrathin SOI n-MOSFETs. IEEE Trans. Electron Dev. **54**, 1156–1164 (2007)
17. Branin, F.H.: Network sensitivity and noise analysis simplified. IEEE Trans. Circ. Theor. **20**, 285–288 (1973)
18. Jungemann, C.: A deterministic approach to RF noise in silicon devices based on the Langevin Boltzmann equation. IEEE Trans. Electron Dev. **54**(5), 1185–1192 (2007)
19. Jungemann, C.: A deterministic solver for the Langevin Boltzmann equation including the Pauli principle. In: SPIE: Fluctuations and Noise, pp. 660007–1–660007–12, **6600** (2007)
20. Gritsch, M., Kosina, H., Grasser, T., Selberherr, S.: Influence of generation/recombination effects in simulations of partially depleted SOI MOSFETs. Solid State Electron. **45**, 621–627 (2001)
21. Gritsch, M., Kosina, H., Grasser, T., Selberherr, S.: Revision of the standard hydrodynamic transport model for SOI simulation. IEEE Trans. Electron Dev. **49**(10), 1814–1820 (2002)
22. Polsky, B., Penzin, O., Sayed, K.E., Schenk, A., Wettstein, A., Fichtner, W.: On negative differential resistance in hydrodynamic simulation of partially depleted SOI transistors. IEEE Trans. Electron Dev. **52**, 500–506 (2005)
23. Jungemann, C., Neinhüs, B., Nguyen, C.D., Meinerzhagen, B.: Impact of the floating body effect on noise in SOI devices investigated by hydrodynamic simulation. In: Proceedings of SISPAD, pp. 235–238 (2004)
24. Thoma, R., Emunds, A., Meinerzhagen, B., Peifer, H.J., Engl, W.L.: Hydrodynamic equations for semiconductors with nonparabolic bandstructures. IEEE Trans. Electron Dev. **38**(6), 1343–1352 (1991)
25. Mains, R.K., Haddad, G.I., Blakey, P.A.: Simulation of GaAs IMPATT diodes including energy and velocity transport equations. IEEE Trans. Electron Dev. **30**(10), 1327–1337 (1983)
26. Jin, W., Chan, P.C., Fung, S.K.H., Ko, P.K.: Shot-noise-induced excess low-frequency noise in floating-body partially depleted SOI MOSFET's. IEEE Trans. Electron Dev. **46**(7), 1180–1185

(1999)
27. Iyer, S.S., Patton, G.L., Stork, J.M.C., Meyerson, B.S., Harame, D.L.: Heterojunction bipolar transistors using Si-Ge alloys. IEEE Trans. Electron Dev. **36**(10), 2043–2064 (1989)
28. Cressler, J.D.: Re-engineering silicon: Si-Ge heterojunction bipolar technology. IEEE Spectrum **3**, 49–55 (1995)
29. Schüppen, A.: SiGe–HBTs for mobile communication. Solid State Electron. **43**, 1373–1381 (1999)
30. Maiti, C.K., Armstrong, G.A., Applications of Silicon–Germanium Heterostructure Devices, Series in Optics and Optoelectronics. Institute of Physics Publishing, Philadelphia (2001)
31. Klaassen, D.B.M., Slotboom, J.W., de Graaf, H.C.: Unified apparent bandgap narrowing in n- and p-type silicon. Solid State Electron. **35**, 125–129 (1992)
32. Hong, S.-M., Jungemann, C.: Electron transport in extremely scaled SiGe HBTs. In: Proceedings of BCTM, pp. 67–74 (2009)
33. Cressler, J.D., Niu, G.: Silicon-Germanium heterojunction bipolar transistors. Artech House, London (2003)
34. Hong, S.-M., Jungemann, C.: Deterministic simulation of SiGe HBTs based on the Boltzmann equation. In: Proceedings of ESSDERC, pp. 170–173 (2008)
35. Gummel, H.K.: A self-consistent iterative scheme for one-dimensional steady state transistor calculations. IEEE Trans. Electron Dev. **11**(10), 455–465 (1964)
36. Park, Y.-J., Navon, D.H., Tang, T.-W.: Monte Carlo simulation of bipolar transistors. IEEE Trans. Electron Dev. **31**(12), 1724–1730 (1984)
37. Gummel, H.K.: On the definition of the cutoff frequency f_T. In: Proceedings of IEEE, p. 2159 (1969)
38. Jungemann, C., Meinerzhagen, B.: Hierarchical Device Simulation: The Monte-Carlo Perspective, Computational Microelectronics. Springer, New York (2003)
39. Neinhüs, B., Decker, S., Graf, P., Bufler, F.M., Meinerzhagen, B.: Consistent hydrodynamic and Monte-Carlo simulation of SiGe HBTs based on table models for the relaxation times. VLSI Des. **8**, 387–391 (1998)
40. Chynoweth, A.G.: Ionization rates for electrons and holes in silicon. Phys. Rev. **109**(5), 1537–1540 (1958)
41. Maes, W., de Meyer, K., Van Overstraeten, R.: Impact ionization in silicon: A review and update. Solid State Electron. **33**, 705–718 (1990)
42. Bork, I., Jungemann, C., Meinerzhagen, B., Engl, W.L.: Influence of heat flux on the accuracy of hydrodynamic models for ultrashort Si MOSFETs. In: NUPAD Tech. Dig., vol. 5. Honolulu (1994)
43. Jungemann, C., Meinerzhagen, B., Decker, S., Keith, S., Yamaguchi, S., Goto, H.: Is physically sound and predictive modeling of NMOS substrate currents possible?. Solid State Electron. **42**, 647–655 (1998)
44. Jungemann, C., Neinhüs, B., Decker, S., Meinerzhagen, B.: Hierarchical 2–D DD and HD noise simulations of Si and SiGe devices: Part II – Results. IEEE Trans. Electron Dev. **49**(7), 1258–1264 (2002)

Part III
Transport in a Quasi-2D Hole Gas

Chapter 6
Coordinate Systems and Systems of Equation

6.1 Coordinate Systems

A device simulator for nanoscale strained SiGe double gate PMOSFETs based on the self-consistent solution of the 1D SE, 2D PE, and multi subband 1D BTE is developed, which is applicable for arbitrary crystallographic orientations and arbitrary channel directions.

In order to handle general surface orientations and channel directions, three different coordinate systems are introduced as shown in Fig. 6.1. These are the crystallographic coordinate system (x_C, y_C, z_C) of the cubic unit cell of the diamond lattice, the primed coordinate system (x', y', z') for which the $\mathbf{k} \cdot \mathbf{p}$ SE is solved, and the device coordinate system (x_D, y_D, z_D) for a planar MOSFET device where z_D is the quantization direction (also the growth direction) and x_D the channel direction. Since the z_D axis and z' axis are aligned, both, the (x_D, y_D) plane and the (x', y') plane are parallel to the wafer plane. For an arbitrary channel direction, the device coordinate system can be transformed to the primed coordinate system with the following transformation:

$$\begin{pmatrix} x_D \\ y_D \\ z_D \end{pmatrix} = \begin{pmatrix} \cos\varphi_D & \sin\varphi_D & 0 \\ -\sin\varphi_D & \cos\varphi_D & 0 \\ 0 & 0 & 1 \end{pmatrix} \begin{pmatrix} x' \\ y' \\ z' \end{pmatrix}, \tag{6.1}$$

where φ_D is the angle between the x_D- and x'-axes. The transformation between the crystallographic coordinate system and the primed coordinate system is given by [1]:

$$\begin{pmatrix} x' \\ y' \\ z' \end{pmatrix} = \hat{\mathbf{U}} \begin{pmatrix} x_C \\ y_C \\ z_C \end{pmatrix}, \qquad \hat{\mathbf{U}} = \begin{pmatrix} \cos\Phi'\cos\theta' & \sin\Phi'\cos\theta' & -\sin\theta' \\ -\sin\Phi' & \cos\Phi' & 0 \\ \cos\Phi'\sin\theta' & \sin\Phi'\sin\theta' & \cos\theta' \end{pmatrix}, \tag{6.2}$$

where θ' and Φ' are the azimuth and polar angles, respectively, of the growth direction z' in the crystallographic coordinate system.

S.-M. Hong et al., *Deterministic Solvers for the Boltzmann Transport Equation*, Computational Microelectronics, DOI 10.1007/978-3-7091-0778-2_6,

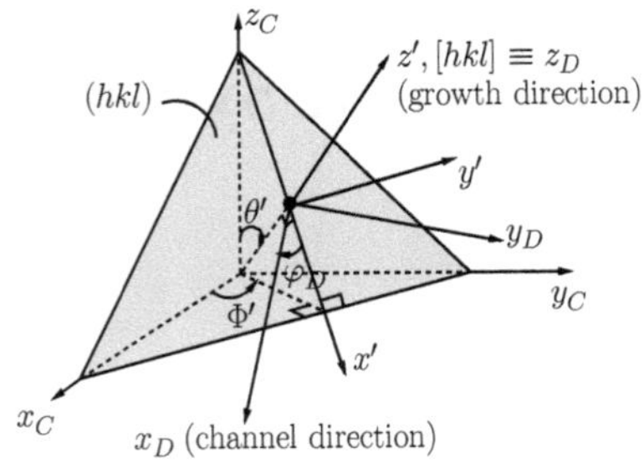

Fig. 6.1 Crystallographic coordinate system (x_C, y_C, z_C), primed coordinate system (x', y', z') for which the $\mathbf{k} \cdot \mathbf{p}$ SE is solved, and device coordinate system (x_D, y_D, z_D)

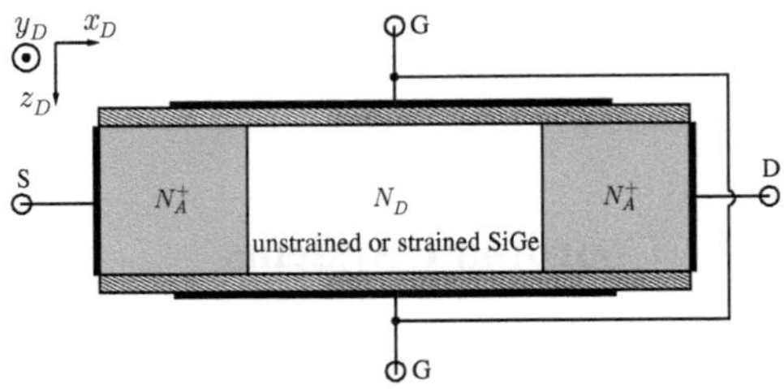

Fig. 6.2 Structure of a double gate PMOSFET with the device coordinate system (x_D, y_D, z_D)

6.2 Systems of Equation

The structure of a double gate PMOSFET is shown in Fig. 6.2 together with the device coordinate system. The device is partitioned into many slices along the transport direction x_D [2]. For each slice at position x_D the subband energies and envelope functions are calculated by solving the 1D SE in z_D direction for the given electrostatic potential $V(x_D, z_D)$. The description of the SE is presented in Chap. 7. The electrostatic potential $V(x_D, z_D)$ entering the SE is the solution of the 2D PE (described in Chap. 10). Based on the resultant subband structure, the stationary 1D BTE is solved to obtain the subband distribution function depending on the wave vector, position x_D, and subband index. The BTE is discussed in Chap. 9. Transport quantities like drain current, transconductance, intrinsic delay, are calculated based on the self-consistent solution of these three equations: SE, BTE, and PE. In Chap. 11 the algorithm to obtain the self-consistent solution of the three equations is described.

References

1. Pham, A.T., Jungemann, C., Meinerzhagen, B.: Microscopic modeling of hole inversion layer mobility in unstrained and uniaxially stressed Si on arbitrarily oriented substrates. Solid State Electron. **52**, 1437–1442 (2008)
2. Widiger, D.J., Kizilyalli, I.C., Coleman, J.J.: Two-dimensional transient simulation of an idealized high electron mobility transistor. IEEE Trans. Electron Dev. **32**(6), 1092–1102 (1985)

Chapter 7
Efficient $\mathbf{k}\cdot\mathbf{p}$ SE Solver

The subband structure is calculated with the $\mathbf{k}\cdot\mathbf{p}$ method. This chapter provides a description of a $\mathbf{k}\cdot\mathbf{p}$ solver which allows to determine accurately the eigen states in an efficient way and to capture the impact of crystallographic orientation and strain.

Based on the time-independent SE for a quantum system with a single particle, the derivation of the N band $\mathbf{k}\cdot\mathbf{p}$ SE is briefly described. An $N\times N$ $\mathbf{k}\cdot\mathbf{p}$ Hamiltonian with arbitrary N is derived firstly. The case $N=6$ which is often used for valence bands in Si or Ge is described in detail. The envelope function approximation $\mathbf{k}\cdot\mathbf{p}$ SE needed for the evaluation of the valence subband structure is described next, assuming an external potential which is non-periodic in the quantization direction. For an arbitrary crystallographic orientation, an efficient rotation of the $\mathbf{k}\cdot\mathbf{p}$ Hamiltonian is proposed which avoids dealing with lengthy formulas due to the presence of trigonometric functions in the rotation matrix of the coordinate system. In this chapter the crystallographic coordinate system $(x_\mathrm{C}, y_\mathrm{C}, z_\mathrm{C})$ is denoted as (x, y, z) for convenience.

7.1 $\mathbf{k}\cdot\mathbf{p}$ Hamiltonian

The motion of a single electron in a semiconductor can be characterized by the wavefunction ψ and the eigen energy ε which fulfill the time-independent Schrödinger equation [1] for a single particle:

$$\mathscr{H}\psi = \varepsilon\psi, \tag{7.1}$$

where the Hamiltonian $\mathscr{H}$ containing the kinetic and potential energy for a single particle system is given by

$$\mathscr{H} = \frac{\mathbf{p}^2}{2m_0} + V_\mathrm{L}(\mathbf{r}). \tag{7.2}$$

S.-M. Hong et al., *Deterministic Solvers for the Boltzmann Transport Equation*, Computational Microelectronics, DOI 10.1007/978-3-7091-0778-2_7,

Here, m_0 is the free electron mass, V_L is the periodic lattice potential, and the momentum operator $\mathbf{p}$ is defined as $\mathbf{p} = -i\hbar\nabla_\mathbf{r}$, where $\hbar$ is the Planck constant and $\mathbf{r}$ the position vector.

Within the $\mathbf{k}\cdot\mathbf{p}$ theory, the wave function is represented in the Bloch form $\psi = u_\mathbf{k}(\mathbf{r})\mathrm{e}^{\mathrm{i}\mathbf{k}^\mathrm{T}\mathbf{r}}$. Here, the Bloch function $u_\mathbf{k}(\mathbf{r})$ is lattice periodic fulfilling $u_\mathbf{k}(\mathbf{r}+\mathbf{R}) = u_\mathbf{k}(\mathbf{r})$ with $\mathbf{R}$ a direct lattice vector. Inserting the Bloch form of the wave function into (7.1) yields the $\mathbf{k}\cdot\mathbf{p}$ Schrödinger equation:

$$\mathscr{H}_{\mathbf{kp}} u_\mathbf{k}(\mathbf{r}) = \varepsilon_\mathbf{k} u_\mathbf{k}(\mathbf{r}), \tag{7.3}$$

where the $\mathbf{k}\cdot\mathbf{p}$ Hamiltonian reads

$$\mathscr{H}_{\mathbf{kp}}(\mathbf{k}) = \frac{\mathbf{p}^2}{2m_0} + V_\mathrm{L}(\mathbf{r}) + \frac{\hbar}{m_0}\mathbf{k}\cdot\mathbf{p} + \frac{\hbar^2k^2}{2m_0}. \tag{7.4}$$

The Bloch function $u_\mathbf{k}(\mathbf{r})$ can be expanded in a basis of Γ-centered[1] Bloch functions $u_\Gamma^\mu(\mathbf{r})$ associated with band μ:$u_\mathbf{k}(\mathbf{r}) = \sum_{\mu=1}^{\infty} a_\mathbf{k}^\mu u_\Gamma^\mu(\mathbf{r})$, where the expansion coefficient $a_\mathbf{k}^\mu$ is position independent and $u_\Gamma^\lambda(\mathbf{r})$ orthonormal: $\int_{\substack{\text{unit}\\\text{cell}}} u_\Gamma^{\mu*}(\mathbf{r})u_\Gamma^\lambda(\mathbf{r})\mathrm{d}^3r = \delta_{\mu\lambda}$. Due to the orthonormality of $u_\Gamma^\lambda(\mathbf{r})$, the projection of the $\mathbf{k}\cdot\mathbf{p}$ SE (7.3) onto $u_\Gamma^\lambda(\mathbf{r})$ defined as $\int_{\substack{\text{unit}\\\text{cell}}} u_\Gamma^{\lambda*}(\mathbf{r})\cdot\mathrm{LHS}\{(7.3)\}\mathrm{d}^3r = \int_{\substack{\text{unit}\\\text{cell}}} u_\Gamma^{\lambda*}(\mathbf{r})\cdot\mathrm{RHS}\{(7.3)\}\mathrm{d}^3r$ leads to an eigen problem for the eigen vectors $a_\mathbf{k}^\mu$:

$$\sum_{\mu=1}^{\infty} \mathscr{H}_{\mathbf{kp}}^{\lambda\mu}(\mathbf{k}) a_\mathbf{k}^\mu = \varepsilon_\mathbf{k} a_\mathbf{k}^\lambda, \tag{7.5}$$

where the matrix elements of the $\mathbf{k}\cdot\mathbf{p}$ Hamiltonian are defined as $\mathscr{H}_{\mathbf{kp}}^{\lambda\mu}(\mathbf{k}) \equiv \int_{\substack{\text{unit}\\\text{cell}}} u_\Gamma^{\lambda*}(\mathbf{r})\mathscr{H}_{\mathbf{kp}}(\mathbf{k})u_\Gamma^\mu(\mathbf{r})\mathrm{d}^3r$. The momentum matrix elements $\mathscr{P}^{\lambda\mu}$ are defined as [2]:

$$\mathscr{P}^{\lambda\mu} \equiv \int_{\substack{\text{unit}\\\text{cell}}} u_\Gamma^{\lambda*}(\mathbf{r})\mathbf{p}u_\Gamma^\mu(\mathbf{r})\mathrm{d}^3r. \tag{7.6}$$

It can be shown that $\mathscr{P}^{\lambda\lambda} = \mathbf{0}$ and $\mathscr{P}^{\mu\lambda} = \mathscr{P}^{\lambda\mu*}$ [2]. In terms of $\mathscr{P}^{\lambda\mu}$, the matrix elements of the $\mathbf{k}\cdot\mathbf{p}$ Hamiltonian can be written as

$$\mathscr{H}_{\mathbf{kp}}^{\lambda\mu}(\mathbf{k}) = \left(\varepsilon_\Gamma^\lambda + \frac{\hbar^2k^2}{2m_0}\right)\delta_{\mu\lambda} + \frac{\hbar}{m_0}\mathbf{k}\cdot\mathscr{P}^{\lambda\mu}. \tag{7.7}$$

Here, $\varepsilon_\Gamma^\lambda$ is the eigen energy of the Γ-centered eigen problem $\left[\frac{\mathbf{p}^2}{2m_0} + V_\mathrm{L}(\mathbf{r})\right]u_\Gamma^\lambda(\mathbf{r}) = \varepsilon_\Gamma^\lambda u_\Gamma^\lambda(\mathbf{r})$.

[1] Γ denotes the point $\mathbf{k} = \mathbf{0}$ in the $\mathbf{k}$-space.

Theoretically, the eigen problem (7.5) should be solved for an infinite number of bands. This is impossible on a computer. In additions, the CPU time increases strongly with the number of bands (N). For the sake of efficiency only N bands (called main bands) which are necessary to describe the valence band structure with sufficient accuracy are explicitly included into the $\mathbf{k}\cdot\mathbf{p}$ Hamiltonian. The other bands are considered as remote bands and their interaction with the main bands is taken into account when calculating the $N \times N$ $\mathbf{k}\cdot\mathbf{p}$ Hamiltonian by the Löwdin perturbation theory [3]. It can be shown that an $N \times N$ eigen problem can be derived from the general eigen problem (7.5):

$$\sum_{\mu=1}^{N}\left\{\mathscr{H}_{\mathbf{kp}}^{\lambda\mu}(\mathbf{k}) + \sum_{\mu'=N+1}^{\infty}\frac{\mathscr{H}_{\mathbf{kp}}^{\lambda\mu'}(\mathbf{k})\mathscr{H}_{\mathbf{kp}}^{\mu'\mu}(\mathbf{k})}{\varepsilon_{\mathbf{k}} - \mathscr{H}_{\mathbf{kp}}^{\mu'\mu'}(\mathbf{k})}\right\} a_{\mathbf{k}}^{\mu} = \varepsilon_{\mathbf{k}} a_{\mathbf{k}}^{\lambda}, \qquad \forall\lambda \le N \tag{7.8}$$

where the inner sum running over $\mu' = (N+1)\ldots\infty$ indicates the interaction of the remote bands $\mu' > N$ with the main bands $\lambda \le N$. The main bands are assumed to be degenerate at the Γ-point, i.e. $\varepsilon_{\Gamma}^{\lambda} = \varepsilon_0, \forall\lambda \le N$. For the main bands it is possible to choose the Γ-centered Bloch functions $u_{\Gamma}^{\lambda}(\mathbf{r})$ so that the momentum matrix elements vanish:

$$\mathscr{P}^{\lambda\mu} = \mathbf{0}, \qquad \mu, \lambda \le N, \tag{7.9}$$

which leads for the Hamiltonian matrix elements in (7.8) to

$$\mathscr{H}_{\mathbf{kp}}^{\lambda\mu}(\mathbf{k}) = \left(\varepsilon_0 + \frac{\hbar^2 k^2}{2m_0}\right)\delta_{\mu\lambda}, \qquad \mu, \lambda \le N, \tag{7.10}$$

$$\mathscr{H}_{\mathbf{kp}}^{\lambda\mu'}(\mathbf{k}) = \frac{\hbar}{m_0}\mathbf{k}\cdot\mathscr{P}^{\lambda\mu'} = \frac{\hbar}{m_0}\sum_{\alpha\in\{x,y,z\}} k_\alpha \mathscr{P}_{\alpha}^{\lambda\mu'}, \quad \mu' > N, \lambda \le N. \tag{7.11}$$

In (7.8) the denominator $\varepsilon_{\mathbf{k}} - \mathscr{H}_{\mathbf{kp}}^{\mu'\mu'}(\mathbf{k})$ of the remote band Hamiltonian part depends on the wave vector $\mathbf{k}$. However, if the main bands are assumed to be isotropic and parabolic $\varepsilon_{\mathbf{k}} \approx \varepsilon_0 + \frac{\hbar^2 k^2}{2m_0}$, then this denominator becomes independent from $\mathbf{k}$:

$$\varepsilon_{\mathbf{k}} - \mathscr{H}_{\mathbf{kp}}^{\mu'\mu'}(\mathbf{k}) \approx \varepsilon_0 + \frac{\hbar^2 k^2}{2m_0} - \left(\varepsilon_{\Gamma}^{\mu'} + \frac{\hbar^2 k^2}{2m_0}\right) = \varepsilon_0 - \varepsilon_{\Gamma}^{\mu'}. \tag{7.12}$$

Introducing all assumptions for main and remote bands and the appropriate choice of Γ-centered Bloch functions fulfilling (7.9) into the eigen problem (7.8) leads to the $N \times N$ $\mathbf{k}\cdot\mathbf{p}$ Schrödinger equation

$$\hat{\mathbf{H}}_{\mathbf{kp}}(\mathbf{k})\left(a_{\mathbf{k}}^{1}\ldots a_{\mathbf{k}}^{\mu}\ldots a_{\mathbf{k}}^{N}\right)^{\mathrm{T}} = (\varepsilon_{\mathbf{k}} - \varepsilon_0)\left(a_{\mathbf{k}}^{1}\ldots a_{\mathbf{k}}^{\mu}\ldots a_{\mathbf{k}}^{N}\right)^{\mathrm{T}}, \tag{7.13}$$

where the matrix elements of the $N \times N$ $\mathbf{k}\cdot\mathbf{p}$ Hamiltonian are given by

$$\hat{\mathbf{H}}^{\lambda\mu}_{\mathbf{kp}}(\mathbf{k}) = \frac{\hbar^2 k^2}{2m_0}\delta_{\mu\lambda} + \sum_{\alpha,\beta\in\{x,y,z\}} \frac{\hbar^2 k_\alpha k_\beta}{m_0^2} \sum_{\mu'=N+1}^{\infty} \frac{\mathscr{P}^{\lambda\mu'}_{\alpha}\mathscr{P}^{\mu'\mu}_{\beta}}{\varepsilon_0 - \varepsilon^{\mu'}_{\Gamma}}. \tag{7.14}$$

Although recently a "full band" (including $N = 30$ bands) $\mathbf{k}\cdot\mathbf{p}$ model has been proposed [4], often 3 bands, namely heavy hole (hh), light hole (lh), and split-off (so) are included into the $\mathbf{k}\cdot\mathbf{p}$ model for Si [5–7] or Ge [7,8] or even III–V materials [8]. The interactions between the spin and orbital angular momentum of p-states are also included. The inclusion of two spin directions in the 3×3 spin-degenerate $\mathbf{k}\cdot\mathbf{p}$ Hamiltonian results in a 6×6 $\mathbf{k}\cdot\mathbf{p}$ Hamiltonian.

Firstly, the 3×3 spin-degenerate $\mathbf{k}\cdot\mathbf{p}$ Hamiltonian $\hat{\mathbf{H}}^{3\times3}_{\mathbf{kp}}(\mathbf{k})$ is described. Very often $|x\rangle = u^X_\Gamma(\mathbf{r})$, $|y\rangle = u^Y_\Gamma(\mathbf{r})$, and $|z\rangle = u^Z_\Gamma(\mathbf{r})$ are chosen as the Γ-centered Bloch functions, where x, y, z are the three axes of the cubic unit cell. This basis set is a special set of degenerate states with strong symmetry [9–11]. With this basis set and an additional argument from the symmetry of the diamond lattice [2,9–11], the 3×3 spin-degenerate $\mathbf{k}\cdot\mathbf{p}$ Hamiltonian takes the following form [2,12]:

$$\hat{\mathbf{H}}^{3\times3}_{\mathbf{kp}}(\mathbf{k}) = \begin{pmatrix} \mathscr{L}k_x^2 + \mathscr{M}(k_y^2+k_z^2) & \mathscr{N}k_xky & \mathscr{N}k_xk_z \\ \mathscr{N}k_xky & \mathscr{L}k_y^2 + \mathscr{M}(k_z^2+k_x^2) & \mathscr{N}k_yk_z \\ \mathscr{N}k_xk_z & \mathscr{N}k_yk_z & \mathscr{L}k_z^2 + \mathscr{M}(k_x^2+k_y^2) \end{pmatrix}, \tag{7.15}$$

where the $\mathbf{k}\cdot\mathbf{p}$ parameters $\mathscr{L}, \mathscr{M}, \mathscr{N}$ given by

$$\mathscr{L} = \frac{\hbar^2}{2m_0} + \frac{\hbar^2}{m_0^2}\sum_{\mu'=N+1}^{\infty} \frac{\mathscr{P}^{X\mu'}_{x}\mathscr{P}^{\mu'X}_{x}}{\varepsilon_0 - \varepsilon^{\mu'}_{\Gamma}}, \tag{7.16}$$

$$\mathscr{M} = \frac{\hbar^2}{2m_0} + \frac{\hbar^2}{m_0^2}\sum_{\mu'=N+1}^{\infty} \frac{\mathscr{P}^{X\mu'}_{y}\mathscr{P}^{\mu'X}_{y}}{\varepsilon_0 - \varepsilon^{\mu'}_{\Gamma}}, \tag{7.17}$$

$$\mathscr{N} = \frac{\hbar^2}{m_0^2}\sum_{\mu'=N+1}^{\infty} \frac{\mathscr{P}^{X\mu'}_{x}\mathscr{P}^{\mu'Y}_{y} + \mathscr{P}^{Y\mu'}_{x}\mathscr{P}^{\mu'X}_{y}}{\varepsilon_0 - \varepsilon^{\mu'}_{\Gamma}} \tag{7.18}$$

are determined by matching certain experiments or simulation results of more fundamental approaches. In this work, the values for $\mathscr{L}, \mathscr{M}, \mathscr{N}$ are consistent with the ones used in the nextnano simulator [13,14].[2]

The Luttinger-Kohn basis states for heavy hole (hh), light hole (lh) and split off (so) bands including the two different spin directions are often used in the literature [2]. The Luttinger-Kohn basis states are given by linear combinations of $|x\rangle$, $|y\rangle$, $|z\rangle$ including the spinor $\uparrow$, $\downarrow$ [2]:

[2]Actually similar to [12], in [13, 14] the term $\hbar^2/(2m_0^2)$ is split from $\mathscr{L}$, $\mathscr{M}$, and $\hat{\mathbf{H}}^{3\times3}_{\mathbf{kp}}(\mathbf{k})$ is written as

$$|\mathbf{u}\rangle^{\mathrm{LK}} = \hat{\mathbf{A}}\,|\mathbf{u}\rangle^{x,y,z} \tag{7.20}$$

with $|\mathbf{u}\rangle^{\mathrm{LK}} = \left(|u_{hh\uparrow}\rangle, |u_{lh1}\rangle, |u_{lh2}\rangle, |u_{hh\downarrow}\rangle, |u_{so1}\rangle, |u_{so2}\rangle\right)^{\mathrm{T}}$ and $|\mathbf{u}\rangle^{x,y,z} = (|x\uparrow\rangle, |y\uparrow\rangle, |z\uparrow\rangle, |x\downarrow\rangle, |y\uparrow\rangle, |z\downarrow\rangle)^T$. The unitary rotation matrix $\hat{\mathbf{A}}$ is given by

$$\hat{\mathbf{A}} = \begin{pmatrix} -\frac{1}{\sqrt{2}} & -\frac{i}{\sqrt{2}} & 0 & 0 & 0 & 0 \\ 0 & 0 & \sqrt{\frac{2}{3}} & -\frac{1}{\sqrt{6}} & -\frac{i}{\sqrt{6}} & 0 \\ \frac{1}{\sqrt{6}} & -\frac{i}{\sqrt{6}} & 0 & 0 & 0 & \sqrt{\frac{2}{3}} \\ 0 & 0 & 0 & \frac{1}{\sqrt{2}} & -\frac{i}{\sqrt{2}} & 0 \\ 0 & 0 & \frac{1}{\sqrt{3}} & \frac{1}{\sqrt{3}} & \frac{i}{\sqrt{3}} & 0 \\ \frac{1}{\sqrt{3}} & -\frac{i}{\sqrt{3}} & 0 & 0 & 0 & -\frac{1}{\sqrt{3}} \end{pmatrix}. \tag{7.21}$$

It can be shown that the Luttinger-Kohn **k · p** Hamiltonian is a rotation of $\hat{\mathbf{H}}^{3\times3}_{\mathbf{kp}}$:

$$\hat{\mathbf{H}}^{\mathrm{LK}}_{\mathbf{kp}} = \hat{\mathbf{A}} \begin{pmatrix} \hat{\mathbf{H}}^{3\times3}_{\mathbf{kp}} & \hat{\mathbf{0}} \\ \hat{\mathbf{0}} & \hat{\mathbf{H}}^{3\times3}_{\mathbf{kp}} \end{pmatrix} \hat{\mathbf{A}}^{\dagger} \tag{7.22}$$

Inserting (7.15) into (7.22), the Luttinger-Kohn 6×6 **k · p** Hamiltonian is established [2, 12]

$$\hat{\mathbf{H}}^{\mathrm{LK}}_{\mathbf{kp}} = \begin{pmatrix} P+Q & -S & R & 0 & -S/\sqrt{2} & \sqrt{2}R \\ -S^* & P-Q & 0 & R & \sqrt{2}Q & \sqrt{3/2}S \\ R^* & 0 & P-Q & S & \sqrt{3/2}S^* & \sqrt{2}Q \\ 0 & R^* & S^* & P+Q & -\sqrt{2}R^* & -S^*/\sqrt{2} \\ -S^*/\sqrt{2} & -\sqrt{2}Q & \sqrt{3/2}S & -\sqrt{2}R & P & 0 \\ \sqrt{2}R^* & \sqrt{3/2}S^* & \sqrt{2}Q & -S/\sqrt{2} & 0 & P \end{pmatrix}. \tag{7.23}$$

P, Q, R, S are determined by the wave vector **k** and the Luttinger-Kohn parameters $\gamma_1, \gamma_2, \gamma_3$

$$\hat{\mathbf{H}}^{3\times3}_{\mathbf{kp}}(\mathbf{k}) = \frac{\hbar^2 k^2}{2m_0^2}\hat{\mathbf{I}} + \begin{pmatrix} \mathscr{L}'k_x^2 + \mathscr{M}'(k_y^2+k_z^2) & \mathscr{N}'k_xky & \mathscr{N}'k_xk_z \\ \mathscr{N}'k_xky & \mathscr{L}'k_y^2 + \mathscr{M}'(k_z^2+k_x^2) & \mathscr{N}'k_yk_z \\ \mathscr{N}'k_xk_z & \mathscr{N}'k_yk_z & \mathscr{L}'k_z^2 + \mathscr{M}'(k_x^2+k_y^2) \end{pmatrix}. \tag{7.19}$$

Thus, the parameters $\mathscr{L}', \mathscr{M}', \mathscr{N}'$ in [13, 14] relate to the parameters $\mathscr{L}, \mathscr{M}, \mathscr{N}$ as follows: $\mathscr{L} = \mathscr{L}' + \frac{\hbar^2}{2m_0^2}$, $\mathscr{M} = \mathscr{M}' + \frac{\hbar^2}{2m_0^2}$, and $\mathscr{N} = \mathscr{N}'$. In [12,13] $\mathscr{L}', \mathscr{M}', \mathscr{N}'$ are negative assuming negative eigen energies. In this work positive eigen energies are considered, thus, positive $\mathscr{L}, \mathscr{M}, \mathscr{N}$ ($\mathscr{L} \leftarrow |\mathscr{L}|, \mathscr{M} \leftarrow |\mathscr{M}|, \mathscr{N} \leftarrow |\mathscr{N}|$) are used.

$$\frac{2m_0}{\hbar^2}P = \gamma_1(k_x^2+k_y^2+k_z^2), \quad \frac{2m_0}{\hbar^2}Q = \gamma_2(k_x^2+k_y^2-2k_z^2),$$
$$\frac{2m_0}{\hbar^2}R = -\gamma_2\sqrt{3}(k_x^2-k_y^2) - i\gamma_3 2\sqrt{3}k_xk_y, \quad \frac{2m_0}{\hbar^2}S = 2\sqrt{3}\gamma_3(k_x+ik_y)k_z. \tag{7.24}$$

The Luttinger-Kohn parameters γ_1, γ_2, γ_3 relate directly to the parameters $\mathscr{L}, \mathscr{M}, \mathscr{N}$

$$\gamma_1 = \frac{2m_0}{\hbar^2}\frac{\mathscr{L}+2\mathscr{M}}{3}, \qquad \gamma_2 = \frac{2m_0}{\hbar^2}\frac{\mathscr{L}-\mathscr{M}}{6}, \qquad \gamma_3 = \frac{2m_0}{\hbar^2}\frac{\mathscr{N}}{6}. \tag{7.25}$$

In order to take into account the interaction between the spin and the orbital angular momentum of the p-states, the spin-orbit Hamiltonian is introduced. For the Luttinger-Kohn basis states given by (7.20), the spin-orbit Hamiltonian is diagonalized and contains only two non-zero elements [2,15]: $\hat{\mathbf{H}}^{\mathrm{LK}}_{\mathrm{so},55} = \hat{\mathbf{H}}^{\mathrm{LK}}_{\mathrm{so},66} = \Delta$. Here, Δ is the spin-orbit split-off potential which is a material dependent parameter and determined as in [13, 14].

In some cases the inversion layer contains multi layers of different materials like Si, Ge or SiGe alloy. The valence band offset of different layers can be taken into account by adding a diagonal band-offset Hamiltonian [13, 14]

$$\hat{\mathbf{H}}^{3\times 3}_{Ev} = E_v\hat{\mathbf{I}}. \tag{7.26}$$

Here, E_v is the valence band offset which depends on the position. The calculation of E_v of strained $Si_{1-y}Ge_y$ layers on a relaxed $Si_{1-x}Ge_x$ layer is described in detail in [14].

The influence of strain on the band structure can be considered by the Pikus-Bir model [16]. The Pikus-Bir strain Hamiltonian can be calculated based on the strain tensor and must be included into the Hamiltonian of the $\mathbf{k}\cdot\mathbf{p}$ SE.

With the $|x\rangle, |y\rangle, |z\rangle$ basis states, the Pikus-Bir strain Hamiltonian is defined as [16]

$$\hat{\mathbf{H}}^{3\times 3}_{\mathrm{strain}} = \begin{pmatrix} l\epsilon_{xx} + m(\epsilon_{yy}+\epsilon_{zz}) & n\epsilon_{xy} & n\epsilon_{xz} \\ n\epsilon_{xy} & l\epsilon_{yy} + m(\epsilon_{zz}+\epsilon_{xx}) & n\epsilon_{yz} \\ n\epsilon_{xz} & n\epsilon_{yz} & l\epsilon_{zz} + m(\epsilon_{xx}+\epsilon_{yy}) \end{pmatrix}, \tag{7.27}$$

where the deformation potentials l, m, n relate to the hydrostatic deformation potential a_v and the shear deformation potentials b, d as follows:

$$l = a_v + 2b, \qquad m = a_v - b, \qquad n = \sqrt{3}d. \tag{7.28}$$

For the Luttinger-Kohn basis states $|\mathbf{u}\rangle^{\mathrm{LK}}$, the 6×6 strain Hamiltonian can be obtained using a unitary rotation similar to (7.22):

$$\hat{\mathbf{H}}^{\mathrm{LK}}_{\mathrm{strain}} = \hat{\mathbf{A}} \begin{pmatrix} \hat{\mathbf{H}}^{3\times 3}_{\mathrm{strain}} & \hat{\mathbf{0}} \\ \hat{\mathbf{0}} & \hat{\mathbf{H}}^{3\times 3}_{\mathrm{strain}} \end{pmatrix} \hat{\mathbf{A}}^{\dagger}. \tag{7.29}$$

The total Hamiltonian $\hat{\mathbf{H}}^{\mathrm{LK}}$ including the effect of spin-orbit coupling and strain is a superposition of the $\mathbf{k} \cdot \mathbf{p}$, spin-orbit, and strain Hamiltonians:

$$\hat{\mathbf{H}}^{\mathrm{LK}}(\mathbf{k}) = \hat{\mathbf{H}}^{\mathrm{LK}}_{\mathbf{kp}}(\mathbf{k}) + \hat{\mathbf{H}}^{\mathrm{LK}}_{\mathrm{so}} + E_v\hat{\mathbf{I}} + \hat{\mathbf{H}}^{\mathrm{LK}}_{\mathrm{strain}} \tag{7.30}$$

which is sometimes denoted as $\hat{\mathbf{H}}$ in this work. Please note that $\hat{\mathbf{H}}^{\mathrm{LK}}_{\mathrm{so}}$ and $\hat{\mathbf{H}}^{\mathrm{LK}}_{\mathrm{strain}}$ do not depend on the wave vector and $\hat{\mathbf{H}}^{\mathrm{LK}}_{\mathbf{kp}}(\mathbf{k})|_{\mathbf{k}=\mathbf{0}} = \hat{\mathbf{0}}$. Therefore, $\hat{\mathbf{H}}^{\mathrm{LK}}(\mathbf{k})|_{\mathbf{k}=\mathbf{0}} = \hat{\mathbf{H}}^{\mathrm{LK}}_{\mathrm{so}} + E_v\hat{\mathbf{I}} + \hat{\mathbf{H}}^{\mathrm{LK}}_{\mathrm{strain}}$. Moreover, the total Hamiltonian is hermitian since the individual Hamiltonians $\hat{\mathbf{H}}^{\mathrm{LK}}_{\mathbf{kp}}(\mathbf{k})$, $\hat{\mathbf{H}}^{\mathrm{LK}}_{\mathrm{so}}$, $E_v\hat{\mathbf{I}}$, and $\hat{\mathbf{H}}^{\mathrm{LK}}_{\mathrm{strain}}$ are hermitian. The hermitian type of the total Hamiltonian ensures a real eigen energy for the $\mathbf{k} \cdot \mathbf{p}$ SE eigen problem.

7.2 Envelope Function Approximation k · p SE

Up to this point, the $\mathbf{k} \cdot \mathbf{p}$ Schrödinger equation is derived only for a bulk system where the lattice potential V_{L} acting on the particle is still periodic. In a confined system like a hole inversion layer, the particle experiences not only V_{L} but also the additional electrostatic potential V of the confining electric field. In contrast to V_{L}, the electrostatic potential V is aperiodic and is assumed to vary slowly in the quantization direction. Under these conditions the $\mathbf{k} \cdot \mathbf{p}$ method can be derived for a hole inversion layer from the bulk case using the envelope function approximation.

The total potential is the superposition of the periodic lattice potential V_{L} and the electrostatic potential V. In the presence of V the time-independent Schrödinger equation (7.1) for a single particle reads:

$$\left[\frac{\mathbf{p}^2}{2m_0} + V_{\mathrm{L}}(\mathbf{r}) + V(\mathbf{r})\right]\psi^{\nu} = \varepsilon^{\nu}\psi^{\nu}. \tag{7.31}$$

The wave function of the eigen state ν is expanded with the Γ-centered Bloch functions $u^{\mu}_{\Gamma}(\mathbf{r})$ and the envelope function $\zeta^{\nu}_{\mu}(\mathbf{r})$

$$\psi^{\nu} = \sum_{\mu=1}^{\infty} \zeta^{\nu}_{\mu}(\mathbf{r})u^{\mu}_{\Gamma}(\mathbf{r}). \tag{7.32}$$

In a confined system, the envelope function $\zeta^{\nu}_{\mu}(\mathbf{r})$ and the wave function ψ^{ν} are both aperiodic. The fundamental assumption for the envelope function approximation (EFA) is that the envelope function is assumed to vary slowly compared to the Γ-centered Bloch functions within the cubic unit cell of the diamond lattice. By taking into account the slow variation of the envelope function, the projection of the Schrödinger equation onto $u^{\lambda}_{\Gamma}(\mathbf{r})$ results in the EFA Schrödinger equation:

$$\sum_{\mu=1}^{\infty}\left[\mathscr{H}^{\lambda\mu}_{\mathbf{kp}}\left(\mathbf{k} \leftarrow \frac{1}{\hbar}\mathbf{p}\right) + V(\mathbf{r})\delta_{\lambda\mu}\right]\zeta^{\nu}_{\mu}(\mathbf{r}) = \varepsilon^{\nu}\zeta^{\nu}_{\lambda}(\mathbf{r}). \tag{7.33}$$

The Löwdin perturbation theory, which has been used to derive the $N \times N$ $\mathbf{k}\cdot\mathbf{p}$ SE for the bulk case, can also be employed for the derivation of the $N \times N$ $\mathbf{k}\cdot\mathbf{p}$ EFA Schrödinger equation:

$$\left[\hat{\mathbf{H}}_{\mathbf{kp}}\left(\mathbf{k} \leftarrow \frac{1}{\hbar}\mathbf{p}\right) + V(\mathbf{r})\hat{\mathbf{I}}\right]\left(\zeta_1^{\nu}(\mathbf{r})\ldots\zeta_N^{\nu}(\mathbf{r})\right)^{\mathrm{T}} = (\varepsilon^{\nu} - \varepsilon_0)\left(\zeta_1^{\nu}(\mathbf{r})\ldots\zeta_N^{\nu}(\mathbf{r})\right)^{\mathrm{T}}. \tag{7.34}$$

The $N \times N$ $\mathbf{k}\cdot\mathbf{p}$ EFA SE eigen problem (7.34) results in an infinite number of eigen states of which each consists of N envelope functions. Each eigen state represents one subband. The number of subbands is theoretically infinite. However, the contribution of an individual subband to the inversion charge or transport quantities like current and average drift velocity is not uniform for all subbands but reduces with increasing subband energy. Therefore, in practice, only a finite number of subbands is taken into account (typically 20–30 subbands).

7.3 Efficient Rotation of the $\mathbf{k}\cdot\mathbf{p}$ Hamiltonian for an Arbitrary Confinement Direction

Up to this point, the 6×6 $\mathbf{k}\cdot\mathbf{p}$ Hamiltonian is derived for the Γ-centered Bloch functions in the crystallographic coordinate system (x, y, z). This means that the derived 6×6 $\mathbf{k}\cdot\mathbf{p}$ Hamiltonian using (7.34) yields only the simple form for a confinement along the main crystal orientations $\mathbf{e}_{x\mathrm{C}}$, $\mathbf{e}_{y\mathrm{C}}$, $\mathbf{e}_{z\mathrm{C}}$. In order to handle an arbitrary confinement direction, an efficient rotation of the $\mathbf{k}\cdot\mathbf{p}$ Hamiltonian to the primed coordinate system (x', y', z') is proposed in this section.

The basis states $|x'\rangle, |y'\rangle, |z'\rangle$ associated with the primed coordinate system are mixtures of the states $|x\rangle, |y\rangle, |z\rangle$ associated with the crystallographic coordinate system. For example, when $z' = z$ ($\theta' = 0°$ in Fig. 6.1) and the x', y' axes are on the (x, y) plane, then the state $|x'\rangle$ is a mixture of the $|x\rangle$ and $|y\rangle$ states. When the angle $\Phi' = \angle\mathbf{e}_{x'}, \mathbf{e}_x$ varies in the range $0 < \Phi' < 45°$, the state $|x'\rangle$ is influenced by $|x\rangle$ more than by $|y\rangle$ and vice versa for $45° < \Phi' < 90°$. When $\Phi' = 90°$, the influence of $|x\rangle$ on $|x'\rangle$ vanishes and $|x'\rangle$ must be exactly $|y\rangle$. The basis states $|x'\rangle, |y'\rangle, |z'\rangle$ are therefore given by a unitary rotation of the states $|x\rangle, |y\rangle, |z\rangle$:[3]

$$(|x'\rangle\,|y'\rangle\,|z'\rangle)^{\mathrm{T}} = \hat{\mathbf{U}}(|x\rangle\,|y\rangle\,|z\rangle)^{\mathrm{T}} \tag{7.35}$$

Here, $\hat{\mathbf{U}}$ is the primed- to crystallographic coordinate system rotation matrix given by (6.2). The basis states $|x'\rangle, |y'\rangle, |z'\rangle$ are therefore linear combinations of the

---

[3]When $z' = z$, $\theta' = 0$ holds and $\hat{\mathbf{U}} = \begin{pmatrix} \cos\Phi' & \sin\Phi' & 0 \\ -\sin\Phi' & \cos\Phi' & 0 \\ 0 & 0 & 1 \end{pmatrix}$. Using (7.35) one obtains $|x'\rangle = \cos\Phi'\,|x\rangle + \sin\Phi'\,|y\rangle$. For the special case $x' = y$, $\Phi' = 90°$ holds and therefore $|x'\rangle = |y\rangle$.

$|x\rangle, |y\rangle, |z\rangle$ basis states. Using the Hamiltonian rotation one can derive the 3×3 $\mathbf{k}\cdot\mathbf{p}$ Hamiltonian (7.15) associated with the primed coordinate system $\hat{\mathbf{H}}'^{3\times3}_{\mathbf{kp}}(\mathbf{k}') = \hat{\bar{\mathbf{U}}}\hat{\mathbf{H}}^{3\times3}_{\mathbf{kp}}(\mathbf{k}')\hat{\bar{\mathbf{U}}}^{\mathrm{T}}$. The Luttinger-Kohn Hamiltonian given by (7.22) takes the following form in the primed coordinate system:

$$\hat{\mathbf{H}}'^{\mathrm{LK}}_{\mathbf{kp}}(\mathbf{k}') = \hat{\mathbf{A}} \begin{pmatrix} \hat{\mathbf{H}}'^{3\times3}_{\mathbf{kp}}(\mathbf{k}') & \hat{\mathbf{0}} \\ \hat{\mathbf{0}} & \hat{\mathbf{H}}'^{3\times3}_{\mathbf{kp}}(\mathbf{k}') \end{pmatrix} \hat{\mathbf{A}}^{\dagger} = \hat{\mathbf{R}}\hat{\mathbf{H}}^{\mathrm{LK}}_{\mathbf{kp}}(\mathbf{k}')\hat{\mathbf{R}}^{\dagger} = \hat{\mathbf{R}}\hat{\mathbf{H}}^{\mathrm{LK}}_{\mathbf{kp}}(\hat{\bar{\mathbf{U}}}\mathbf{k})\hat{\mathbf{R}}^{\dagger} \tag{7.36}$$

where $\hat{\mathbf{R}}$ is a unitary rotation matrix

$$\hat{\mathbf{R}} = \hat{\mathbf{A}} \begin{pmatrix} \hat{\bar{\mathbf{U}}} & \hat{\mathbf{0}} \\ \hat{\mathbf{0}} & \hat{\bar{\mathbf{U}}} \end{pmatrix} \hat{\mathbf{A}}^{\dagger} \tag{7.37}$$

The Luttinger-Kohn Hamiltonian in the $\mathbf{k}'$-space given by (7.36) has the same form as the Luttinger-Kohn Hamiltonian $\hat{\mathbf{H}}^{\mathrm{LK}}_{\mathbf{kp}}(\mathbf{k})$ shown in (7.23) [15]. However the dependence of P, Q, R, S on $\mathbf{k}'$ for an arbitrary surface orientation is different from the one for (001) surface orientation (given by (7.24)).

In this work quantization is considered only in z' direction. With this constrain the envelope function is therefore not periodic in z' but its periodicity in x' and y' still remains. This motivates the following ansatz for $\zeta'^{\nu}_{\mu}(\mathbf{r})$:

$$\zeta'^{\nu}_{\mu}(\mathbf{r}) = \zeta'^{\nu}_{\mu}(z', k'_x, k'_y)\mathrm{e}^{\mathrm{i}k'_x x'}\mathrm{e}^{\mathrm{i}k'_y y'} \tag{7.38}$$

For convenience, the three components of the wave vector $\mathbf{k}'$ along the primed coordinate axes x', y', z' are denoted as k'_x, k'_y, and k'_z, respectively.

Using the ansatz (7.38) for the envelope function, the 6×6 EFA Schrödinger equation can be derived:

$$\left[\hat{\mathbf{H}}'^{\mathrm{LK}}_{\mathbf{kp}}\left(k'_x, k'_y, k'_z \leftarrow -i\frac{\partial}{\partial z'}\right) + V(z')\hat{\mathbf{I}}\right]\boldsymbol{\zeta}'^{\nu}(z', k'_x, k'_y) = \left(\varepsilon^{\nu}(k'_x, k'_y) - \varepsilon_0\right) \times\boldsymbol{\zeta}'^{\nu}(z', k'_x, k'_y) \tag{7.39}$$

Here, $\boldsymbol{\zeta}'^{\nu}(z', k'_x, k'_y) = \left(\zeta'^{\nu}_1(z', k'_x, k'_y)\ldots\zeta'^{\nu}_6(z', k'_x, k'_y)\right)^{\mathrm{T}}$. In order to develop a simple method for implementing the substitution $k'_z \leftarrow -i\frac{\partial}{\partial z'}$ in (7.39), the 6×6 $\mathbf{k}\cdot\mathbf{p}$ Hamiltonian $\hat{\mathbf{H}}'^{\mathrm{LK}}_{\mathbf{kp}}$ is expanded into a second order polynomial of k'_z:

$$\hat{\mathbf{H}}'^{\mathrm{LK}}_{\mathbf{kp}} = \hat{\mathscr{M}}^{(2)}_{\mathbf{kp}}k'^2_z + \hat{\mathscr{M}}^{(1)}_{\mathbf{kp}}k'_z + \hat{\mathscr{M}}^{(0)}_{\mathbf{kp}}, \tag{7.40}$$

where $\hat{\mathscr{M}}^{(2)}_{\mathbf{kp}}$, $\hat{\mathscr{M}}^{(1)}_{\mathbf{kp}}(k'_x, k'_y)$, $\hat{\mathscr{M}}^{(0)}_{\mathbf{kp}}(k'_x, k'_y)$ are 6×6 matrices. With this expansion the substitution $k'_z \leftarrow -i\frac{\partial}{\partial z'}$ in (7.39) can be easily performed:

$$\left[-\hat{\mathcal{M}}_{\mathbf{kp}}^{(2)}\frac{\partial^2}{\partial z'^2} - i\hat{\mathcal{M}}_{\mathbf{kp}}^{(1)}\frac{\partial}{\partial z'} + \hat{\mathcal{M}}_{\mathbf{kp}}^{(0)} + V(z')\hat{\mathbf{I}}\right]\boldsymbol{\zeta}'^{\nu}(z',k'_x,k'_y)$$
$$= (\varepsilon^{\nu}(k'_x,k'_y) - \varepsilon_0)\boldsymbol{\zeta}'^{\nu}(z',k'_x,k'_y) \tag{7.41}$$

The analytical formulas of the 6 × 6 matrices $\hat{\mathcal{M}}_{\mathbf{kp}}^{(0),(1),(2)}$, which can be theoretically derived from (7.36), are very lengthy even for the $(11n)$ substrates as shown in [15]. In addition the form of $\hat{\mathcal{M}}_{\mathbf{kp}}^{(0),(1),(2)}$ is completely different from one surface orientation to the other. This approach therefore limits the generalization of the $\mathbf{k}\cdot\mathbf{p}$ solver for an arbitrary orientation.

In this work an alternative approach is proposed which avoids dealing with such complicated formulas and which can be generalized for an arbitrary surface orientation by expanding $\hat{\mathbf{H}}_{\mathbf{kp}}^{\prime\mathrm{LK}}$ in k'_x, k'_y, k'_z:

$$\hat{\mathbf{H}}_{\mathbf{kp}}^{\prime\mathrm{LK}} = \hat{\mathcal{M}}_{\mathbf{kp}}^{xx}k_x'^2 + \hat{\mathcal{M}}_{\mathbf{kp}}^{yy}k_y'^2 + \hat{\mathcal{M}}_{\mathbf{kp}}^{zz}k_z'^2 + \hat{\mathcal{M}}_{\mathbf{kp}}^{xy}k'_xk'_y + \hat{\mathcal{M}}_{\mathbf{kp}}^{yz}k'_yk'_z + \hat{\mathcal{M}}_{\mathbf{kp}}^{zx}k'_zk'_x. \tag{7.42}$$

All $\mathbf{k}$ independent components $\hat{\mathcal{M}}_{\mathbf{kp}}^{\alpha\beta}$, $\alpha,\beta \in \{x,y,z\}$ of this expansion can be determined by calculating $\hat{\mathbf{H}}_{\mathbf{kp}}^{\prime\mathrm{LK}}$ for a few specified wave vectors using (7.36):

$$\hat{\mathcal{M}}_{\mathbf{kp}}^{\alpha\alpha} = \hat{\mathbf{H}}_{\mathbf{kp}}^{\prime\mathrm{LK}}\Big|_{\mathbf{k}'=\mathbf{e}_{\alpha'}} \tag{7.43}$$

$$\hat{\mathcal{M}}_{\mathbf{kp}}^{\alpha\beta} = \hat{\mathbf{H}}_{\mathbf{kp}}^{\prime\mathrm{LK}}\Big|_{\mathbf{k}'=\mathbf{e}_{\alpha'}+\mathbf{e}_{\beta'}} - \hat{\mathbf{H}}_{\mathbf{kp}}^{\prime\mathrm{LK}}\Big|_{\mathbf{k}'=\mathbf{e}_{\alpha'}} - \hat{\mathbf{H}}_{\mathbf{kp}}^{\prime\mathrm{LK}}\Big|_{\mathbf{k}'=\mathbf{e}_{\beta'}}, \quad \alpha \neq \beta \tag{7.44}$$

where $\mathbf{e}_{\alpha'}$ is the unit vector in direction α'. For a specific wave vector and a given surface orientation, the calculation of $\hat{\mathbf{H}}_{\mathbf{kp}}^{\prime\mathrm{LK}}$ using (7.36) is straightforward avoiding lengthy formulas due to the trigonometric functions contained in the rotation matrix $\hat{\mathbf{U}}$ (6.2). As soon as $\hat{\mathcal{M}}_{\mathbf{kp}}^{\alpha\alpha}$ and $\hat{\mathcal{M}}_{\mathbf{kp}}^{\alpha\beta}$ are determined, $\hat{\mathcal{M}}_{\mathbf{kp}}^{(2),(1),(0)}$ can be evaluated directly for given k'_x, k'_y:

$$\hat{\mathcal{M}}_{\mathbf{kp}}^{(2)} = \hat{\mathcal{M}}_{\mathbf{kp}}^{zz}, \tag{7.45}$$

$$\hat{\mathcal{M}}_{\mathbf{kp}}^{(1)}(k'_x,k'_y) = \hat{\mathcal{M}}_{\mathbf{kp}}^{yz}k'_y + \hat{\mathcal{M}}_{\mathbf{kp}}^{zx}k'_x, \tag{7.46}$$

$$\hat{\mathcal{M}}_{\mathbf{kp}}^{(0)}(k'_x,k'_y) = \hat{\mathcal{M}}_{\mathbf{kp}}^{xx}{k'_x}^2 + \hat{\mathcal{M}}_{\mathbf{kp}}^{yy}{k'_y}^2 + \hat{\mathcal{M}}_{\mathbf{kp}}^{xy}k'_xk'_y. \tag{7.47}$$

In order to verify the proposed approach, the coefficients P, Q, R, S of the Luttinger-Kohn Hamiltonian $\hat{\mathbf{H}}'_{\mathbf{kp}}$ are calculated and compared with the results published in [15] for $(11n)$ surface orientations. As shown in Fig. 7.1 the calculated coefficients P, Q, R, S reproduce the results of [15].

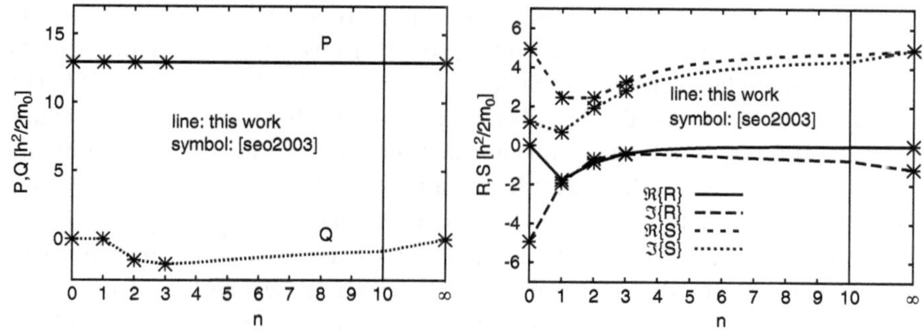

Fig. 7.1 Coefficients P, Q, R, S of the Luttinger-Kohn Hamiltonian $\hat{\mathbf{H}}'_{\mathbf{kp}}$ of this work compared with the results of [15] for $(11n)$ surfaces with $k'_x = k'_y = k'_z = 1$

7.4 Discretization of EFA k · p SE in the Quantization Direction

Including the strain and split-off Hamiltonians in the EFA $\mathbf{k}\cdot\mathbf{p}$ Schrödinger equation (7.41) and assuming $\varepsilon_0 = 0$ leads to the following form, where the subband index ν and the wave numbers k'_x, k'_y are not shown for the sake of brevity:

$$\left(-\hat{\mathscr{M}}^{(2)}_{\mathbf{kp}}(z')\frac{\partial^2}{\partial z'^2} - i\hat{\mathscr{M}}^{(1)}_{\mathbf{kp}}(z')\frac{\partial}{\partial z'} + \hat{\mathbf{H}}'^{(0)}(z')\right)\boldsymbol{\zeta}(z') = \varepsilon\boldsymbol{\zeta}(z'). \qquad (7.48)$$

The $\mathbf{k}$-independent Hamiltonian $\hat{\mathbf{H}}'^{(0)}$ is given by

$$\hat{\mathbf{H}}'^{(0)} = \hat{\mathscr{M}}^{(0)}_{\mathbf{kp}} + \hat{\mathbf{H}}'_{\mathrm{so}} + \hat{\mathbf{H}}'_{\mathrm{strain}} + E_v(z')\hat{\mathbf{I}} + V(z')\hat{\mathbf{I}}. \qquad (7.49)$$

The Pikus-Bir strain Hamiltonian $\hat{\mathbf{H}}'_{\mathrm{strain}}$ and split-off Hamiltonian $\hat{\mathbf{H}}'_{\mathrm{so}}$ for the Luttinger-Kohn basis states in the (x', y', z') system are obtained by using the same rotation applied to the $\mathbf{k}\cdot\mathbf{p}$ Hamiltonian $\hat{\mathbf{H}}'_{\mathrm{strain}} = \hat{\mathbf{R}}\hat{\mathbf{H}}^{\mathrm{LK}}_{\mathrm{strain}}\hat{\mathbf{R}}^{\dagger}$, $\hat{\mathbf{H}}'_{\mathrm{so}} = \hat{\mathbf{R}}\hat{\mathbf{H}}^{\mathrm{LK}}_{\mathrm{so}}\hat{\mathbf{R}}^{\dagger}$

For a conventional PMOS structure containing only a single Si, SiGe or Ge layer in the channel, the 3 point finite difference method can be used to discretize the EFA $\mathbf{k}\cdot\mathbf{p}$ SE (7.48), because the matrices $\hat{\mathscr{M}}^{(2)}_{\mathbf{kp}}$ and $\hat{\mathscr{M}}^{(1)}_{\mathbf{kp}}$ do not depend on z' and the envelope functions are assumed to vanish at the interface between the channel and the oxide. However, an advanced PMOS structure can contain multiple layers of Si, SiGe or Ge in the channel [17]. For a multi layer PMOS structure (or a PMOS heterostructure), it is difficult to capture the abrupt change of $\hat{\mathscr{M}}^{(2)}_{\mathbf{kp}}$ and $\hat{\mathscr{M}}^{(1)}_{\mathbf{kp}}$ at the interface between two layers if the finite difference method is used.

Actually the difficulties with a heterostructure not only occur in the discretization of the $\mathbf{k}\cdot\mathbf{p}$ SE, but also in the derivation of the $\mathbf{k}\cdot\mathbf{p}$ theory. The derivation of the $\mathbf{k}\cdot\mathbf{p}$ theory for a heterostructure is much more complicated than for a single layer structure due to the nature of the problem: the periodicity of the lattice is destroyed in a heterostructure. In principle, this aperiodicity of the lattice should be included in the derivation of the $\mathbf{k} \cdot \mathbf{p}$ SE [10, 11]. However, the aperiodicity of the lattice for multiple layers in a Si–SiGe heterostructure is often neglected [5]. This assumption allows to use the described $\mathbf{k} \cdot \mathbf{p}$ theory for both single layer structures and heterostructures. Along the quantization direction z' of a heterostructure, the variation of the material dependent $\mathbf{k} \cdot \mathbf{p}$ parameters like P, Q, R, S or the valence band-offset E_v is taken into account [5]. Each matrix $\hat{\mathscr{M}}^{(2)}_{\mathbf{kp}}$, $\hat{\mathscr{M}}^{(1)}_{\mathbf{kp}}$, and $\hat{\mathbf{H}}'^{(0)}$ is therefore a function of z'. The $\mathbf{k} \cdot \mathbf{p}$ theory is assumed to be unchanged for heterostructures. The discretization of the $\mathbf{k} \cdot \mathbf{p}$ SE at the interface between two layers of a heterostructure is not trivial in order to obtain a hermitian eigen problem, which ensures real values for the eigen energies. To solve this problem, the box integration method is used in this work for the discretization of the EFA $\mathbf{k} \cdot \mathbf{p}$ SE (7.48) on a non-equidistant z' grid. Appendix A.1 shows the details of discretization and the discretized equations of all grid points can be casted into matrix form as follows:

$$\begin{pmatrix} \ddots & & & & \\ \hat{\mathbf{D}}^{-}_{j-1} & \hat{\mathbf{D}}_{j-1} & \hat{\mathbf{D}}^{+}_{j-1} & & \\ & \hat{\mathbf{D}}^{-}_{j} & \hat{\mathbf{D}}_{j} & \hat{\mathbf{D}}^{+}_{j} & \\ & & \hat{\mathbf{D}}^{-}_{j+1} & \hat{\mathbf{D}}_{j+1} & \hat{\mathbf{D}}^{+}_{j+1} \\ & & & & \ddots \end{pmatrix} \begin{pmatrix} \cdots \\ \zeta_{j-1} \\ \zeta_{j} \\ \zeta_{j+1} \\ \cdots \end{pmatrix} = \varepsilon \hat{\mathbf{C}} \begin{pmatrix} \cdots \\ \zeta_{j-1} \\ \zeta_{j} \\ \zeta_{j+1} \\ \cdots \end{pmatrix}, \tag{7.50}$$

where $\hat{\mathbf{C}}$ is a positive-definite and diagonal matrix. In Appendix A.1 it is demonstrated that the discretization results in $\hat{\mathbf{D}}^{\dagger}_{j} = \hat{\mathbf{D}}_{j}$ and $(\hat{\mathbf{D}}^{-}_{j})^{\dagger} = \hat{\mathbf{D}}^{+}_{j-1}$. The eigen problem (7.50) is therefore hermitian which ensures real valued eigen energies.

7.5 Symmetry of Subband Structure

The symmetry of the subband structure is often analyzed based on the subband equi energy line $k'^{\nu}(\varepsilon, \phi')$ for a fixed energy level ε in the polar coordinate system (k', ϕ') of the 2D $\mathbf{k}'$-space. The subband equi energy line $k'^{\nu}(\varepsilon, \phi')$ is defined based on the eigen energies $\varepsilon^{\nu}(k', \phi')$ resulting from the eigen energy of the $\mathbf{k} \cdot \mathbf{p}$ SE such that $\varepsilon^{\nu}(k', \phi') = \varepsilon$. The symmetry of the projection of the unit cell of the diamond lattice onto the wafer surface depends on the surface orientation. Consequently, within the 2D $\mathbf{k}'$-space, the equi energy line $k'^{\nu}(\varepsilon, \phi')$ of subband ν resulting from the eigen energy of the $\mathbf{k} \cdot \mathbf{p}$ SE has different symmetries in the polar angle ϕ'. If the symmetry of the subband structure is correctly determined, the $\mathbf{k} \cdot \mathbf{p}$ SE eigen problem has to be solved only for an irreducible ϕ' range [6, 18]. The eigen energies outside of this ϕ' range can be obtained by exploiting the symmetry of subbands.

The symmetry of the subband structure is a property of the Hamiltonian and does not depend on the subband index or the electrostatic potential. The symmetry is determined by the crystallographic orientation and stress/strain conditions. This implies that the symmetry of the subband structure in the 2D $\mathbf{k}'$-space is the same as the symmetry that a bulk band structure possesses in the 2D $\mathbf{k}'$ transport plane under the same crystallographic orientation and stress/strain conditions. Therefore, to investigate the symmetry of the subband structure, one can analyze the eigen energies of a simplified eigen problem for a bulk system

$$\hat{\mathbf{H}}'(\mathbf{k}' = (k'_x, k'_y, 0)^{\mathrm{T}})\boldsymbol{\zeta} = \varepsilon\boldsymbol{\zeta}, \tag{7.51}$$

where the total Hamiltonian $\hat{\mathbf{H}}'$ is calculated for a wave vector in the transport plane with $k'_z = 0$. For $k'_z = 0$, the total Hamiltonian $\hat{\mathbf{H}}'$ including the $\mathbf{k}\cdot\mathbf{p}$, split-off and strain Hamiltonians reads:

$$\hat{\mathbf{H}}'(\mathbf{k}' = (k'_x, k'_y, 0)^{\mathrm{T}}) = k'^2 \hat{\mathscr{M}}^{\phi'}_{\mathbf{kp}} + \hat{\mathbf{H}}'_{\mathrm{so}} + \hat{\mathbf{H}}'_{\mathrm{strain}} + E_v \hat{\mathbf{I}}, \tag{7.52}$$

where only the $\mathbf{k}\cdot\mathbf{p}$ Hamiltonian given by (7.42) depends on $\mathbf{k}'$, and the polar angle dependent matrix $\hat{\mathscr{M}}^{\phi'}_{\mathbf{kp}}$ is defined as

$$\hat{\mathscr{M}}^{\phi'}_{\mathbf{kp}} = \hat{\mathscr{M}}^{xx}_{\mathbf{kp}} \cos^2\phi' + \hat{\mathscr{M}}^{yy}_{\mathbf{kp}} \sin^2\phi' + \frac{1}{2}\hat{\mathscr{M}}^{xy}_{\mathbf{kp}} \sin(2\phi'). \tag{7.53}$$

In order to obtain the equi energy line $k'(\varepsilon, \phi')$ which provides information about the symmetry of the subband structure, it is more convenient to reformulate the above eigen problem with the eigen value k'^2 for a given ε and ϕ':

$$\left(\varepsilon\hat{\mathbf{I}} - \hat{\mathbf{H}}'_{\mathrm{so}} - \hat{\mathbf{H}}'_{\mathrm{strain}} - E_v\hat{\mathbf{I}}\right)\boldsymbol{\zeta} = k'^2 \hat{\mathscr{M}}^{\phi'}_{\mathbf{kp}}\boldsymbol{\zeta}. \tag{7.54}$$

One important property of the matrix $\hat{\mathscr{M}}^{\phi'}_{\mathbf{kp}}$ given by (7.53) should be noted:

$$\hat{\mathscr{M}}^{\phi'}_{\mathbf{kp}} = \hat{\mathscr{M}}^{\pi+\phi'}_{\mathbf{kp}}. \tag{7.55}$$

This symmetry holds for all crystallographic orientations and all stress/strain conditions. Therefore, the subband structure in the $\mathbf{k}'$-space for an arbitrary surface orientation combined with any stress/strain configuration always has the fundamental symmetry:

$$k'(\varepsilon, \pi + \phi') = k'(\varepsilon, \phi'). \tag{7.56}$$

In addition, depending on the surface orientation and stress/strain configuration, the subband structure may have the following symmetries:

$$k'(\varepsilon, \phi') = k'(\varepsilon, \pi - \phi'), \qquad \forall\phi' \in [\pi/2, \pi], \tag{7.57}$$

$$k'(\varepsilon, \phi') = k'(\varepsilon, \pi/2 - \phi'), \qquad \forall\phi' \in [\pi/4, \pi/2], \tag{7.58}$$

$$k'(\varepsilon, \phi') = k'(\varepsilon, 2\pi/3 - \phi'), \qquad \forall\phi' \in [\pi/3, 2\pi/3]. \tag{7.59}$$

All these additional symmetries can be easily checked by solving the 6×6 eigen problem (7.54) with minor numerical effort. This provides an efficient method to classify the symmetry of the subband structure.

References

1. Schrödinger, E.: An undulatory theory of the mechanics of atoms and molecules. Phys. Rev. **28**(6), 1049–1070 (1926)
2. Luttinger, J.M., Kohn, W.: Motion of electron and holes in perturbes periodic fields. Phys. Rev. **97**, 869–883 (1955)
3. Löwdin, P-O.: Perturbation Theory and Its Applications in Quantum Mechanics. Wiley, New York (1966)
4. Rideau, D., Feraille, M., Michaillat, M., Niquet, Y.M., Tavernier, C., Jaouen, H.: On the validity of the effective mass approximation and the Luttinger k.p model in fully depleted SOI MOSFETs. Solid State Electron. **53**, 452–461 (2009)
5. Oberhuber, R., Zandler, G., Vogl, P.: Subband structure and mobility of two-dimensional holes in strained Si/SiGe MOSFETs. Phys. Rev. B **58**, 9941–9948 (1998)
6. Fischetti, M.V., Ren, Z., Solomon, P.M., Yang, M., Rim, K.: Six-band $k \cdot p$ calculation of the hole mobility in silicon inversion layers: Dependence on surface orientation, strain, and silicon thickness. J. Appl. Phys. **94**, 1079–1095 (2003)
7. Pham, A.T., Jungemann, C., Meinerzhagen, B.: Physics-based modeling of hole inversion layer mobility in strained SiGe on insulator. IEEE Trans. Electron Dev. **54**(9), 2174–2182 (2007)
8. Zhang, Y., Fischetti, M.V.: Calculation of hole mobility in Ge and III-V p-channels. In: Proceedings of IWCE, pp. 41–44 (2009)
9. Yu, P.Y., Cardona, M.: Fundamentals of Semiconductors. Springer, Berlin (1997)
10. Foreman, B.A.: Analytical envelope-function theory of interface band mixing. Phys. Rev. Lett. **81**(2), 425–428 (1998)
11. Foreman, B.A.: Theory of the effective Hamiltonian for degenerate bands in an electric field. J. Phys. Condens. Matter **12**, R435–R461 (2000)
12. Dresselhaus, G., Kip, A.F., Kittel, C.: Cyclotron resonance of electrons and holes in silicon and germanium crystals. Phys. Rev. **98**, 368–384 (1955)
13. TU-München, `nextnano` 3D device simulator, http://www.nextnano.de (2008)
14. Rieger, M.M., Vogl, P.: Electronic-band parameters in strained $Si_{1-x}Ge_x$ alloys on $Si_{1-y}Ge_y$ substrates. Phys. Rev. B **48**, 14276–14287 (1993)
15. Seo, W.H., Donegan, J.F.: 6×6 effective mass Hamiltonian for heterostructures grown on (11N)-oriented substrates. Phys. Rev. B **68**, 0753181–0753188 (2003)
16. Bir, G.L., Pikus, G.E.: Symmetry and Strain-Induced Effects in Semiconductors. Wiley, New York (1974)
17. Aberg, I., Chleirigh, C.N., Hoyt, J.L.: Ultrathin-body strained-Si and SiGe heterostructure-on-insulator MOSFETs. IEEE Trans. Electron Dev. **53**(5), 1021–1029 (2006)
18. Pham, A.T., Jungemann, C., Meinerzhagen, B.: A fast k*p solver for hole inversion layers with an efficient 2D k-space discretization. J. Compu. Electr. **7**(3), 99–102 (2008)

Chapter 8
Efficient 2D k-Space Discretization and Non-Linear Interpolation Schemes

Within this chapter the primed coordinate system is denoted as (x, y, z) for convenience. The 2D **k**-space is discretized in polar coordinates, which are denoted as k, ϕ. The foci of this section are the efficient 2D **k**-space discretization and non-linear interpolation schemes. The $\mathbf{k} \cdot \mathbf{p}$ solver consumes typically the largest part of the CPU time during transport simulations including the $\mathbf{k} \cdot \mathbf{p}$ SE [1–3]. An efficient discretization of the 2D **k**-space combined with an efficient interpolation is therefore required in order to reduce the number of grid points in the 2D **k**-space and thus the CPU time spent on the $\mathbf{k} \cdot \mathbf{p}$ solver. The $\mathbf{k} \cdot \mathbf{p}$ SE is solved in this work on an orthogonal grid in the 2D (k, ϕ)-space. The discretization of the 2D **k**-space with polar coordinates is actually not new and has been used by Fischetti et al. [3] before. However, in [3] a linear interpolation of the subband energy based on the coordinates k, ϕ has been used, which requires to solve the $\mathbf{k} \cdot \mathbf{p}$ SE for many (k, ϕ) grid points in order to get an accurate energy interpolation. In this work, the subband energy and the probability density function for an arbitrary in-plane wave vector are determined by an efficient interpolation which requires a much smaller number of (k, ϕ) grid points. Firstly, the dispersion relation for each radial angle ϕ_n is interpolated with monotonic cubic splines. Secondly, the resultant interpolation coefficients for constant ϕ are interpolated for a constant modulus of the wave vector k_l with harmonics in ϕ.

8.1 Efficient 2D k-Space Discretization

The discretization of the 2D **k**-space using polar coordinates is briefly described as follows. Due to the symmetry of the subband structure the $\mathbf{k} \cdot \mathbf{p}$ SE only needs to be solved for an irreducible ϕ range [2–4]. The irreducible ϕ range is partitioned into slices at predefined angles ϕ_n:

$$\phi_n = \frac{2\pi}{N} n + \varphi, \qquad \forall n = 0, \dots, N_{\text{irr}}. \tag{8.1}$$

S.-M. Hong et al., *Deterministic Solvers for the Boltzmann Transport Equation*, Computational Microelectronics, DOI 10.1007/978-3-7091-0778-2_8,

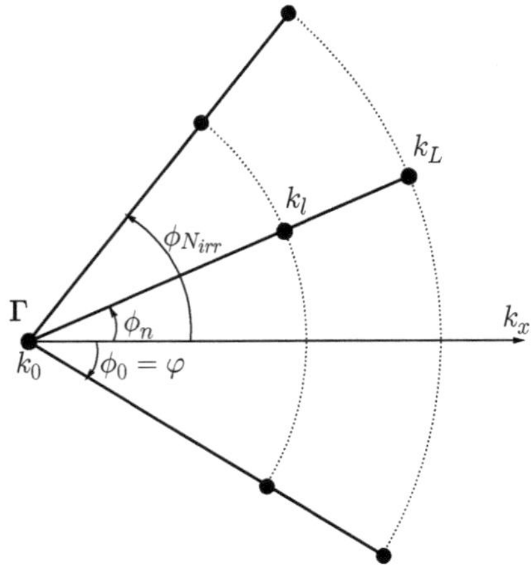

Fig. 8.1 Discretization of the 2D **k**-space within the irreducible (k, ϕ) range, $\varphi = \phi_0 \le \phi \le \phi_{N_{\mathrm{irr}}}$ and $0 \le k \le k_L$, where k, ϕ represent the polar coordinate system. The primed coordinate system (x', y', z') is denoted as (x, y, z), for convenience

N is a positive integer number. The constants φ and $\frac{2\pi}{N} N_{\mathrm{irr}} + \varphi$ are the lower and upper bound of the irreducible ϕ range, respectively. On each slice ϕ_n, the magnitude of the wave vector k is discretized into $k_0 = 0, k_1, \ldots, k_L$. The discrete values of the wave vector $k_0, k_1, \ldots, k_L$ are the same for all slices ϕ_n. Therefore, for all slices ϕ_n the wave vector $(k_0, \phi_n) = \mathbf{0}$ indicates the Γ point in the 2D **k**-space. A non-equidistant discretization of k is employed: a fine grid near the Γ point and a coarser grid for larger k. The subband structure is determined for wave vectors **k** up to $k = k_L$ assuming that the contribution of eigen energies for $k > k_L$ to transport can be neglected (e.g. $k_L = 0.6\pi/a_{Si}$ is used in [2] for low-field mobility calculations). In Fig. 8.1 this grid is shown.

8.2 Monotonic Cubic Spline Interpolation of Subband Energy

The numerical approach to solve the BTE presented in Chap. 9 relies on a dispersion relation $\varepsilon^\nu(k, \phi)$ that is for each subband ν monotonically increasing for growing k and constant ϕ:

$$\varepsilon^\nu(k, \phi) < \varepsilon^\nu(k', \phi), \qquad k < k'. \tag{8.2}$$

A cubic spline interpolation conserving this monotonicity condition (8.2) is therefore required.

For a given polar angle ϕ a cubic spline interpolation of the eigen energy $\varepsilon^\nu(k, \phi)$ based on the eigen energies $\varepsilon^\nu(k_l, \phi)$ is given by:

$$\begin{aligned}\varepsilon^\nu(k, \phi) &= a^\nu(k_l, \phi)(k - k_l)^3 + b^\nu(k_l, \phi)(k - k_l)^2 \\ &\quad + \hbar v_k^\nu(k_l, \phi)(k - k_l) + \varepsilon^\nu(k_l, \phi), \quad k \in [k_l, k_{l+1}).\end{aligned} \tag{8.3}$$

Here, a, b, and v_k are interpolation coefficients. In principle, the cubic spline interpolation (8.3) should conserve the continuity of the eigen energy and its derivative w.r.t k

$$\lim_{k\to k_{l+1}^-} \varepsilon^\nu(k,\phi) = \lim_{k\to k_{l+1}^+} \varepsilon^\nu(k,\phi) = \varepsilon^\nu(k_{l+1},\phi), \tag{8.4}$$

$$\lim_{k\to k_{l+1}^-} \frac{\partial \varepsilon^\nu(k,\phi)}{\partial k} = \lim_{k\to k_{l+1}^+} \frac{\partial \varepsilon^\nu(k,\phi)}{\partial k} = \hbar v_k^\nu(k_{l+1},\phi), \tag{8.5}$$

which requires a fixed relation between a,b and $\varepsilon^\nu(k_l,\phi)$, $v_k^\nu(k_l,\phi)$, $\varepsilon^\nu(k_{l+1},\phi)$, $v_k^\nu(k_{l+1},\phi)$.

Often the natural cubic spline interpolation is employed with its additional continuity condition for the second order derivative:

$$\lim_{k\to k_{l+1}^-} \frac{\partial^2 \varepsilon^\nu(k,\phi)}{\partial k^2} = \lim_{k\to k_{l+1}^+} \frac{\partial^2 \varepsilon^\nu(k,\phi)}{\partial k^2}. \tag{8.6}$$

A big advantage of the natural cubic spline is the maximum smoothness (C^2) of the interpolation (i.e. fulfilling the continuity conditions (8.4)–(8.6) at the same time). The natural cubic spline interpolation results in a unique solution for all interpolation coefficients. However, a dispersion relation which is interpolated with natural cubic splines does not always fulfill (8.2) for $k, k' \in [k_l, k_{l+1})$. A monotonic cubic spline interpolation is therefore preferred.

The monotonicity condition (8.2) leads to the condition $\varepsilon^\nu(k_l,\phi) < \varepsilon^\nu(k_{l+1},\phi)$ for all l. Let us assume that $\varepsilon^\nu(k_l,\phi) < \varepsilon^\nu(k_{l+1},\phi)$ is fulfilled for all l. Even under this condition the monotonicity of the interpolation is not guaranteed and the continuity of the second derivative has to be sacrificed to solve this problem. Disregarding (8.6) on the other hand provides the degree of freedom to adjust $v_k^\nu(k_l,\phi)$ so that monotonicity is achieved. Such a scheme was proposed by Fritsch and Carlson [5] based on hermite spline interpolation.

Using the relation between a,b and $\varepsilon^\nu(k_l,\phi)$, $v_k^\nu(k_l,\phi)$, the interpolation (8.3) can be rewritten in the hermite spline interpolation form:

$$\begin{aligned}
\varepsilon^\nu(k,\phi) = {} & h_0\left(\frac{k_{l+1}-k}{k_{l+1}-k_l}\right)\varepsilon^\nu(k_l,\phi) + h_0\left(\frac{k-k_l}{k_{l+1}-k_l}\right)\varepsilon^\nu(k_{l+1},\phi) \\
& + h_1\left(\frac{k_{l+1}-k}{k_{l+1}-k_l}\right)(k_l-k_{l+1})\hbar v_k^\nu(k_l,\phi) \\
& + h_1\left(\frac{k-k_l}{k_{l+1}-k_l}\right)(k_{l+1}-k_l)\hbar v_k^\nu(k_{l+1},\phi),
\end{aligned} \tag{8.7}$$

where the hermite functions h are defined as: $h_0(x) = -2x^3 + 3x^2, h_1(x) = x^3 - x^2, x \in [0,1]$. For convenience, $v_k^\nu(k_l,\phi)$ and $v_k^\nu(k_{l+1},\phi)$ are normalized and the normalization results are denoted as α, β:

$$\alpha \equiv \alpha^{\nu}(k_l, \phi) = \frac{k_{l+1} - k_l}{\varepsilon^{\nu}(k_{l+1}, \phi) - \varepsilon^{\nu}(k_l, \phi)} \hbar v_k^{\nu}(k_l, \phi), \tag{8.8}$$

$$\beta \equiv \beta^{\nu}(k_l, \phi) = \frac{k_{l+1} - k_l}{\varepsilon^{\nu}(k_{l+1}, \phi) - \varepsilon^{\nu}(k_l, \phi)} \hbar v_k^{\nu}(k_{l+1}, \phi). \tag{8.9}$$

According to the derivation in [5], the interpolation (8.7) fulfills condition (8.2) when α, β are located within a region $\mathscr{M}$ defined as:

$$\mathscr{M} = \left\{ \frac{(\alpha + \beta - 4)^2}{2^2} + \frac{(\alpha - \beta)^2}{(2\sqrt{3})^2} \leq 1 \right\} \cup \mathscr{C}, \tag{8.10}$$

where $\mathscr{C}$ is given by

$$\mathscr{C} = \left\{ \alpha \geq 0, \beta \geq 0, \frac{\alpha^2}{3^2} + \frac{\beta^2}{3^2} \leq 1 \right\}. \tag{8.11}$$

The region $\mathscr{M}$ for monotonic interpolation and its subset $\mathscr{C}$ are plotted in Fig. 8.2.

By iteration the normalized $v_k^{\nu}(k_l, \phi)$ and $v_k^{\nu}(k_{l+1}, \phi)$ can be adjusted to be within $\mathscr{M}$ [5]. At first, $v_k^{\nu}(k_l, \phi)$ is initialized such that $v_k^{\nu}(k_l, \phi) = \mathring{v}_k^{\nu}(k_l, \phi) > 0$. The initialization of $\mathring{v}_k^{\nu}(k_l, \phi)$ is performed as follows:

$$\mathring{v}_k^{\nu}(k_l, \phi) = \begin{cases} v_{k,\mathrm{ncs}}^{\nu}(k_l, \phi), & v_{k,\mathrm{ncs}}^{\nu}(k_l, \phi) > 0 \\ \underline{v}_k^{\nu}(k_l, \phi), & v_{k,\mathrm{ncs}}^{\nu}(k_l, \phi) < 0 \end{cases}. \tag{8.12}$$

Here, $v_{k,\mathrm{ncs}}^{\nu}(k_l, \phi)$ is the interpolation coefficient resulting from the natural cubic spline based on $\varepsilon^{\nu}(k_l, \phi)$. The reason for this initialization is that if the normalized

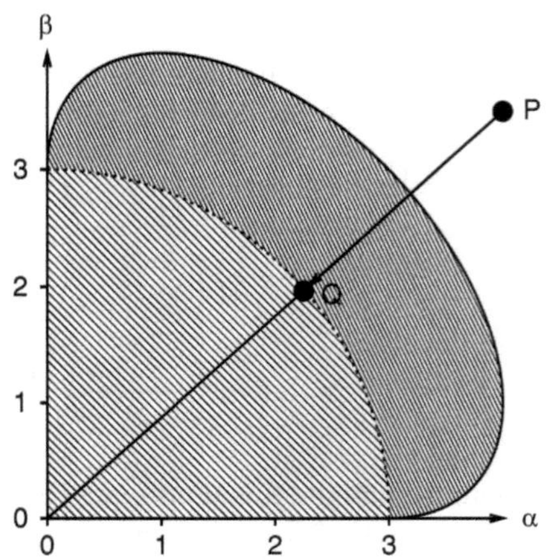

Fig. 8.2 The region $\mathscr{M}$ for monotonic interpolation is bounded by the *continuous lines*. The subset $\mathscr{C}$ is the region bounded by the *circle* (*dashed line*) and the lines $\alpha = 0, \beta = 0$. Any point P outside $\mathscr{M}$ is adjusted to a point Q on the circular boundary of subset $\mathscr{C}$

$v^{\nu}_{k,\mathrm{ncs}}(k_l,\phi)$ and $v^{\nu}_{k,\mathrm{ncs}}(k_{l+1},\phi)$ are within region $\mathscr{M}$, then the interpolation coefficients of the monotonic cubic spline are the same as the ones of the natural cubic spline scheme, and consequently the cubic spline interpolation has maximum order of smoothness C^2. For a negative $v^{\nu}_{k,\mathrm{ncs}}(k_l,\phi)$, a positive number is assigned to $\mathring{v}^{\nu}_k(k_l,\phi)$ by setting it to $\underline{v}^{\nu}_k(k_l,\phi)$ given by the three point finite difference:

$$\underline{v}^{\nu}_k(k_l,\phi) = \frac{1}{2\hbar}\frac{\varepsilon^{\nu}(k_{l+1},\phi)-\varepsilon^{\nu}(k_l,\phi)}{k_{l+1}-k_l} + \frac{1}{2\hbar}\frac{\varepsilon^{\nu}(k_l,\phi)-\varepsilon^{\nu}(k_{l-1},\phi)}{k_l-k_{l-1}}. \quad (8.13)$$

After initialization, $v^{\nu}_k(k_l,\phi)$ and $v^{\nu}_k(k_{l+1},\phi)$ are normalized using (8.8) and (8.9). The resulting α, β are checked if they are within region $\mathscr{M}$. If a point P with the coordinates $(\alpha,\ \beta)$ is outside of $\mathscr{M}$, it is replaced by the point Q at the intersection between the line OP and the circular boundary of subset $\mathscr{C}$ [5] as shown in Fig. 8.2. This implies an adjustment for $v^{\nu}_k(k_l,\phi)$ and $v^{\nu}_k(k_{l+1},\phi)$:

$$v^{\nu}_k(k_l,\phi) \leftarrow \frac{3}{\sqrt{\alpha^2+\beta^2}}v^{\nu}_k(k_l,\phi) \text{ and } v^{\nu}_k(k_{l+1},\phi) \leftarrow \frac{3}{\sqrt{\alpha^2+\beta^2}}v^{\nu}_k(k_{l+1},\phi). \quad (8.14)$$

The adjusted $v^{\nu}_k(k_{l+1},\phi)$ for node k_l enters the calculation of $\alpha^{\nu}(k_{l+1},\phi)$ for the next node k_{l+1}, which is treated similarly. The described procedure is repeated for all indices l.

Please note that the adjusted $v^{\nu}_k(k_{l+1},\phi)$ for node k_{l+1} may again cause the point $(\alpha^{\nu}(k_l,\phi), \beta^{\nu}(k_l,\phi))$ to be out of region $\mathscr{M}$. Therefore, an iteration is required, which repeats the adjustment procedure until there is no need for further adjustment. In practice, it takes a few iteration steps to ensure monotonicity for the interpolated dispersion relation. After interpolation the monotonicity of the subband energy dispersion relation for each polar angle ϕ is ensured. The continuity of the subband energy dispersion relation and its derivative (or group velocity) is also conserved. However, the continuity of the second order derivative of the dispersion relation is not always ensured.

In reality, the dispersion relation $\varepsilon^{\nu}(k,\phi)$ resulting from the SE does not always monotonically increase with increasing k. This is due to the interaction of the bands included in the $\mathbf{k}\cdot\mathbf{p}$ Hamiltonian. It leads to dispersion relations which are bended downwards and electron-like for some k. Consequently, there exists the situation that $\varepsilon^{\nu}(k_{l'},\phi) \geq \varepsilon^{\nu}(k_{l'+1},\phi)$ at certain nodes $k_{l'}$, violating the monotonicity condition for the dispersion relation.

In order to enforce the monotonicity condition $\varepsilon^{\nu}(k_l,\phi) < \varepsilon^{\nu}(k_{l+1},\phi)$ for all l, all samples $\varepsilon^{\nu}(k_{l'},\phi)$ resulting from the $\mathbf{k}\cdot\mathbf{p}$ SE are pre-processed by dropping all "bad" eigen energies, as shown in Fig. 8.3. The eigen energy at the Γ point of the 2D **k**-space ($\varepsilon^{\nu}(k_0,\phi) = \varepsilon^{\nu}(0,\phi)$) is always considered as "good". With increasing l', the dropping procedure compares the eigen energy $\varepsilon^{\nu}(k_{l'},\phi)$ with the previous "good" eigen energy for the same radial angle ϕ. If the considered eigen energy is higher, then it is "good". Otherwise the eigen energy is "bad" and is dropped by ignoring the discrete value $k_{l'}$ and renumbering the index $l''+1 \rightarrow l''$ for all $l'' \geq l'$.

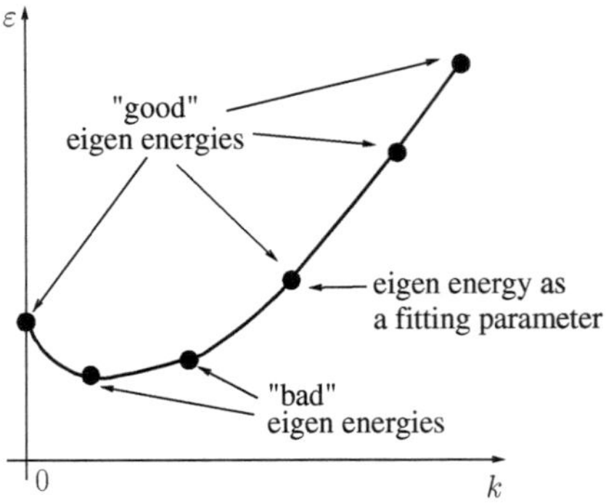

Fig. 8.3 "Good" and "bad" eigen energies of the dispersion relation $\varepsilon = \varepsilon^{\nu}(k, \phi)$. All "bad" eigen energies are removed. The remaining "Good" eigen energies $\varepsilon^{\nu}(k_l, \phi)$ monotonically increase with increasing l. Monotonic cubic spline interpolation of the dispersion relation is finally performed based on the "good" eigen energies only

After the dropping procedure, the energy of the remaining nodes monotonically increases with increasing l. The described monotonic spline interpolation is finally performed for the "good" samples $\varepsilon^{\nu}(k_l, \phi)$ only.

The "bad" eigen energy dropping procedure described above on the one hand helps to obtain a monotonic dispersion relation after spline interpolation, but on the other hand reshapes the dispersion relation. If "bad" eigen energies have been ignored in the range $k_l < k < k_{l+1}$ where $\varepsilon^{\nu}(k_l, \phi)$ and $\varepsilon^{\nu}(k_{l+1}, \phi)$ are "good" eigen energies, then the eigen energy $\varepsilon^{\nu}(k_{l+1}, \phi)$ is considered as a fitting parameter (see Fig. 8.3). In case that there is another range $k_{\hat{l}'} < k < k_{\hat{l}'+1}$ containing "bad" eigen energies where $k_{l+1} < k_{\hat{l}'}$, the fitting parameter is still $\varepsilon^{\nu}(k_{l+1}, \phi)$. In order to improve the quality of the approximated dispersion relation the fitting parameter for each discrete value of ϕ is evaluated such that the mobility related quantity shown below is as accurate as possible in comparison to the value for the same quantity resulting from the exact dispersion relation without any approximation.

$$\int \left[v^{\nu}_{k,\mathrm{mono}}(k, \phi)\right]^2 f_{\mathrm{eq}}(\varepsilon^{\nu}_{\mathrm{mono}}(k, \phi))[1 - f_{\mathrm{eq}}(\varepsilon^{\nu}_{\mathrm{mono}}(k, \phi))]k\mathrm{d}k$$
$$\rightarrow \int \left[v^{\nu}_{k,\mathrm{exact}}(k, \phi)\right]^2 f_{\mathrm{eq}}(\varepsilon^{\nu}_{\mathrm{exact}}(k, \phi))[1 - f_{\mathrm{eq}}(\varepsilon^{\nu}_{\mathrm{exact}}(k, \phi))]k\mathrm{d}k. \quad (8.15)$$

f_{eq} is the Fermi-Dirac distribution function. The Levenberg-Marquardt non-linear least squares fitting algorithm [6, 7] with constraint to good eigen energy is employed to obtain the optimized fitting parameter.

The approximations resulting from the "bad" eigen energy dropping scheme including the non-linear least squares fit are studied in Fig. 8.4 showing dispersion relations of the first 12 valence subbands for $\phi = 0$ (along [100]) (top) and $\phi = 45°$ (along [110]) (bottom) in an unstrained Si double gate hole inversion layer with

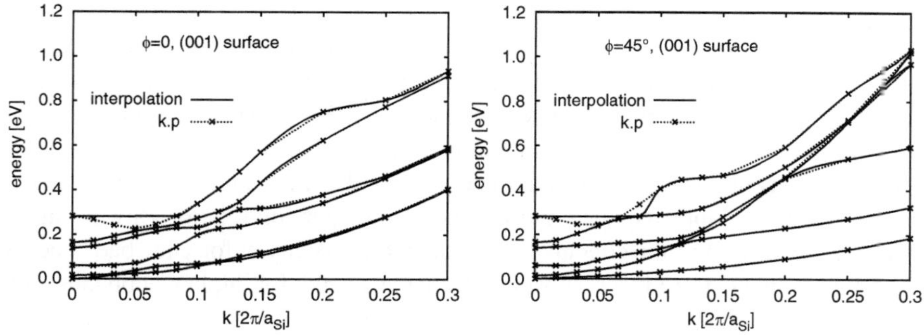

Fig. 8.4 Dispersion relation $\varepsilon = \varepsilon^{\nu}(k, \phi)$ of the first 12 valence subbands for $\phi = 0$ (along [100]) (*left*) and $\phi = 45°$ (along [110]) (*right*) in an unstrained Si double gate hole inversion layer without (*dotted lines* + *symbols*) and with (*solid lines*) monotonic interpolation. Si thickness is 5 nm, (001) surface orientation, $N_{\text{inv}} = 10^{13}\,\text{cm}^{-2}$ and $T = 300\,\text{K}$

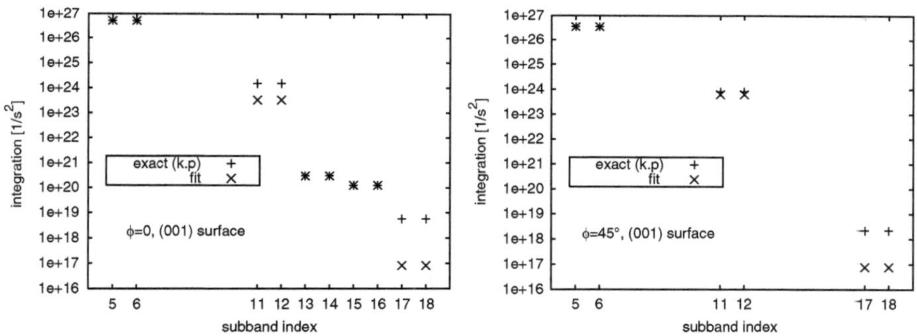

Fig. 8.5 Accuracy of fitting quantity (8.15) for the first 18 subbands with $\phi = 0$ (along [100]) (*left*) and $\phi = 45°$ (along [110]) (*right*). Only subbands with "bad" eigen energies are shown. Simulation conditions are the same as in Fig. 8.4

(001) surface orientation. The eigen energies are shifted such that $\varepsilon(k = 0, \phi) = 0$ for the first subband. For a small Si thickness (5 nm in this case) the non-degeneracy has a minor impact on the subband structure, and the dispersion curves of the subbands $1, 3, 5, \ldots$ nearly match the dispersion curves of the subbands $2, 4, 6, \ldots$. In Fig. 8.4 the eigen energies are evaluated by the $\mathbf{k} \cdot \mathbf{p}$ solver only for the discrete k_l (symbols). A non-monotonicity of the exact dispersion curve with discrete eigen energies exists for the subbands 5, 6 and 11, 12, though it is only clearly visible for 11, 12. However, as shown in Fig. 8.4 the dispersion relation after "bad" eigen energy dropping and interpolation is again monotonic for all subbands. For the Γ point the interpolated eigen energy of each subband reproduces exactly the eigen energy from the $\mathbf{k} \cdot \mathbf{p}$ solver of the same subband index in order to retain the degeneracy of the subband structure at the Γ point. Moreover, the subband structure remains isotropic at the Γ point, i.e. $\varepsilon^{\nu}(k = 0, \phi) = \varepsilon^{\nu}(k = 0)$.

The accuracy of the fitting procedure for the quantity (8.15) is shown in Fig. 8.5. Only subbands with "bad" eigen energies are shown in Fig. 8.5. The exact function is reproduced excellently for the subbands 5, 6, 13, 14, 15, and 16. However, there is still a deviation between the fit and the exact result for the subbands 11, 12 and 17, 18, because for these subbands the minimum eigen energies are much smaller than the eigen energies at the Γ point (visible in Fig. 8.4 for subbands 11, 12).

For the other cases with different combinations between surface orientation and stress/strain configuration bad eigen energies hardly exist and as a consequence the exact and the approximated and interpolated dispersion relations always agree very well. This allows the conclusion that the monotonicity enforcing "bad" energy dropping scheme introduces substantial changes only for higher subbands (11, 12). Therefore, this approximation scheme is much more flexible and much more accurate than the alternative approximation used in [8,9].

To evaluate the subband energy for a wave vector $\mathbf{k} = (k, \phi)$ using the cubic spline interpolation (8.7), the coefficients $\varepsilon^{\nu}(k_l, \phi)$ and $v_k^{\nu}(k_l, \phi)$ being consistent with (8.7) so that the monotonicity of dispersion relation is ensured have to be calculated for the given angle ϕ. For an arbitrary angle ϕ, the coefficients $\varepsilon^{\nu}(k_l, \phi)$ and $v_k^{\nu}(k_l, \phi)$ can be determined by a Fourier interpolation based on the discrete values of these coefficients at the polar angles $\phi = \phi_n$ evaluated by solving the $\mathbf{k} \cdot \mathbf{p}$ SE.

The Fourier interpolation of the coefficient $X^{\nu}(k_l, \phi)$ representing $v_k^{\nu}(k_l, \phi)$, $\varepsilon^{\nu}(k_l, \phi)$ is given by:

$$X^{\nu}(k_l, \phi) = \sum_{m=-M+1}^{M} X_{m,l}^{\nu} Y_m(\phi), \tag{8.16}$$

where the harmonic function $Y_m(\phi)$ is defined such that $\int_0^{2\pi} Y_m(\phi) Y_n(\phi) \mathrm{d}\phi = \delta_{m,n}$ (see Appendices A.2 and A.3).

Because the expansion (8.16) reproduces[1] $X^{\nu}(k_l, \phi)$ at $\phi = \phi_n$, the $X_{m,l}^{\nu}$ are the unknowns of a linear system with the system matrix $(Y_m(\phi_n))_{m=-M+1,\ldots,M}^{n=0,\ldots,N-1}$ and the right hand side vector $(X^{\nu}(k_l, \phi_n))_{n=0,\ldots,N-1}$ (see (8.1) for the definition of N). A unique solution for this linear system requires that the number of equations equals the number of unknowns, i.e. $N = 2M$. Due to the orthogonality of Y_m the solution of the linear system can be shown to be [10]:

$$X_{m,l}^{\nu} = \frac{1}{1 + \delta_{m,M} \cos(2M\varphi)} \sum_{n=0}^{N-1} X^{\nu}(k_l, \phi_n) Y_m(\phi_n) \frac{2\pi}{N}, \tag{8.17}$$

where φ is defined as in (8.1). Inserting the Fourier expansion of $\varepsilon^{\nu}(k_l, \phi)$ and $v_k^{\nu}(k_l, \phi)$ with (8.16) into (8.3) yields an interpolation of the subband energy for

[1]For dropped "bad" eigen energies $\varepsilon^{\nu}(k_l, \phi)$, the "bad" values are updated with the new value resulting from the interpolation (8.7). The group velocity $v_k^{\nu}(k_l, \phi)$ is also assigned by the derivative of the dispersion relation at node k_l using (8.5).

an arbitrary wave vector (k, ϕ) in the 2D **k**-space:[2]

$$\varepsilon^{\nu}(k,\phi) = \sum_{m=-M+1}^{M} \varepsilon_m^{\nu}(k) Y_m(\phi), \tag{8.18}$$

where

$$\begin{aligned}\varepsilon_m^{\nu}(k) &= h_0\left(\frac{k_{l+1}-k}{k_{l+1}-k_l}\right)\varepsilon_{m,l}^{\nu} + h_0\left(\frac{k-k_l}{k_{l+1}-k_l}\right)\varepsilon_{m,l+1}^{\nu} \\ &\quad + h_1\left(\frac{k_{l+1}-k}{k_{l+1}-k_l}\right)(k_l-k_{l+1})\hbar\{v_k\}_{m,l}^{\nu} \\ &\quad + h_1\left(\frac{k-k_l}{k_{l+1}-k_l}\right)(k_{l+1}-k_l)\hbar\{v_k\}_{m,l+1}^{\nu}. \end{aligned} \tag{8.19}$$

Here, $\{v_k\}_{m,l}^{\nu}$ and $\varepsilon_{m,l}^{\nu}$ are determined by (8.17) and h_0 and h_1 are hermit functions. As soon as $\varepsilon_m^{\nu}(k)$ has been determined, the anisotropy of any type of subband structure can described by (8.18) up to a given harmonic order.

In the literature, the warped structure of the hole subbands has been interpolated as a function of the polar angle in the 2D **k**-space [8]. However, the analytical formula presented in [8] is restricted to the subband structure of unstrained Si and includes only positive order harmonics up to 8th order. All these limitations existing in [8] can be avoided with the Fourier interpolation scheme presented in this work.

Figure 8.6 shows the equi energy lines of the first subband in unstrained (left) or biaxially strained 1.5 GPa (center) or uniaxially stressed −1.5 GPa [110] (right) Si double gate hole inversion layers with (001) surface orientation. The irreducible ϕ range for which the SE is solved is $[0, \pi/4]$ for unstrained and biaxially strained Si and is enlarged to $[-\pi/4, \pi/4]$ for uniaxially stressed Si. The symmetry of the subband structure is eight-fold for unstrained and biaxially strained Si or four-fold for the uniaxial stress case, because the uniaxial stress reduces the symmetry. For the unstrained and biaxially strained cases, the eigen states are determined for 6 or 11 ϕ-slices ($N_{\text{irr}} = 5$ or $N_{\text{irr}} = 10$) within the irreducible ϕ range. This implies that the subband energy interpolation (8.18) includes Fourier harmonics up to order $M = 20$ or $M = 40$, respectively, due to $M = N/2 = 4N_{\text{irr}}$. It can be clearly observed that the differences between the equi energy lines for $M = 20$ and $M = 40$ are very small. For uniaxially stressed Si, the irreducible ϕ range increases by a factor of two compared to the unstrained case. Consequently, the relation between M and N_{irr} is changed: $M = N/2 = 2N_{\text{irr}}$. In order to include Fourier harmonics up to order $M = 20$ (or $M = 40$), the irreducible ϕ range must be partitioned into 11 (or 21) ϕ-slices. Again, the difference of the equi energy lines for $M = 20$ and $M = 40$ is very small. It turns out that the interpolation based on eigen energies for

[2]Please note that using the Fourier interpolation (8.18) the monotonicity of $\varepsilon^{\nu}(k,\phi)$ for a fixed ϕ between two discrete polar angles may be violated. However, for solving the BTE, the monotonicity is only needed for discrete polar angles.

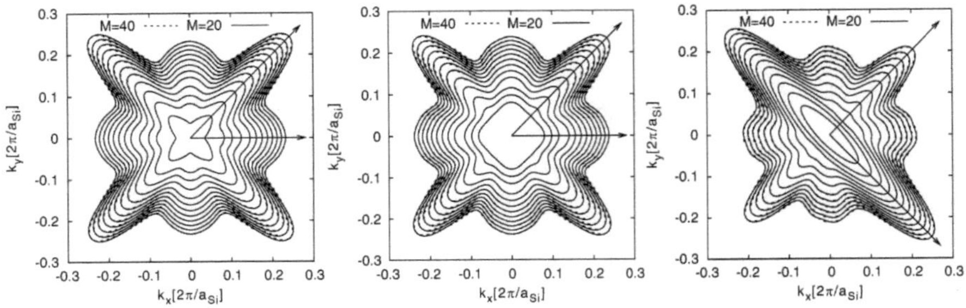

Fig. 8.6 Equienergy lines of the first subband in unstrained (*left*) or biaxially strained 1.5 GPa (*center*) or uniaxially stressed −1.5 GPa ([110] direction) (*right*) Si double gate hole inversion layers interpolated with up to $M = 20$ or $M = 40$ Fourier harmonics. The bottom of the first subband is shifted to 0 eV. The equi energy lines are for multiples of 20 meV. The two *arrows* indicate the lower and upper bound of the irreducible ϕ range for which the SE is solved. Si thickness is 5 nm, (001) surface orientation, $N_{\text{inv}} = 10^{13}\,\text{cm}^{-2}$ and $T = 300\,\text{K}$

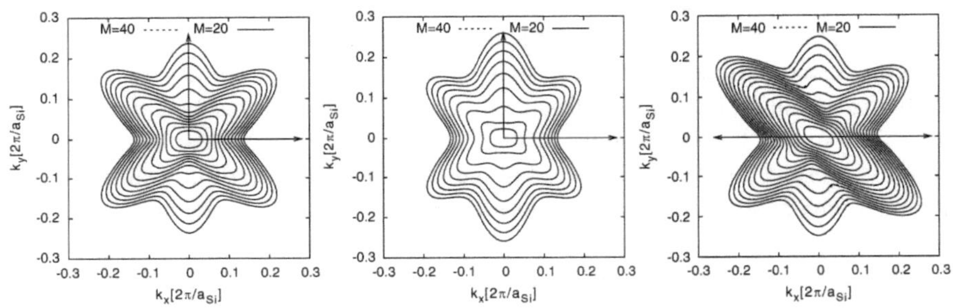

Fig. 8.7 Same as Fig. 8.6 but for (110) surface orientation. The uniaxial stress direction is $[\bar{1}1\overline{\sqrt{2}}]$ resulting in an angle of 45° with k_x-axis

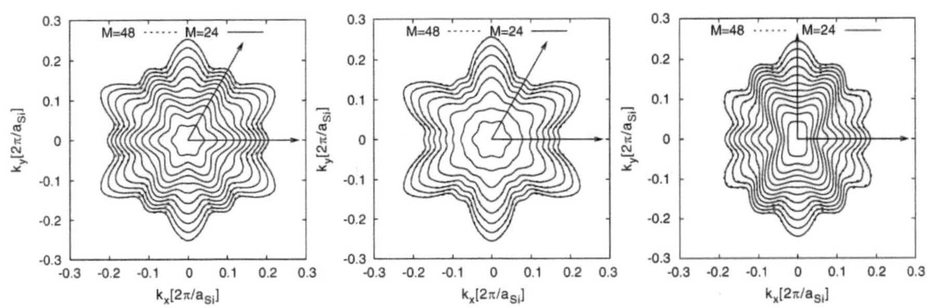

Fig. 8.8 Same as Fig. 8.6 but for (111) surface orientation. The uniaxial stress direction is $[11\bar{2}] \parallel k_x$-axis

6 ϕ-slices for unstrained and biaxially strained Si or 11 ϕ-slices for uniaxially stressed Si accurately reproduces the subband structure.

Similar cases of the equi energy lines for (110) and (111) surface orientations are shown in Figs. 8.7 and 8.8, respectively. It is shown that the subband structure symmetry changes due to the variation of the surface orientation or stress/strain conditions, leading to the change in the irreducible ϕ range. The uniaxial stress reduces the symmetry of the subband structure for (110) or (111) surfaces as well. For the (110) surface, the difference of the equi energy lines for $M = 20$ and $M = 40$ is very small. For the (111) surface, the difference of the equi energy lines for $M = 24$ and $M = 48$ is also very small.

8.3 Calculation of Group Velocity

The group velocity is a fundamental subband quantity which directly influences the transport in the device. Due to the warped subband structure the group velocity is often calculated numerically (e.g. [3]). This can cause systematic numerical errors. In this work a closed form equation for the group velocity is derived consistent with the interpolation of the subband energy given by (8.18). Using the closed form of the group velocity, additional numerical errors are avoided.

Inserting (8.18) into the definition of the group velocity gives:

$$\mathbf{v}^{\nu}(\mathbf{k}) \equiv \frac{1}{\hbar}\nabla_{\mathbf{k}}\varepsilon^{\nu}(\mathbf{k}) = \frac{1}{\hbar}\sum_{m=-M+1}^{M}\nabla_{\mathbf{k}}\left(\varepsilon_m^{\nu}(k)Y_m(\phi)\right) = v_k^{\nu}(\mathbf{k})\mathbf{e}_k + v_{\phi}^{\nu}(\mathbf{k})\mathbf{e}_{\phi}, \tag{8.20}$$

where

$$v_k^{\nu}(\mathbf{k}) = \frac{1}{\hbar}\sum_{m=-M+1}^{M} Y_m(\phi)\left[\, 3a_{m,l}^{\nu}(k-k_l)^2 + 2b_{m,l}^{\nu}(k-k_l) + \hbar\{v_k\}_{m,l}^{\nu}\,\right] \tag{8.21}$$

and

$$v_{\phi}^{\nu}(\mathbf{k}) = \frac{1}{\hbar k}\sum_{m=-M+1}^{M} -mY_{-m}(\phi)\Big[\, a_{m,l}^{\nu}(k-k_l)^3 + b_{m,l}^{\nu}(k-k_l)^2 + \hbar\{v_k\}_{m,l}^{\nu}(k-k_l) + \varepsilon_{m,l}^{\nu}\,\Big]. \tag{8.22}$$

Within the primed coordinate system (x, y, z) in which the SE is solved, the components x and y of the group velocity are determined based on the rotation

$$\begin{pmatrix} v_x^{\nu}(\mathbf{k}) \\ v_y^{\nu}(\mathbf{k}) \end{pmatrix} = \begin{pmatrix} \cos\phi & -\sin\phi \\ \sin\phi & \cos\phi \end{pmatrix}\begin{pmatrix} v_k^{\nu}(\mathbf{k}) \\ v_{\phi}^{\nu}(\mathbf{k}) \end{pmatrix}. \tag{8.23}$$

If the channel of a device is oriented in the direction x_{D} creating an angle φ_{D} with the x-axis of the primed coordinate system, then the x_{D}- and y_{D}-components of the group velocity can also be determined by a rotation of the vector $(v_x^\nu(\mathbf{k}), v_y^\nu(\mathbf{k}))^\mathrm{T}$:

$$\begin{pmatrix} v_{x_\mathrm{D}}^\nu(\mathbf{k}) \\ v_{y_\mathrm{D}}^\nu(\mathbf{k}) \end{pmatrix} = \begin{pmatrix} \cos\varphi_\mathrm{D} & \sin\varphi_\mathrm{D} \\ -\sin\varphi_\mathrm{D} & \cos\varphi_\mathrm{D} \end{pmatrix} \begin{pmatrix} v_x^\nu(\mathbf{k}) \\ v_y^\nu(\mathbf{k}) \end{pmatrix}. \tag{8.24}$$

8.4 Interpolation of the Probability Density Function

For the calculation of the hole density in Chap. 9, the probability density function is required which is defined as follows:

$$\mathscr{P}^\nu(z, \mathbf{k}) = \boldsymbol{\zeta}^{\dagger\nu}(z, \mathbf{k})\boldsymbol{\zeta}^\nu(z, \mathbf{k}). \tag{8.25}$$

$\boldsymbol{\zeta}$ is the envelope function which is the eigen vector of the $\mathbf{k}\cdot\mathbf{p}$ eigen problem (7.41). At the discrete grid points (k, ϕ) in the 2D $\mathbf{k}$-space for which the $\mathbf{k}\cdot\mathbf{p}$ SE is solved, the probability density function $\mathscr{P}^\nu(z, \mathbf{k})$ can be evaluated directly from the resultant envelope function $\boldsymbol{\zeta}$ using (8.25). For other wave vectors $\mathscr{P}^\nu(z, \mathbf{k})$ is determined by interpolation. The interpolation of $\mathscr{P}^\nu(z, \mathbf{k})$ in the 2D $\mathbf{k}$-space is done independently for each position z. The interpolation of $\mathscr{P}^\nu(z, \mathbf{k})$ for each position z contains two steps. First, $\mathscr{P}^\nu(z, \mathbf{k})$ is interpolated with the Fourier interpolation for a constant k. Second, the resultant Fourier expansion coefficients are interpolated with the cubic spline interpolation scheme. For each position z, the Fourier interpolation of $\mathscr{P}$ is given by:

$$\mathscr{P}^\nu(z, k, \phi) = \sum_{m=-M+1}^{M} \mathscr{P}_m^\nu(z, k) Y_m(\phi), \tag{8.26}$$

which is similar to the interpolation of the subband energy given by (8.18). The Fourier expansion coefficients $\mathscr{P}_m^\nu(z, k)$ can be therefore evaluated

$$\mathscr{P}_m^\nu(z, k) = \frac{1}{1 + \delta_{m,M}\cos(2M\varphi)} \sum_{n=0}^{N-1} \mathscr{P}^\nu(z, k, \phi_n) Y_m(\phi_n) \frac{2\pi}{N} \tag{8.27}$$

similar to (8.17). For each order m, the Fourier expansion coefficient $\mathscr{P}_m^\nu(z, k)$ is interpolated with a cubic spline interpolation scheme

$$\begin{aligned} \mathscr{P}_m^\nu(z, k) = {} & h_0\left(\frac{k_{l+1} - k}{k_{l+1} - k_l}\right) \mathscr{P}_m^\nu(z, k_l) + h_0\left(\frac{k - k_l}{k_{l+1} - k_l}\right) \mathscr{P}_m^\nu(z, k_{l+1}) \\ & + h_1\left(\frac{k_{l+1} - k}{k_{l+1} - k_l}\right)(k_l - k_{l+1})\mathscr{Q}_m^\nu(z, k_l) \\ & + h_1\left(\frac{k - k_l}{k_{l+1} - k_l}\right)(k_{l+1} - k_l)\mathscr{Q}_m^\nu(z, k_{l+1}) \end{aligned} \tag{8.28}$$

similar to (8.7) which ensures the continuity conditions for $\mathscr{P}_m^\nu(z,k)$ and $\partial\mathscr{P}_m^\nu(z,k)/\partial k$ at each node k_l. Here, $\mathscr{Q}_m^\nu(z,k_l) = \partial\mathscr{P}_m^\nu(z,k)/\partial k|_{k=k_l}$ is the cubic spline interpolation coefficient. Since the probability density function is not required to be monotonic with increasing k for each radial angle ϕ, the interpolation (8.28) can reach the maximum smoothness (C^2) without any problems. Therefore, the continuity condition for $\partial^2\mathscr{P}_m^\nu(z,k)/\partial k^2$ at each node k_l similar to (8.6) can be achieved, which yields a unique solution for $\mathscr{Q}_m^\nu(z,k_l)$.

References

1. Oberhuber, R., Zandler, G., Vogl, P.: Subband structure and mobility of two-dimensional holes in strained Si/SiGe MOSFETs. Phys. Rev. B **58**, 9941–9948 (1998)
2. Wang, E.X., Matagne, P., Shifren, L., Obradovic, B., Kotlyar, R., Cea, S., Stettler, M., Giles, M.D.: Physics of hole transport in strained silicon MOSFET inversion layers. IEEE Trans. Electron Dev. **53**(8), 1840–1851 (2006)
3. Fischetti, M.V., Ren, Z., Solomon, P.M., Yang, M., Rim, K.: Six-band $k \cdot p$ calculation of the hole mobility in silicon inversion layers: Dependence on surface orientation, strain, and silicon thickness. J. Appl. Phys. **94**, 1079–1095 (2003)
4. Pham, A.T., Jungemann, C., Meinerzhagen, B.: Physics-based modeling of hole inversion layer mobility in strained SiGe on insulator. IEEE Trans. Electron Dev. **54**(9), 2174–2182 (2007)
5. Fritsch, F.N., Carlson, R.E.: Monotone piecewise cubic interpolation. SIAM J. Num. Anal. **17**(2), 238–246 (1980)
6. Levenberg, K.: A method for the solution of certain problems in least squares. Quart. Appl. Math. **2**, 164–168 (1944)
7. Marquardt, D.: An algorithm for least-squares estimation of nonlinear parameters. SIAM J. Appl. Math. **11**, 431–441 (1963)
8. De Michielis, M., Esseni, D., Palestri, P., Selmi, L.: A new analytical model for the energy dispersion in two-dimensional hole inversion layers. Solid State Electron. **51**, 598–603 (2007)
9. De Michielis, M., Esseni, D., Palestri, P., Selmi, L.: A new multi subband monte carlo simulator for nano p-MOSFETs. In: Proceedings of ULIS, pp. 67–70 (2008)
10. Pham, A.T., Jungemann, C., Meinerzhagen, B.: A fast k*p solver for hole inversion layers with an efficient 2D k-space discretization. J. Comp. Electr. **7**(3), 99–102 (2008)

Chapter 9
Deterministic Solver for the Multisubband Stationary BTE

In this chapter a deterministic method used to solve the multisubband BTE is presented. The method is based on the Fourier expansion of the distribution function in the 2D **k**-space, which is analogous to the spherical harmonic expansion method for the bulk case [1].

In the semiclassical framework stationary transport of carriers within a device can be described by a distribution function $f^{\nu}(\mathbf{r},\mathbf{k})$, which is a function of position in the phase space $(\mathbf{r},\mathbf{k})$ and depends also on the subband index ν. Within this section the device coordinate system $(x_{\rm D}, y_{\rm D}, z_{\rm D})$ (shown in Fig. 6.1) is denoted as (x, y, z), for convenience. Having a 2D **k**-space the wave vector on which the distribution function depends is given by $\mathbf{k} = (k_x, k_y)$. In real space the PMOSFET is assumed to be homogenous in y direction. Therefore, the distribution function depends only on x in the real space. The distribution function $f^{\nu}(x,\mathbf{k})$ fulfills the stationary BTE [2, 3] for subband ν:

$$\frac{1}{\hbar}\mathbf{F}^{\nu\mathrm{T}}\nabla_{\mathbf{k}} f^{\nu} + \mathbf{v}^{\nu\mathrm{T}}\nabla_{\mathbf{r}} f^{\nu} = \hat{S}^{\nu}\{f\}. \tag{9.1}$$

Here, $\hat{S}^{\nu}\{f\}$ is the scattering integral. The total force $\mathbf{F}$ due to a gradient in the total energy and a magnetic field is given by:

$$\mathbf{F}^{\nu}(x,\mathbf{k}) = \mathbf{F}^{\nu}_{\mathrm{E}}(x,\mathbf{k}) + \mathbf{F}^{\nu}_{\mathrm{B}}(x,\mathbf{k}). \tag{9.2}$$

In terms of the total subband energy $\varepsilon^{\nu}(x,\mathbf{k})$, which is the eigen energy of the $\mathbf{k}\cdot\mathbf{p}$ SE, $\mathbf{F}^{\nu}_{\mathrm{E}}(x,\mathbf{k})$ is given by:

$$\mathbf{F}^{\nu}_{\mathrm{E}}(x,\mathbf{k}) = -\nabla_{\mathbf{r}}\left[\varepsilon^{\nu}(x,\mathbf{k})\right] = -\frac{\partial}{\partial x}\left[\varepsilon^{\nu}(x,\mathbf{k})\right]\mathbf{e}_x. \tag{9.3}$$

The force caused by the magnetic field $\mathbf{F}^{\nu}_{\mathrm{B}}(x,\mathbf{k})$ is given by:

$$\mathbf{F}^{\nu}_{\mathrm{B}}(x,\mathbf{k}) = q\mathbf{v}^{\nu}(x,\mathbf{k}) \times \mathbf{B}, \tag{9.4}$$

S.-M. Hong et al., *Deterministic Solvers for the Boltzmann Transport Equation*, Computational Microelectronics, DOI 10.1007/978-3-7091-0778-2_9,

where $\mathbf{B}$ is assumed to be constant and perpendicular to the 2DHG surface, i.e. $\mathbf{B} = B\mathbf{e}_z$.

The total subband energy $\varepsilon^\nu(x, \mathbf{k})$ is decomposed into the potential subband energy $\varepsilon^\nu_{\text{pot}}(x)$ and the kinetic subband energy $\tilde{\varepsilon}^\nu(x, \mathbf{k})$ [4]:

$$\varepsilon^\nu(x, \mathbf{k}) = \varepsilon^\nu_{\text{pot}}(x) + \tilde{\varepsilon}^\nu(x, \mathbf{k}) \tag{9.5}$$

with $\tilde{\varepsilon}^\nu(x, \mathbf{k} = \mathbf{0}) = 0$.

The force given by (9.3) is accordingly decomposed into two parts

$$\mathbf{F}^\nu_{\text{E}}(x, \mathbf{k}) = \left[F^\nu_{\text{E,pot}}(x) + \tilde{F}^\nu_{\text{E}}(x, \mathbf{k})\right]\mathbf{e}_x \tag{9.6}$$

with

$$F^\nu_{\text{E,pot}}(x) = -\frac{\partial}{\partial x}\varepsilon^\nu_{\text{pot}}(x) \quad \text{and} \quad \tilde{F}^\nu_{\text{E}}(x, \mathbf{k}) = -\frac{\partial}{\partial x}\tilde{\varepsilon}^\nu(x, \mathbf{k}). \tag{9.7}$$

The force due to the magnetic field can be decomposed into the two components $\mathbf{e}_k$ and $\mathbf{e}_\phi$:

$$\mathbf{F}^\nu_{\text{B}}(x, \mathbf{k}) = -qBv^\nu_k(x, \mathbf{k})\mathbf{e}_\phi + qBv^\nu_\phi(x, \mathbf{k})\mathbf{e}_k. \tag{9.8}$$

The components $v^\nu_k(x, \mathbf{k})$ and $v^\nu_\phi(x, \mathbf{k})$ of the group velocity are calculated in the manner described in Sect. 8.3. $v^\nu_k(x, \mathbf{k})$ and $v^\nu_\phi(x, \mathbf{k})$ are determined by (8.21) and (8.22) directly from the interpolation of the kinetic subband energy for position x. The group velocity components $v^\nu_x(x, \mathbf{k})$ and $v^\nu_y(x, \mathbf{k})$ are then determined using (8.23) and (8.24).

The monotonicity of the dispersion relation (8.2) at position x together with the total subband energy decomposition (9.5) leads directly to the following important properties:

$$\tilde{\varepsilon}^\nu(x, k, \phi) > \tilde{\varepsilon}^\nu(x, k', \phi), \text{ if } k > k' \tag{9.9}$$

$$\text{and therefore, } \tilde{\varepsilon}^\nu(x, k, \phi) \geq \tilde{\varepsilon}^\nu(x, 0, \phi) = 0, \quad \forall\nu, \forall x, \forall k. \tag{9.10}$$

For each discrete angle ϕ on which the monotonic interpolation of the dispersion relation has been done, the transformation $\tilde{\varepsilon} = \tilde{\varepsilon}^\nu(x, k, \phi) \leftrightarrow k = k^\nu(x, \tilde{\varepsilon}, \phi)$ is always unique in both directions due to the monotonic increase of $\tilde{\varepsilon}^\nu(x, k, \phi)$ with increasing k resulting from (9.9). Moreover, (9.10) implies that for a given positive kinetic energy $\tilde{\varepsilon}$ the distribution function of any subband at any position x can always be defined as a function of energy and angle ϕ in a unique way. Firstly, this makes it possible to discretize the distribution function on an $\tilde{\varepsilon}$ grid, and secondly to use the same kinetic energy grid $\tilde{\varepsilon}$ for all subbands and positions x when discretizing the BTE in the next steps.

The BTE is solved for a modified distribution function g [1, 5], which is defined as

$$g^{\nu}(x,\tilde{\varepsilon},\phi) = f^{\nu}(x,\mathbf{k}^{\nu}(x,\tilde{\varepsilon},\phi))Z^{\nu}(x,\tilde{\varepsilon},\phi) \tag{9.11}$$

with the generalized density of states for a single spin direction Z [1,5]:

$$Z^{\nu}(x,\tilde{\varepsilon},\phi) = \frac{1}{(2\pi)^2}k^{\nu}(x,\tilde{\varepsilon},\phi)\frac{\partial k^{\nu}(x,\tilde{\varepsilon},\phi)}{\partial\tilde{\varepsilon}}. \tag{9.12}$$

At each position x and kinetic energy $\tilde{\varepsilon}$, the modified distribution function g is expanded into a Fourier series

$$g^{\nu}(x,\tilde{\varepsilon},\phi) = \sum_{m=-M'}^{M'} g^{\nu}_m(x,\tilde{\varepsilon})Y_m(\phi), \tag{9.13}$$

where the relation between the Fourier expansion coefficient g_m and the original distribution function f is given by:

$$\begin{aligned} g^{\nu}_m(x,\tilde{\varepsilon}) &= \frac{1}{(2\pi)^2}\int f^{\nu}(x,\mathbf{k})\delta(\tilde{\varepsilon}-\tilde{\varepsilon}^{\nu}(x,\mathbf{k}))Y_m(\phi)\mathrm{d}^2k \\ &= \int_0^{2\pi} g^{\nu}(x,\tilde{\varepsilon},\phi)Y_m(\phi)\mathrm{d}\phi. \end{aligned} \tag{9.14}$$

The final equality is obtained by using (A.21) (see Appendix A.4). M' indicates the maximum order of harmonics up to which the modified distribution function g is expanded. Actually, M' is not necessarily the same as M, which is the maximum harmonic order of the Fourier expansion of the subband energy given by (8.18). Theoretically, M' must be infinitely large in order to obtain an exact solution of the BTE. However, an M' in the range of 9–13 is often used (see Sect. 12.2.1).

The macroscopic quantity $\overline{X}(x)$ corresponding to a microscopic quantity $X^{\nu}(x,\mathbf{k})$ (like kinetic energy, group velocity etc.) is defined via the distribution function f as follows

$$\overline{X}(x) = \sum_{\nu}\frac{1}{(2\pi)^2}\int X^{\nu}(x,\mathbf{k})f^{\nu}(x,\mathbf{k})\mathrm{d}^2k \tag{9.15}$$

and can be easily calculated as well in terms of the Fourier expansion coefficients g_m

$$\begin{aligned} \overline{X}(x) &= \sum_{\nu}\int\frac{1}{(2\pi)^2}X^{\nu}(x,\mathbf{k})f^{\nu}(x,\mathbf{k})\mathrm{d}^2k \\ &= \sum_{\nu}\int\int_0^{2\pi} X^{\nu}(x,\tilde{\varepsilon}^{\nu}(x,\mathbf{k}),\phi)g^{\nu}(x,\tilde{\varepsilon}^{\nu}(x,\mathbf{k}),\phi)\mathrm{d}\phi\mathrm{d}\tilde{\varepsilon}^{\nu} \end{aligned}$$

$$
\begin{aligned}
&= \sum_{\nu} \int \int_0^{2\pi} X^{\nu}(x, \tilde{\varepsilon}, \phi) g^{\nu}(x, \tilde{\varepsilon}, \phi) \mathrm{d}\phi \mathrm{d}\tilde{\varepsilon} \\
&= \sum_{\nu} \int \sum_{m=-M'}^{M'} X_m^{\nu}(x, \tilde{\varepsilon}) g_m^{\nu}(x, \tilde{\varepsilon}) \mathrm{d}\tilde{\varepsilon}.
\end{aligned} \tag{9.16}
$$

The final result (9.16) is obtained by expanding X^{ν} into Fourier harmonics on a constant kinetic equi energy line

$$
X^{\nu}(x, \mathbf{k}^{\nu}(x, \tilde{\varepsilon}, \phi)) = \sum_{m=-M''}^{M''} X_m^{\nu}(x, \tilde{\varepsilon}) Y_m(\phi), \tag{9.17}
$$

where X_m^{ν} is the Fourier expansion coefficient of X^{ν} given by

$$
X_m^{\nu}(x, \tilde{\varepsilon}) = \int_0^{2\pi} X^{\nu}(x, \mathbf{k}^{\nu}(x, \tilde{\varepsilon}, \phi)) Y_m(\phi) \mathrm{d}\phi. \tag{9.18}
$$

The microscopic quantity X mentioned above can be a scalar, a vector or a matrix. The maximum harmonic order M'' in the expansion (9.17) does not have to be equal to M but can be larger. Nevertheless, the average (9.16) involves only X_m^{ν} for $-M' \leq m \leq M'$ because the Fourier expansion of g includes only harmonics in that range.

The calculation of the inversion charge along the x direction is straightforward with (9.15) and (9.16) for $X = 1$

$$
N_{\mathrm{inv}}(x) \equiv \sum_{\nu} \frac{1}{(2\pi)^2} \int f^{\nu}(x, \mathbf{k}) \mathrm{d}^2 k = \frac{1}{Y_0} \sum_{\nu} \int g_0^{\nu}(x, \tilde{\varepsilon}) \mathrm{d}\tilde{\varepsilon}. \tag{9.19}
$$

It turns out that only g_m with $m = 0$ contributes to the inversion charge due to the definition of g.

The current density is defined with $\mathbf{X}^{\nu}(x, \mathbf{k}) = \mathbf{v}^{\nu}(x, \mathbf{k})$ by applying (9.15) to each component of the vector-valued microscopic quantity. The current density along the transport direction x reads

$$
j_x(x) = \sum_{\nu} \int \sum_{m=-M'}^{M'} \{v_x^{\nu}\}_m(x, \tilde{\varepsilon}) g_m^{\nu}(x, \tilde{\varepsilon}) \mathrm{d}\tilde{\varepsilon} \tag{9.20}
$$

due to (9.16). It can be recognized that only g_m with odd m contributes to $j_x(x)$ because $\{v_x\}_m = 0$ for all even m due to the symmetry of the subband structure.

9.1 Projection of the BTE onto Fourier Harmonics

The calculation of any macroscopic quantity (9.16) requires the Fourier expansion coefficient g_m of the modified distribution function. In order to determine g_m the BTE is projected onto the harmonic Y_m:

$$\frac{1}{(2\pi)^2}\int\{\mathrm{BTE}\}\delta(\tilde{\varepsilon}-\tilde{\varepsilon}^{\nu}(x,\mathbf{k}))Y_m(\phi)\mathrm{d}^2k. \tag{9.21}$$

As soon as the solution of the projected BTE for the unknowns g_m is available, a continuous solution of the distribution g over energy and angle can be obtained by (9.13).

The projection (9.21) results in a system of partial differential equations for the unknowns g_m:

$$\sum_{m'=-M'}^{M'}\left\{\tilde{\nabla}^{\mathrm{T}}\left[\tilde{\mathbf{j}}^{\nu}_{m,m'}(x,\tilde{\varepsilon})\right]-\left[b^{\mathrm{E}\nu}_{m,m'}(x,\tilde{\varepsilon})+b^{\mathrm{B}\nu}_{m,m'}(x,\tilde{\varepsilon})\right]g^{\nu}_{m'}(x,\tilde{\varepsilon})\right\}=\hat{S}^{\nu}_{m}\{g\}(x,\tilde{\varepsilon}) \tag{9.22}$$

where the operator $\tilde{\nabla}^{\mathrm{T}}$ is defined as $\tilde{\nabla}^{\mathrm{T}}=\left(\frac{\partial}{\partial x},\frac{\partial}{\partial\tilde{\varepsilon}}\right)$ and the vector $\tilde{\mathbf{j}}$ is given by

$$\tilde{\mathbf{j}}^{\nu}_{m,m'}(x,\tilde{\varepsilon})=\begin{pmatrix}1\\ F^{\nu}_{\mathrm{E,pot}}(x)\end{pmatrix}d^{\nu}_{m,m'}(x,\tilde{\varepsilon})g^{\nu}_{m'}(x,\tilde{\varepsilon}). \tag{9.23}$$

The derivation of the LHS of (9.22) can be found in Appendix A.5. For this derivation the integrals with a delta function mentioned in Appendix A.4 should be noted. The magnetic field, which enters the drift term of the BTE, appears only in the b^B term as long as the magnetic field is perpendicular to the transport plane.

For the calculation of the coefficients $d^{\nu}_{m,m'}$, $b^{\mathrm{E}\nu}_{m,m'}$ and $b^{\mathrm{B}\nu}_{m,m'}$ given by (A.33), (A.34), and (A.35), respectively, the microscopic quantities v_x, $F_{\mathrm{E}}/(\hbar k)$ and $v_k/(\hbar k)$ are expanded into Fourier harmonics using (9.17) with $X\leftarrow v_x$, $X\leftarrow F_{\mathrm{E}}/(\hbar k)$, and $X\leftarrow v_k/(\hbar k)$, respectively. The Fourier expansion coefficients $\{v^{\nu}_x\}_{m''}$, $\left\{\frac{F^{\nu}_{\mathrm{E}}}{\hbar k^{\nu}}\right\}_{m''}$, $\left\{\frac{qv^{\nu}_k}{\hbar k^{\nu}}\right\}_{m''}$ are calculated numerically using (9.18) with $m\leftarrow m''$. The symmetry of the subband structure results in $\{v^{\nu}_x\}_{m''}\neq 0$ only for odd m'' and $\left\{\frac{F^{\nu}_{\mathrm{E}}}{\hbar k^{\nu}}\right\}_{m''}\neq 0$, $\left\{\frac{qv^{\nu}_k}{\hbar k^{\nu}}\right\}_{m''}\neq 0$ only for even m''. In terms of these Fourier expansion coefficients the coefficients $d_{m,m'}$, $b^{\mathrm{E}}_{m,m'}$, and $b^{\mathrm{B}}_{m,m'}$ can be expressed as:

$$d^{\nu}_{m,m'}(x,\tilde{\varepsilon})=\sum_{m''=-M''}^{M''}\{v^{\nu}_x\}_{m''}(x,\tilde{\varepsilon})\Upsilon_{m,m',m''}, \tag{9.24}$$

$$b^{\mathrm{E}\nu}_{m,m'}(x,\tilde{\varepsilon}) = m \sum_{m''=-M''}^{M''} \left\{ \frac{F^{\nu}_{\mathrm{E}}}{\hbar k^{\nu}} \right\}_{m''} (x,\tilde{\varepsilon}) \times \left[\cos\varphi_{\mathrm{D}} \{\Upsilon S\}_{m,m',m''} - \sin\varphi_{\mathrm{D}} \{\Upsilon C\}_{m,m',m''} \right], \tag{9.25}$$

$$b^{\mathrm{B}\nu}_{m,m'}(x,\tilde{\varepsilon}) = mB \sum_{m''=-M''}^{M''} \left\{ \frac{q v^{\nu}_{k}}{\hbar k^{\nu}} \right\}_{m''} (x,\tilde{\varepsilon}) \Upsilon_{-m,m',m''} \tag{9.26}$$

with

$$\Upsilon_{m,m',m''} = \int_0^{2\pi} Y_m(\phi) Y_{m'}(\phi) Y_{m''}(\phi) \mathrm{d}\phi, \tag{9.27}$$

$$\{\Upsilon C\}_{m,m',m''} = \int_0^{2\pi} Y_{-m}(\phi) Y_{m'}(\phi) Y_{m''}(\phi) \cos\phi \mathrm{d}\phi, \tag{9.28}$$

$$\{\Upsilon S\}_{m,m',m''} = \int_0^{2\pi} Y_{-m}(\phi) Y_{m'}(\phi) Y_{m''}(\phi) \sin\phi \mathrm{d}\phi. \tag{9.29}$$

All three expressions can be evaluated analytically as described in Appendix A.3. The following important properties of the symbols Υ, $\{\Upsilon C\}$, $\{\Upsilon S\}$ should be noted:

$$\Upsilon_{m,m',0} = Y_0 \delta_{m,m'}, \tag{9.30}$$

$$\Upsilon_{m,m',m''} = 0 \text{ if } (m \pm m' \pm m'') \neq 0, \tag{9.31}$$

$$\{\Upsilon C \text{ or } \Upsilon S\}_{m,m',m''} = 0 \text{ if } (m \pm m' \pm m'' \pm 1) \neq 0. \tag{9.32}$$

Taking into account (9.31) and (9.32) the coefficients $d_{m,m'}$, $b^{\mathrm{E}}_{m,m'}$ given by (9.24) and (9.25), respectively, vanish if $(m+m')$ is an even number. However, if $(m+m')$ has an odd parity then $b^{\mathrm{B}}_{m,m'}$ given by (9.26) is zero due to (9.31). For an accurate calculation of d, b^{E}, and b^{B} with (9.24)–(9.26), respectively, M'' should be as large as possible. In practice, $M'' = 30$ is chosen, which is much larger than $M = 20$ or $M = 24$ (see section on subband energy interpolation).

$\hat{S}^{\nu}_{m}\{g\}(x,\tilde{\varepsilon})$ on the RHS of (9.22) is the projection of the scattering integral. The Pauli exclusion principle is included in the scattering integral. Taking into account the assumption that scattering events are local in the real space, the scattering integral reads [6,7]:

$$\hat{S}\{f\} = \frac{1}{(2\pi)^2} \sum_{\nu'} \left\{ \int (1 - f^{\nu}(x,\mathbf{k})) S^{\nu,\nu'}(x,\mathbf{k}|\mathbf{k}') f^{\nu'}(x,\mathbf{k}') \mathrm{d}^2 k' - \int (1 - f^{\nu'}(x,\mathbf{k}')) S^{\nu',\nu}(x,\mathbf{k}'|\mathbf{k}) f^{\nu}(x,\mathbf{k}) \mathrm{d}^2 k' \right\} \tag{9.33}$$

where the transition rate $S^{\nu',\nu}(x,\mathbf{k}'|\mathbf{k})$ is given by:

$$S^{\nu',\nu}(x,\mathbf{k}'|\mathbf{k}) = \sum_{\eta,\mp} c_{\eta,\mp}^{\nu',\nu}(x,\mathbf{k}'|\mathbf{k})\delta(\varepsilon^{\nu'}(x,\mathbf{k}') - \varepsilon^{\nu}(x,\mathbf{k}) \mp \hbar\omega_\eta). \tag{9.34}$$

η is the scattering mechanism index, c the transition coefficient from the initial state $(\nu',\mathbf{k}')$ to the final state $(\nu,\mathbf{k})$, and $\hbar\omega_\eta$ the energy transfer due to scattering. For an elastic scattering mechanism η, $\hbar\omega_\eta = 0$, $c_{\eta,+} = c_\eta$, and $c_{\eta,-} = 0$. Since a constant energy transfer in (9.34) is assumed, it is advantageous to expand the angular dependence of the transition rate on kinetic equi energy lines into a Fourier series [1]

$$c_{\eta,\mp}^{\nu',\nu}(x,\mathbf{k}'|\mathbf{k}) = \sum_{l,l'=-L}^{L} c_{\eta,\mp,l',l}^{\nu',\nu}(x,\tilde{\varepsilon}^{\nu'}(x,\mathbf{k}'),\tilde{\varepsilon}^{\nu}(x,\mathbf{k}))Y_{l'}(\phi')Y_l(\phi) \tag{9.35}$$

where the maximum order of the expansion is L. The projection of the scattering integral onto harmonics $Y_m(\phi)$ can be derived in a manner similar to [5]. The projected $\hat{S}_m^{\nu}\{g\}(x,\tilde{\varepsilon})$ is non-linear due to the Pauli principle (see Appendix A 6):

$$\begin{aligned}\hat{S}_m^{\nu}\{g\}(x,\tilde{\varepsilon}) = &\sum_{m'=-M}^{M}\sum_{l=-L}^{L}\Upsilon_{m,m',l}\sum_{l'=-L}^{L}\sum_{\nu',\eta,\mp}\\ &\Big\{\Big[Z_{m'}^{\nu}(x,\tilde{\varepsilon})c_{\eta,\mp,l,l'}^{\nu,\nu'}(x,\tilde{\varepsilon},\tilde{\varepsilon}')g_{l'}^{\nu'}(x,\tilde{\varepsilon}') - Z_{l'}^{\nu'}(x,\tilde{\varepsilon}')c_{\eta,\pm,l',l}^{\nu',\nu}(x,\tilde{\varepsilon}',\tilde{\varepsilon})g_{m'}^{\nu}(x,\tilde{\varepsilon})\Big]\\ &-\Big[c_{\eta,\mp,l,l'}^{\nu,\nu'}(x,\tilde{\varepsilon},\tilde{\varepsilon}') - c_{\eta,\pm,l',l}^{\nu',\nu}(x,\tilde{\varepsilon}',\tilde{\varepsilon})\Big]g_{l'}^{\nu'}(x,\tilde{\varepsilon}')g_{m'}^{\nu}(x,\tilde{\varepsilon})\Big\}.\end{aligned} \tag{9.36}$$

Here, $\tilde{\varepsilon}' = \tilde{\varepsilon}^{\nu'}(x,\mathbf{k}')$ denotes the kinetic subband energy of the initial state in Fermi's Golden Rule [2,7,8]:

$$\tilde{\varepsilon}^{\nu'}(x,\mathbf{k}') + \varepsilon_{\text{pot}}^{\nu'}(x) = \tilde{\varepsilon}^{\nu}(x,\mathbf{k}) + \varepsilon_{\text{pot}}^{\nu}(x) \mp \hbar\omega_\eta \tag{9.37}$$

and is therefore given by

$$\tilde{\varepsilon}' = \tilde{\varepsilon} + \varepsilon_{\text{pot}}^{\nu}(x) - \varepsilon_{\text{pot}}^{\nu'}(x) \mp \hbar\omega_\eta. \tag{9.38}$$

Four important scattering mechanisms are considered: acoustic phonon scattering, optical phonon scattering, alloy scattering, and surface roughness scattering. The scattering matrix elements of phonon scattering and alloy scattering [7, 9] are approximated such that they can be considered as velocity randomizing (see Chap. A.7).

The projection of the LHS of the BTE for $m = 0$ ensures current continuity

$$\partial j_x(x)/\partial x = 0 \tag{9.39}$$

based on the assumption that particle number conservation holds for the scattering integral

$$\frac{1}{(2\pi)^2}\sum_{\nu}\int \hat{S}\{f\}\mathrm{d}^2k = 0 \tag{9.40}$$

and the following boundary conditions w.r.t. energy are valid:

$$g^{\nu}_{m'}(x,\tilde{\varepsilon})|^{m'\mathrm{odd}}_{\tilde{\varepsilon}=0} = g^{\nu}_{m'}(x,\tilde{\varepsilon})|^{m'\mathrm{odd}}_{\tilde{\varepsilon}=\infty} = 0, \quad \forall\nu, \forall x. \tag{9.41}$$

As long as current continuity holds, the current density along the channel direction j_x is a constant. It should be noted that the b terms in (9.22) do not influence the current continuity due to $b^{\mathrm{E}\nu}_{m,m'} = b^{\mathrm{B}\nu}_{m,m'}\Big|_{m=0} = 0$.

9.2 Discretization of the Projected BTE

For the discretization of the projected BTE, a mesh of x and kinetic energy $\tilde{\varepsilon}$ is required. x is discretized with x_i and the kinetic energy $\tilde{\varepsilon}$ with $\tilde{\varepsilon}_j$. The 2D grid of $(x,\tilde{\varepsilon})$ is given by the tensor product of the two grids, which do not have to be equidistant. The $\tilde{\varepsilon}$ axis is discretized with 5 meV spacing within a kinetic energy range of [0,100 meV] in order to capture the significant change of the distribution function within this energy range. For higher energies $\tilde{\varepsilon}$ the spacing increases in order to reduce the number of grid points. Depending on the drain-to-source biases, the maximum level of the kinetic energy, up to which the distribution function is calculated, is adjusted (from 0.5 eV for $V_{\mathrm{DS}} = -1$ mV to 2–3 eV for $V_{\mathrm{DS}} = -1$ V). The number of grid points for the x axis is about 50–100. Again, a non-equidistant grid is used. The spacing of the x grid around the junctions between source or drain and the channel should be dense in order to capture the large gradient w.r.t to x of the potential subband energy in these regions. Within the channel and source and drain regions, the spacing of the x grid can be larger in order to reduce the number of slices for which the $\mathbf{k}\cdot\mathbf{p}$ SE must be solved.

The discretization of the projected BTE requires also an adjoint grid which is generated directly based on the main grid [1]. The adjoint grid points $i+1/2$ and $j+1/2$ are located on the x- and $\tilde{\varepsilon}$ axes at the positions

$$x_{i+1/2} = \frac{x_i + x_{i+1}}{2} \text{ and } \tilde{\varepsilon}_{j+1/2} = \frac{\tilde{\varepsilon}_j + \tilde{\varepsilon}_{j+1}}{2}, \tag{9.42}$$

respectively. Figure 9.1 shows the direct grid together with the adjoint grid for the x and $\tilde{\varepsilon}$ axes. Direct and adjoint grid lines are shown in solid and dashed lines, respectively. Intersections of the direct grid lines are the direct grid nodes, marked by full circles, whereas intersections of the adjoint grid lines are the adjoint grid nodes, marked by open circles.

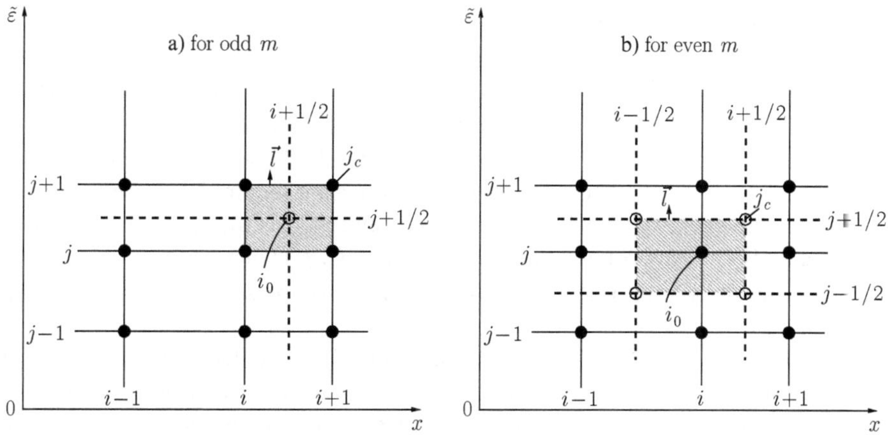

Fig. 9.1 Energy/real space discretization with dual grids. Boxes are centered around node i_0 for odd or even m and are given by the shaded areas. j_c denotes a corner node of the box

In classical drift-diffusion based TCAD device simulators (e.g. [10, 11]), the carrier density is defined on the direct grid and the current density on the adjoint grid. Similarly the unknowns g_m of the projected BTE (9.22) are defined either on the direct or adjoint grid depending on the parity of m. The "densities" (m even) are defined on the direct grid and the "fluxes" (m odd) on the adjoint.

The box-integration method is used for the discretization of the differential equation (9.22), and current continuity is conserved [1, 12]. The box for the integration of the differential equation (9.22) for the center node i_0 is the shaded area in Fig. 9.1 for either odd or even parity of m. The box-integration couples the unknowns at the four corner nodes j_c with the unknown at the center node i_0. A coupling between different subbands does not exist in the discretized drift and diffusion terms of the BTE. However, due to the inclusion of intersubband scattering mechanisms the unknowns for different subband indices are coupled in the discretized scattering integral of the BTE. Moreover, the discretization of the full BTE results in a non-linear system of equations for the unknowns $g_m^{\nu}(i_0)$ due to the Pauli principle in the scattering integral [5].

The fact that there is no subband coupling in the drift and diffusion terms of BTE motivates to parallelize the transport code in the subband index. The parallel computation involves N_b processes where N_b is the number of the included subbands (N_b is practically in the range from 20 to 30). The system matrix due to drift and diffusion terms associated for each subband is set up simultaneously. The scattering matrix elements for all subbands are broadcasted to each process in order to set up in parallel the system matrix due to the scattering integral for each subband. The full system matrix of all subbands is then gathered by the main process. Using such a parallel scheme, a few minutes of CPU time are typically sufficient to setup the system matrix.

9.3 Boundary Conditions

The discretized BTE can only be solved for x and $\tilde{\varepsilon}$ grid points from 0 to a finite value $N_x - 1$ and from 0 up to a finite value $N_{\tilde{\varepsilon}} - 1$, respectively. Due to the $\tilde{\mathbf{j}}$ term or the scattering integral, $g_m(i, j)$ can couple with $g_{m'}(i', j')$ where $i' < 0$ or $i' \geq N_x$ and $j' < 0$ or $j' \geq N_{\tilde{\varepsilon}}$. Thus, boundary conditions are required. For the first point $j = 0$ of the energy grid $\tilde{\varepsilon}(j, m) = 0$ holds and $g_m^{\nu}(i, j)|_{m \text{ odd}} = 0$ is assumed because of (9.41). Any coupling of $g_m^{\nu}(i, j)$ to $g_{m'}^{\nu'}(i', j')$ where $i' \geq N_x$ or $j' \geq N_{\tilde{\varepsilon}}$ due to either the $\tilde{\mathbf{j}}$ term or the scattering integral is suppressed by neglecting such terms.

The boundary conditions for the source and drain contacts are discussed in the following. The source and drain terminals are treated as carrier reservoirs in thermodynamic equilibrium. At the interface between the active device area and the reservoir, the interface distribution function contains two parts:

$$f_s^{\nu}(x, \mathbf{k}) = \Theta\left(\mathbf{v}^{\nu\mathrm{T}}(x, \mathbf{k})\mathbf{n}(x)\right) f_{\mathrm{eq}}(\varepsilon^{\nu}(x, \mathbf{k})) + \Theta\left(-\mathbf{v}^{\nu\mathrm{T}}(x, \mathbf{k})\mathbf{n}(x)\right) f^{\nu}(x, \mathbf{k}) \tag{9.43}$$

The first part $\Theta\left(\mathbf{v}^{\mathrm{T}}\mathbf{n}\right) f_{\mathrm{eq}}$ describes the particles flowing into the active area of the device from the reservoir with an equilibrium distribution. The second part $\Theta\left(-\mathbf{v}^{\mathrm{T}}\mathbf{n}\right) f$ describes the particles leaving the active area of the device (see Fig. 9.2). Here, $\mathbf{n}(x) = \pm\mathbf{e}_x$ is the surface vector pointing into the device. f_{eq} is the Fermi-Dirac distribution function, which is the solution of the BTE at equilibrium. $\Theta(t)$ is the Heaviside step function given by:

$$\Theta(t) = \begin{cases} 0 & , t < 0 \\ 1 & , t > 0 \end{cases}. \tag{9.44}$$

The interface generation rate is the product of the interface distribution function and the group velocity $\mathbf{v}^{\mathrm{T}}\mathbf{n}$:

$$\gamma_{\mathrm{s}}^{\nu}(x, \mathbf{k}) = f_{\mathrm{s}}^{\nu}(x, \mathbf{k})\mathbf{v}^{\nu\mathrm{T}}(x, \mathbf{k})\mathbf{n}(x). \tag{9.45}$$

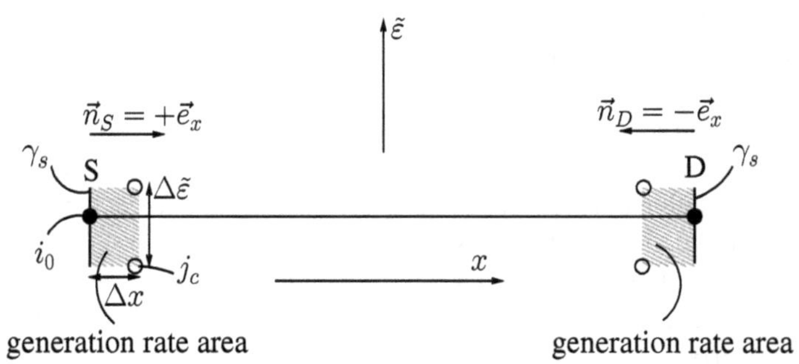

Fig. 9.2 Interface generation rate γ_s and the area for the equivalent area generation rate at the source and drain terminals

The interface generation rate γ_s is implemented by an equivalent area generation rate γ within the boxes of the source and drain terminals shown in Fig. 9.2. Within the generation area, the area generation rate γ is assumed to be constant. The number of generated particles over the total generation area is therefore given by $\gamma \Delta x \Delta \tilde{\varepsilon}$, where Δx and $\Delta \tilde{\varepsilon}$ are the width and height of the area, respectively. The number of generated particles $\gamma \Delta x \Delta \tilde{\varepsilon}$ is consistent with the number of particles generated by the interface generation rate $\gamma_s \Delta \tilde{\varepsilon}$, when the following identity holds:

$$\gamma^{\nu}(x, \mathbf{k}) = \frac{1}{\Delta x(x)} \gamma_{\mathrm{s}}^{\nu}(x, \mathbf{k}) = \frac{1}{\Delta x(x)} f_{\mathrm{s}}^{\nu}(x, \mathbf{k}) \mathbf{v}^{\nu \mathrm{T}}(x, \mathbf{k}) \mathbf{n}(x). \quad (9.46)$$

At the source and drain terminals the area generation rate is added to the scattering integral of the BTE.

References

1. Jungemann, C., Pham, A.T., Meinerzhagen, B., Ringhofer, C., Bollhöfer, M.: Stable discretization of the Boltzmann equation based on spherical harmonics, box integration, and a maximum entropy dissipation principle. J. Appl. Phys. **100**, 024502.1–024502.13 (2006)
2. Madelung, O.: Introduction to Solid State Theory. Springer, Berlin (1978)
3. Jacoboni, C., Lugli, P.: The Monte Carlo Method for Semiconductor Device Simulation. Springer, New York (1989)
4. Jin, S., Tang, T., Fischetti, M.V.: Simulation of silicon nanowire transistors using Boltzmann Transport Equation under relaxation time approximation. IEEE Trans. Electron Dev. **55**(3), 727–736 (2008)
5. Pham, A.T., Jungemann, C., Klawitter, M., Meinerzhagen, B.: Efficient simulation of hole transport in strained Si and SiGe on insulator inversion layers. Solid State Electron. **52**, 1660–1668 (2008)
6. Jungemann, C., Emunds, A., Engl, W.L.: Simulation of linear and nonlinear electron transport in homogeneous silicon inversion layers. Solid State Electron. **36**, 1529–1540 (1993)
7. Fischetti, M.V., Ren, Z., Solomon, P.M., Yang, M., Rim, K.: Six-band $k \cdot p$ calculation of the hole mobility in silicon inversion layers: Dependence on surface orientation, strain, and silicon thickness. J. Appl. Phys. **94**, 1079–1095 (2003)
8. Jungemann, C., Meinerzhagen, B.: Hierarchical Device Simulation: The Monte-Carlo Perspective. Computational Microelectronics. Springer, New York (2003)
9. Harrison, J.W., Hauser, J.R.: Alloy scattering in ternary III-V compounds. Phys. Rev. B **13**, 5347–5350 (1976)
10. Engl, W.L., Dirks, H.K., Meinerzhagen, B.: Device modeling. Proc. IEEE **71**, 10–33 (1983)
11. Meinerzhagen, B., Dirks, H.K., Engl, W.L.: Quasi-simultaneous solution method: A new highly efficient strategy for numerical MOST simulations. IEEE Trans. Comp. Aided Des. **4**, 575–582 (1985)
12. Varga, R.S.: Matrix Iterative Analysis, Series in Automatic Computation. Prentice-Hall, New Jersey (1962)

Chapter 10
Poisson Equation

The electrostatic potential $V(x,z)$ in the device is the solution of the PE

$$\nabla_{\mathbf{r}}^{\mathrm{T}} \kappa \nabla_{\mathbf{r}} V(x,z) = -q(N_{\mathrm{D}}(x,z) - N_{\mathrm{A}}(x,z) - n(x,z) + p(x,z)), \tag{10.1}$$

where κ is dielectric constant, q is the positive elementary charge, and N_{A} and N_{D} are the ionized acceptor and donor concentrations,[1] respectively. The hole density p is calculated with the probability density function $\mathscr{P}$ and the distribution function

$$\begin{aligned} p(x,z) &= \frac{1}{(2\pi)^2} \int \sum_{\nu} \mathscr{P}^{\nu}(x,z,\mathbf{k}) f^{\nu}(x,\mathbf{k}) \mathrm{d}^2 k \\ &= \sum_{\nu} \int \sum_{m=-M'}^{M'} \mathscr{P}_m^{\nu}(x,z,\tilde{\varepsilon}) g_m^{\nu}(x,\tilde{\varepsilon}) \mathrm{d}\tilde{\varepsilon}. \end{aligned} \tag{10.2}$$

The Fourier expansion coefficient $\mathscr{P}_m^{\nu}(x,z,\tilde{\varepsilon})$ of the probability density function is evaluated numerically by

$$\mathscr{P}^{\nu}(x,z,\tilde{\varepsilon}) = \int_0^{2\pi} \mathscr{P}^{\nu}(x,z,k^{\nu}(\tilde{\varepsilon},\phi),\phi) Y_m(\phi) \mathrm{d}\phi. \tag{10.3}$$

Here, $\mathscr{P}^{\nu}(x,z,k^{\nu}(\tilde{\varepsilon},\phi),\phi)$ is the probability density function for the wave vector $\mathbf{k} = (k^{\nu}(\tilde{\varepsilon},\phi),\phi)$ which is determined by the efficient interpolation in the 2D $\mathbf{k}$ space (see Sect. 8.4). Due to the symmetry of the subband structure, the Fourier expansion of the probability density function includes only even harmonics.

For a bulk PMOSFET the electron density $n(x,z)$ needs to be taken into account. For a bulk PMOSFET $n(x,z)$ is calculated with the classical expression

[1] In this work the acceptors and donors are assumed to be completely ionized.

S.-M. Hong et al., *Deterministic Solvers for the Boltzmann Transport Equation*, Computational Microelectronics, DOI 10.1007/978-3-7091-0778-2_10,

$$n(x,z) = N_c(x)\exp\left(-\frac{E_c(x) + qV(x,z) - \varepsilon_F(x)}{k_B T}\right). \tag{10.4}$$

N_c and E_c are the effective density of state and the energy of the conduction band edge, respectively. The electrostatic potential $V(x,z)$, the doping concentrations $N_A(x,z), N_D(x,z)$, the hole density $p(x,z)$, and the electron density $n(x,z)$ are defined only on the nodes of the direct grid. The x grid for the PE is therefore the same as the direct grid used for the BTE. The z grid is consistent with the one used for the SE. Box-integration is also used for the discretization of the PE. Dirichlet boundary conditions are applied to the gate contacts

$$V(x_{\text{gate}}, z_{\text{gate}}) = V_{\text{gate,bias}} + \frac{1}{q}\Phi_{ms}. \tag{10.5}$$

Here, $V_{\text{gate,bias}}$ is the applied gate bias and Φ_{ms} is the work function difference. Neumann boundary conditions are applied for the electrostatic potential at the remaining boundaries of the device and especially at the source and drain contacts [1], assuming that the electric field component, which is perpendicular to the contact interface, vanishes:

$$\frac{\partial V(x_{\text{source}}, z_{\text{source}})}{\partial x} = 0, \quad \frac{\partial V(x_{\text{drain}}, z_{\text{drain}})}{\partial x} = 0. \tag{10.6}$$

In order to improve the convergence of the nonlinear relaxation method for the solution of the SE-PE-BTE system the hole density is assumed to depend on the electrostatic potential during the solution of the PE similar to [2,3]. For the solution of the PE the hole density is assumed to have the following form:

$$\tilde{p}(x,z) = \exp\left(-\frac{V(x,z) - V^{(n')}(x,z)}{k_B T}\right) p(x,z). \tag{10.7}$$

Here, $p(x,z)$ is calculated with (10.2) and $V^{(n')}$ is the electrostatic potential of the previous iteration step. When the nonlinear relaxation method converges, $V^{(n')}(x,z) \rightarrow V(x,z)$ holds. Therefore, the transformed hole density $\tilde{p}(x,z)$ is finally equal to the original hole density $p(x,z)$ due to (10.7).

References

1. Ren, Z., Venugopal, R., Goasguen, S., Datta, S., Lundstrom, M.S.: NanoMOS 2.5: A two-dimensional simulator for quantum transport in double-gate MOSFETs. IEEE Trans. Electron Dev. **50**(9), 1914–1925 (2003)
2. Jungemann, C., Emunds, A., Engl, W.L.: Simulation of linear and nonlinear electron transport in homogeneous silicon inversion layers. Solid State Electron. **36**, 1529–1540 (1993)
3. De Michielis, M., Esseni, D., Palestri, P., Selmi, L.: A new multi subband monte carlo simulator for nano p-MOSFETs. In: Proceedings of ULIS, pp. 67–70 (2008)

Chapter 11
Iteration Methods

11.1 Gummel Type Iteration Scheme

The set of equations solved in this work is similar to [1–3]. One possible strategy for the solution of the SE-BTE-PE system is to solve each equation independently and successively and to consider the coupling between the equations by an outer loop. This constitutes a nonlinear relaxation scheme comparable to the classical Gummel loop [4, 5]. The outer loops are repeated until a self-consistent solution is obtained. In literature the multisubband Monte Carlo (MC) method is often used to solve the BTE [1–3] (Fig. 11.1 (left)). However, the nonlinear relaxation schemes with the MC method converge slowly. Moreover, a true DC solution of BTE is impossible with the transient MC method. In contrast, the BTE in this work is solved with a new deterministic method based on the Fourier expansion of the distribution function (see Chap. 9). Such nonlinear relaxation methods (Fig. 11.1 (right)) with the new deterministic method to solve BTE converge linearly with simulation time similar to the Gummel loop in the classical drift-diffusion based TCAD device simulators [4, 6], and thus much faster than an MC algorithm with its square root dependence [7]. In addition, the deterministic relaxation method yields a true DC solution.

11.2 Convergence Enhancement Method for the Iteration Loop

The convergence of the Gummel type iteration scheme used in [5] can be enhanced further by solving the BTE and PE simultaneously with a Newton approach (Fig. 11.2). Self-consistency with the SE is again obtained by a Gummel type nonlinear relaxation scheme. The new approach converges faster than the previous nonlinear relaxation scheme because solving BTE and PE simultaneously by a Newton approach takes already care of the predominant part of the coupling between these two equations.

S.-M. Hong et al., *Deterministic Solvers for the Boltzmann Transport Equation*, Computational Microelectronics, DOI 10.1007/978-3-7091-0778-2_11,

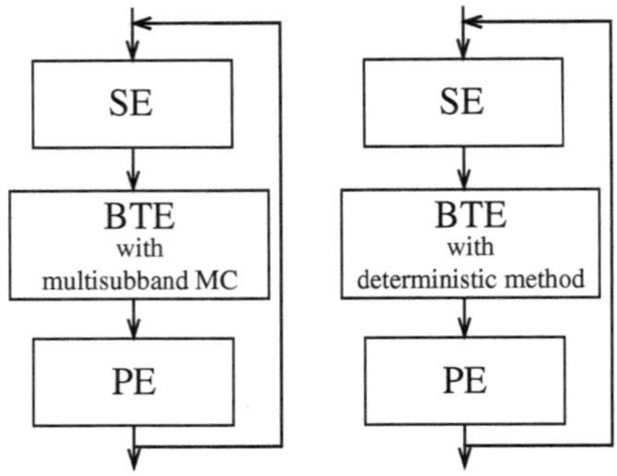

Fig. 11.1 Traditional Gummel type iteration loop to solve self-consistently the system of the three equations SE, BTE, and PE. *Left:* BTE is solved with a multisubband MC method [1–3]. *Right:* BTE is solved with a deterministic method in this work

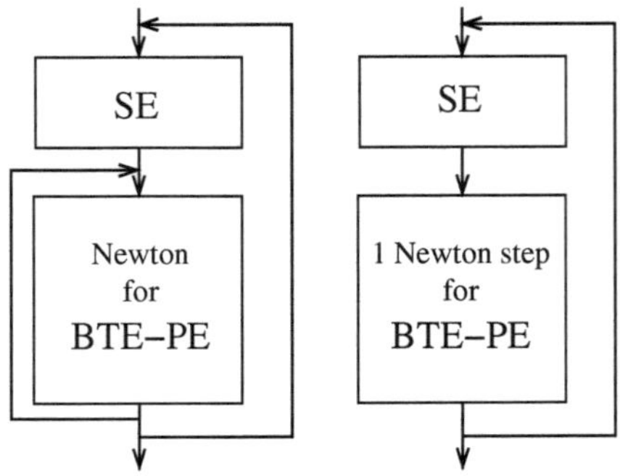

Fig. 11.2 Improved Gummel type iteration loop to solve self-consistently the SE-BTE-PE system with multi- (*left*) or single-step (*right*) Newton algorithm for subsystem BTE-PE. Here, the BTE is solved with a deterministic method

In the following the details of the method are described.

The coupled system of PE and BTE which is intended to be solved by a Newton–Raphson iteration scheme is written as:

$$F_{\mathrm{PE}}(V(\underline{x},\underline{z}), g_m^{\nu}(\underline{x},\underline{\tilde{\varepsilon}})) = 0 \tag{11.1}$$

$$F_{\mathrm{BTE}}(V(\underline{x},\underline{z}), g_m^{\nu}(\underline{x},\underline{\tilde{\varepsilon}})) = 0, \tag{11.2}$$

where $\underline{x},\underline{z}$ are the positions at a discrete space grid point along the transport and quantization directions, respectively. $\underline{\tilde{\varepsilon}}$ is the kinetic subband energy at a discrete energy grid point, V the electrostatic potential, and g_m^{ν} the Fourier expansion coefficient of order m for the generalized distribution function belonging to subband ν. Please note that $V(\underline{x},\underline{z})$ and $g_m^{\nu}(\underline{x},\underline{\tilde{\varepsilon}})$ are vectors containing discrete V and discrete g at a discrete $(\underline{x},\underline{z})$ grid point or a discrete $(\underline{x},\underline{\tilde{\varepsilon}})$ grid point.

For the Newton–Raphson iteration step $(n+1)$, the V and g_m^{ν} corrections, denoted as δV and δg_m^{ν}, are the solution of the following coupled system of equations resulting from a Taylor series expansion of (11.1) and (11.2) up to first order:

$$-F_{\mathrm{PE}}^{(n)} = \frac{\partial F_{\mathrm{PE}}}{\partial V(\underline{x},\underline{z})}\delta V(\underline{x},\underline{z}) + \frac{\partial F_{\mathrm{PE}}}{\partial g_m^{\nu}(\underline{x},\underline{\tilde{\varepsilon}})}\delta g_m^{\nu}(\underline{x},\underline{\tilde{\varepsilon}}), \tag{11.3}$$

$$-F_{\mathrm{BTE}}^{(n)} = \frac{\partial F_{\mathrm{BTE}}}{\partial V(\underline{x},\underline{z})}\delta V(\underline{x},\underline{z}) + \frac{\partial F_{\mathrm{BTE}}}{\partial g_m^{\nu}(\underline{x},\underline{\tilde{\varepsilon}})}\delta g_m^{\nu}(\underline{x},\underline{\tilde{\varepsilon}}). \tag{11.4}$$

As soon as the solutions for δV and δg_m^ν are available, the unknowns of the system (11.1) and (11.2) can be updated:

$$V^{(n+1)}(\underline{x}, \underline{z}) = V^{(n)}(\underline{x}, \underline{z}) + \delta V(\underline{x}, \underline{z}), \tag{11.5}$$

$$g_m^{\nu(n+1)}(\underline{x}, \underline{\tilde{\varepsilon}}) = g_m^{\nu(n)}(\underline{x}, \underline{\tilde{\varepsilon}}) + \delta g_m^\nu(\underline{x}, \underline{\tilde{\varepsilon}}). \tag{11.6}$$

With the updated V and g_m^ν, the correction procedure is repeated for the next iteration step. The iteration converges when the corrections $|\delta V|$ and $|\delta g_m^\nu|$ get smaller than a preset criterion.

Except $\frac{\partial F_{\mathrm{BTE}}}{\partial V(\underline{x},\underline{z})}$, the derivatives appearing in (11.3) and (11.4) can be calculated in a straightforward manner. A direct calculation of $\frac{\partial F_{\mathrm{BTE}}}{\partial V(\underline{x},\underline{z})}$ is impossible because the multisubband BTE does not depend explicitly on V. Actually, the electrostatic potential V entering the 6×6 $\mathbf{k} \cdot \mathbf{p}$ Schrödinger eigen problem is captured by the multisubband BTE implicitly through the total subband energy, which is the eigen energy of the eigen problem. In order to obtain an explicit coupling between the BTE and the PE, the variation of the total subband energy $\delta\varepsilon^\nu(\underline{x}, \mathbf{k}) \equiv \varepsilon^{\nu,(n+1)}(\underline{x}, \mathbf{k}) - \varepsilon^{\nu,(n)}(\underline{x}, \mathbf{k})$ is evaluated in terms of the variation of the electrostatic potential $\delta V(\underline{x}, z)$ using first order perturbation theory and $\delta\hat{\mathbf{H}} = \delta V(\underline{x}, z)\hat{\mathbf{I}}$:

$$\delta\varepsilon^\nu(\underline{x}, \mathbf{k}) = \int \mathscr{P}^\nu(\underline{x}, z, \mathbf{k})\delta V(\underline{x}, z)\mathrm{d}z. \tag{11.7}$$

The wave function $\boldsymbol{\zeta}$ (the eigen vector of the eigen problem) is assumed to be unchanged during the Newton–Raphson iteration process. Consequently, the probability density function $\mathscr{P} = \boldsymbol{\zeta}^\dagger\boldsymbol{\zeta}$ is the same for all Newton–Raphson iteration steps.

In order to simplify the derivative w.r.t V of the scattering integral of the multisubband BTE, $\delta\varepsilon^\nu(\underline{x}, \mathbf{k})$ is further approximated so that it does not depend on the subband index or wave vector. Multiplying the aforementioned $\delta\varepsilon^\nu(\underline{x}, \mathbf{k})$ with the distribution function $f^{\nu,(n)}(\underline{x}, \mathbf{k})$ and then integrating over $\mathbf{k}$-space and summing over all subband indices gives:

$$\sum_\nu \frac{1}{(2\pi)^2} \int \delta\varepsilon^\nu(\underline{x}, \mathbf{k}) f^{\nu,(n)}(\underline{x}, \mathbf{k})\mathrm{d}^2k = \int \delta V(\underline{x}, z) p^{(n)}(\underline{x}, z)\mathrm{d}z, \tag{11.8}$$

where $p^{(n)}(\underline{x}, z)$ is the hole density after iteration step n given by (10.2) with $f^\nu(\underline{x}, \mathbf{k}) \leftarrow f^{\nu,(n)}(\underline{x}, \mathbf{k})$.

If the total subband energy variation is assumed to be dominated by the variation of the subband potential energy $\delta\varepsilon_{\mathrm{pot}}^\nu(\underline{x})$, which in turn is assumed to be the same for all subbands:

$$\delta\tilde{\varepsilon}^\nu(\underline{x}, \mathbf{k}) \approx 0, \tag{11.9}$$

$$\delta\varepsilon_{\mathrm{pot}}^\nu(\underline{x}) \approx \delta\varepsilon_{\mathrm{pot}}(\underline{x}), \tag{11.10}$$

then the LHS of (11.8) can be simplified:

$$\sum_{\nu} \frac{1}{(2\pi)^2} \int \delta\varepsilon^{\nu}(\underline{x}, \mathbf{k}) f^{\nu,(n)}(\underline{x}, \mathbf{k}) \mathrm{d}^2 k \approx \delta\varepsilon_{\mathrm{pot}}(\underline{x}) \sum_{\nu} \frac{1}{(2\pi)^2} \int f^{\nu,(n)}(\underline{x}, \mathbf{k}) \mathrm{d}^2 k$$

$$= \delta\varepsilon_{\mathrm{pot}}(\underline{x}) \int p^{(n)}(\underline{x}, z) \mathrm{d}z. \tag{11.11}$$

The final equality in (11.11) is obtained directly from (10.2) with $\int \mathscr{P}^{\nu}(\underline{x}, z, \mathbf{k})\mathrm{d}z{=}1$.

Thus, the variation of the potential energy, which does not depend on the subband index, is obtained:

$$\delta\varepsilon_{\mathrm{pot}}(\underline{x}) = \frac{\displaystyle\int \delta V(\underline{x}, z) p^{(n)}(\underline{x}, z) \mathrm{d}z}{\displaystyle\int p^{(n)}(\underline{x}, z) \mathrm{d}z}. \tag{11.12}$$

With (11.9) and (11.12) the derivative w.r.t V of the drift operator (LHS of multisubband BTE) can be evaluated. Moreover, the derivative of the drift operator using (11.9) and (11.12) is much simpler than the calculation based on (11.7), which depends on the subband index and wave vector.

The derivative w.r.t V of the scattering integral (the RHS of the multisubband BTE) involves only the derivative w.r.t V of the kinetic subband energy of the initial and final states in Fermi's Golden Rule given by (9.37). The variations of the initial and final energies cancel each other

$$\delta\varepsilon^{\nu}_{\mathrm{pot}}(\underline{x}) = \delta\varepsilon^{\nu'}_{\mathrm{pot}}(\underline{x}) \tag{11.13}$$

in the energy conserving delta function of Fermi's Golden Rule due to approximation (11.10). Therefore, the derivative of the scattering integral w.r.t V vanishes.

Within each Gummel type iteration, only one single Newton–Raphson step is performed for the self-consistent solution of BTE and PE (Fig. 11.2 (right)) instead of multi Newton–Raphson steps which are often used in the usual Newton method (Fig. 11.2 (left)). The electrostatic potential after this single step is immediately used for the SE in the next Gummel type iteration. Solving the BTE and PE simultaneously by a single step Newton–Raphson scheme even reduces the CPU time per nonlinear iteration compared to the case where BTE and PE are solved successively.

In summary the multisubband BTE (9.1) does not depend explicitly on the electrostatic potential V. In order to capture most of the coupling between the BTE and PE the variation of the total subband energy appearing in the BTE is evaluated in terms of the variation of V between the Newton iteration steps n and $n+1$ by applying first order perturbation theory to the $6{\times}6$ $\mathbf{k}{\cdot}\mathbf{p}$ eigen problem. The variation of the subband kinetic energy is neglected and the variation of the total subband energy is further approximated such that it does not depend on the subband index. Using this approximation the derivative of the scattering integral w.r.t. V vanishes.

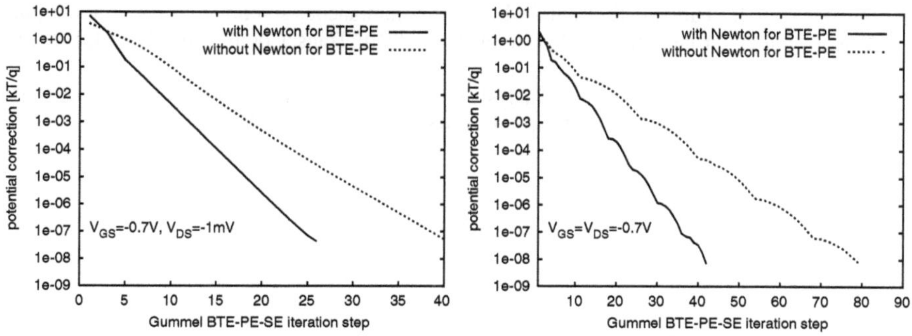

Fig. 11.3 Convergence of the BTE-PE-SE Gummel loop for $V_{GS} = -0.7\,\text{V}$, $V_{DS} = -1\,\text{mV}$ (*left*) and $V_{GS} = V_{DS} = -0.7\,\text{V}$ (*right*). A 16 nm gate length Si double gate PMOSFET with 5 nm body thickness is simulated. More details about simulation conditions (geometry, doping concentration, surface, channel orientations, etc.) can be found in Sect. 12.2.1 of Chap. 12

The derivative of the BTE w.r.t V is therefore simple and involves only the drift term of the BTE.

Convergence of the Gummel type loops is shown in Fig. 11.3 for a 16 nm gate length Si double gate PMOSFET with 5 nm body thickness for a low $V_{DS} = -1\,\text{mV}$ or a higher $V_{DS} = -0.7\,\text{V}$. For both V_{DS} cases an electrostatic potential correction of less than $10^{-7}\,kT/q$ is achievable in 20–80 iteration steps with or even without Newton algorithm. It is very difficult to achieve such a high accuracy if the MC method is used to solve the BTE. Here, this high accuracy is feasible because the BTE is solved with a deterministic algorithm. Moreover, the results in Fig. 11.3 clearly show that the introduction of the Newton approach improves the convergence speed of the nonlinear relaxation loop very much for both low and higher V_{DS} cases. For $V_{DS} = -1\,\text{mV}$ in order to achieve an electrostatic potential correction of less than $10^{-7}\,kT/q$, the quasi Newton approach requires only 24 iteration steps, which is much less than 39 iteration steps required by the traditional approach without the Newton algorithm.

References

1. Krishnan, S., Vasileska, D., Fischetti, M.V.: Modeling p-channel SiGe MOSFETs by taking into account the band-structure and the size quantization effects self-consistently. J. Compu. Electr. **5**(4), 435–438 (2006)
2. De Michielis, M., Esseni, D., Palestri, P., Selmi, L.: A new multi subband monte carlo simulator for nano p-MOSFETs. In: Proceedings of ULIS, pp. 67–70 (2008)
3. Lu, T., Du, G., Jiang, H., Liu, X., Zhang, P.: Multi subband deterministic simulation of an ultra-thin double gate MOSFET with 2D electron gas. In: Proceedings of IWCE, pp. 200–203 (2009)
4. Engl, W.L., Dirks, H. K., Meinerzhagen, B.: Device modeling. Proc. IEEE **71**, 10–33 (1983)
5. Pham, A.T., Jungemann, C., Meinerzhagen, B.: Deterministic multisubband device simulations for strained double gate PMOSFETs including magnetotransport. In: IEDM Tech. Dig. (2008), pp. 895–898

6. Meinerzhagen, B., Dirks, H.K., Engl, W.L.: Quasi-simultaneous solution method: A new highly efficient strategy for numerical MOST simulations. IEEE Trans. Comp. Aided Des. **4**, 575–582 (1985)
7. Jungemann, C., Meinerzhagen, B.: Analysis of the stochastic error of stationary Monte Carlo device simulations. IEEE Trans. Electron Dev. **48**(5), 985–992 (2001)

Chapter 12
Results

12.1 Low Field Mobility in a Homogenous Channel PMOS

For a Si hole inversion layer the parameters for phonon and surface roughness scattering, which are shown in Table 12.1, have been obtained by matching the simulation results to the long-channel low-field mobility measurements for bulk Si PMOSFETs from Takagi [5] at three lattice temperatures 223, 300, and 443 K shown in Fig. 12.1 (top). As in [3,4], a phonon energy of $\hbar\omega = 61.2\,\text{meV}$ was used and kept fixed in this work. The other four scattering parameters were considered as adjustable parameters and their values were extracted by the fitting procedure. The Levenberg-Marquardt least square fitting algorithm [6, 7] was employed and the parameter set of Fischetti in [2] was chosen for initial values. The optimum scattering parameters after fitting are still in good agreement with the ones of Fischetti in [2]. After calibration, all scattering parameters are kept fixed for all simulations. Figure 12.1 (top) shows that the calculated channel mobility fits well the long channel low-field mobility measurements of Takagi [5].

It can be observed that the simulated phonon limited mobility $\mu_{\text{phon}}(N_{\text{inv}})$ obeys the typical functional form for the phonon limited mobility degradation $\mu_{\text{phon}}(N_{\text{inv}}) \propto N_{\text{inv}}^{-1/3}$ for different lattice temperatures as shown in Fig. 12.1 (middle). It can also be seen that the simulation results for μ_{phon} in this work are in good agreement with the simulation results of Fischetti [2]. As shown in Fig. 12.1 (bottom) the surface roughness limited mobility is much larger than the phonon limited mobility, meaning that moving particles in the p-channel quantum well are predominantly influenced by lattice vibration or phonon scattering mechanisms and *not* by surface roughness scattering. With increasing inversion charge, the surface roughness limited mobility decreases much faster than the phonon limited mobility, which implies that the impact of surface roughness scattering on the transport is stronger for stronger inversion. Matthiessen's rule for calculating the total mobility based on the mobilities of the individual scattering mechanisms

S.-M. Hong et al., *Deterministic Solvers for the Boltzmann Transport Equation*, Computational Microelectronics, DOI 10.1007/978-3-7091-0778-2_12,

Table 12.1 Calibrated scattering parameters for a Si hole inversion layer

Parameter	Present work	Literature	Unit
Ξ	6.94	9.64,[a] 7.12,[b] 5.0[c]	eV
$D_t K$	13.06	12.72,[a] 13.24,[b] 7.63[c]	10^8 eV cm^{-1}
$\hbar\omega$	61.2	65,[a] 61.2[c,d]	meV
Δ	0.36	0.758,[a] 0.4,[b] 0.5[c]	nm
L	2.46	2.0,[a] 2.6,[b] 2.0[c]	nm

[a][1] [b][2] [c][3] [d][4]

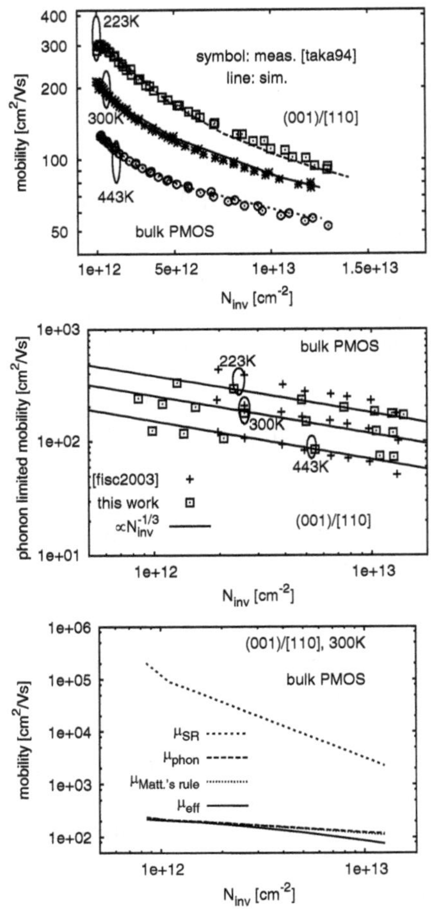

Fig. 12.1 *Top:* Simulated and measured low-field mobilities at different temperatures. Experimental data are from Takagi [5]. *Middle:* Phonon limited mobility at different temperatures compared to the simulation results of Fischetti [2]. Best approximation $\propto N_{\mathrm{inv}}^{-1/3}$ is also shown. *Bottom:* Phonon (μ_{phon}) and surface roughness (μ_{SR}) limited mobility, effective mobility (μ_{eff}), and Matthiessen's rule mobility ($\mu_{\mathrm{Matt.'s\ rule}}$) defined as $1/\mu_{\mathrm{Matt.'s\ rule}} = 1/\mu_{\mathrm{phon}} + 1/\mu_{\mathrm{SR}}$ are shown

$$\frac{1}{\mu_{\mathrm{Matt.'s\ rule}}} = \frac{1}{\mu_{\mathrm{phon}}} + \frac{1}{\mu_{\mathrm{SR}}} \tag{12.1}$$

fails especially for high inversion charge as shown in Fig. 12.1 (bottom).

The validation of the transport model is now performed for SG and DG SOI PMOS structures. No further adjustment for the Si phonon and surface roughness

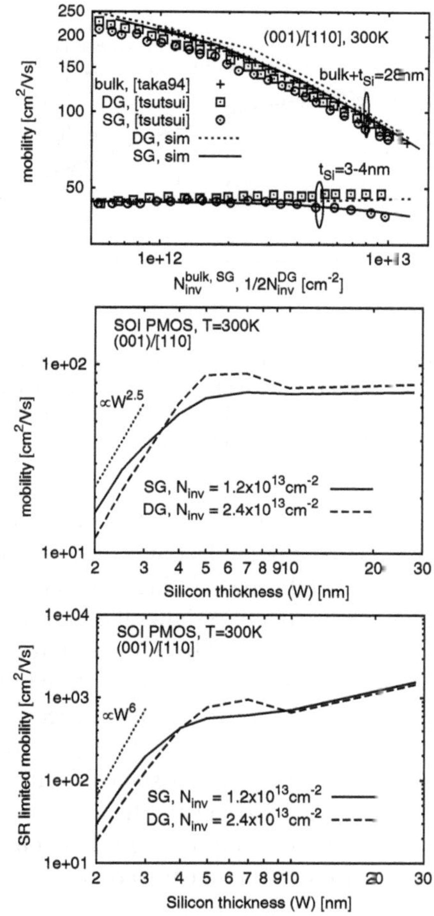

Fig. 12.2 *Top:* Calculated low-field mobility for SG and DG SOI and bulk MOS structures compared to the measurements of Tsutsui and Hiramoto [8] (SOI case) and Takagi [5] (bulk MOS case). *Middle:* Effective mobility (μ_{eff}) versus Si thickness for SG or DG PMOS structures. *Bottom:* Surface roughness (μ_{SR}) limited mobility vs. Si thickness for SG or DG PMOS structures

scattering parameters is made. Mobility curves for SG and DG SOI structures with scaled Si thickness are calculated and compared with experimental data. Figure 12.2 (top) shows that the calculated channel mobility reproduces long channel low-field mobility measurements of Tsutsui and Hiramoto [8] for both SG and DG PMOS structures with thin (3–4 nm) or thick (28 nm) SOI layers. The degradation of the mobility due to the scaled Si thickness shown by the experiments is accurately predicted by the simulator. For a fixed inversion charge of $1.2 \times 10^{13}\,\text{cm}^{-2}$ (SG SOI structure) or $2.4 \times 10^{13}\,\text{cm}^{-2}$ (DG case), the effective mobility μ_{eff} as a function of Si thickness W is shown in Fig. 12.2 (middle). It can be observed that for small Si thickness ($W \leq 3\,\text{nm}$), the simulated $\mu_{\text{eff}}(W) \propto W^{2.6}$, which is consistent with experimental results (from $W^{1.8}$ to $W^{3.3}$) [9]. For a thicker Si layer ($W > 4\,\text{nm}$), the $\mu_{\text{eff}}(W)$ characteristic of a DG structure is always higher than the $\mu_{\text{eff}}(W)$ characteristic of a SG structure (Fig. 12.2 (middle)). This indicates a transport advantage of the DG structure compared to the SG structure. Please note

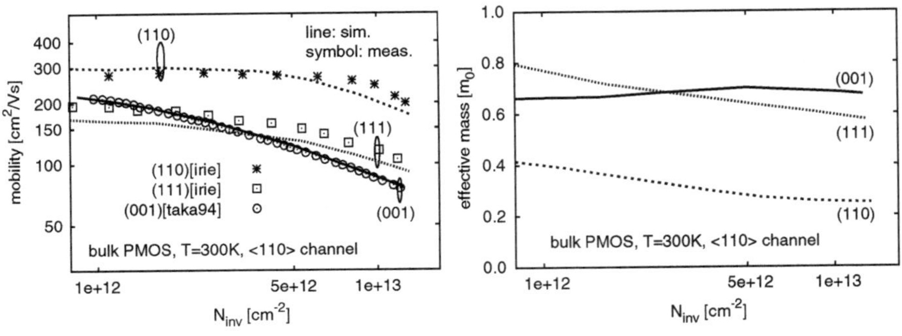

Fig. 12.3 Low-field mobility (*left*) and effective mass in the transport direction (*right*) for (001), (110), and (111) crystallographic surface orientations vs. N_{inv}. Experimental data are from Takagi [5] for (001) and Irie [13] for (110) and (111) surfaces. The channel direction is oriented in $\langle 110 \rangle$ for all three surface orientations

the saturation of the $\mu_{\mathrm{eff}}(W)$ characteristic for $W > 10\,\mathrm{nm}$. Within the $\mu_{\mathrm{eff}}(W)$ characteristic of the DG SOI structure, a hump appears for $5\,\mathrm{nm} \leq W \leq 7\,\mathrm{nm}$. This implies that the ideal Si thickness for a DG PMOS structure is between 5 and 7 nm. Figure 12.2 (bottom) shows the SR limited mobility (μ_{SR}) versus Si thickness. For an ultrathin body (UTB) SOI structure and small $W < 3\,\mathrm{nm}$, the typical asymptotic behavior $\mu_{\mathrm{SR}}(W) \propto W^6$ is confirmed by the simulation results. Please note that $\mu_{\mathrm{SR}}(W) \propto W^6$ for $W < 4\mathrm{nm}$ has been reported frequently in previous publications, e.g. [10, 11] for SOI NMOS or [12] for GaAs/AlAs quantum wells. The relation $\mu_{\mathrm{SR}}(W) \propto W^6$ is also assumed for the macroscopic mobility model for UTB SOI PMOS by Chleirigh et al. [9].

The next mobility model validation is for a bulk PMOS with different surface orientations (001), (110), and (111). For a $\langle 110 \rangle$ channel the simulated mobility curves are shown together with the low-field mobility measurements by Takagi [5] and Irie [13] in Fig. 12.3 (left). There is a good agreement between the simulation results and the experimental data. The simulator accurately predicts the enhancement of the mobility due to the non-traditional (110) surface orientation. Figure 12.3 (right) shows the effective mass in the transport direction $\langle 110 \rangle$ versus the inversion charge. For the (110) surface the effective mass is significantly smaller than the effective mass for the (001) or the (111) surface, which is consistent with a higher mobility for the (110) surface observed in Fig. 12.3 (left). The enhancement of the mobility for the (110) surface compared to mobility for the (001) surface is therefore mainly due to the reduction of the effective mass in the transport direction, which is confirmed by the study of Saitoh [14]. Moreover, the point of intersection of the effective mass curves for (001) and (111) surfaces is consistent with the point of intersection of the mobility curves for these two surfaces.

It is fair to note that the mobility model for the Ge hole inversion layer in this work is not as well calibrated and validated as the model for the Si hole inversion layer. State-of-the-art Ge technologies employ a large variety of gate dielectrics (e.g. HfO_2 ($\kappa = 25$) [15], ZrO_2 ($\kappa = 25$) [16], Al_2O_3 ($\kappa = 9$) [17], $LaAlO_3$ ($\kappa = 25$)

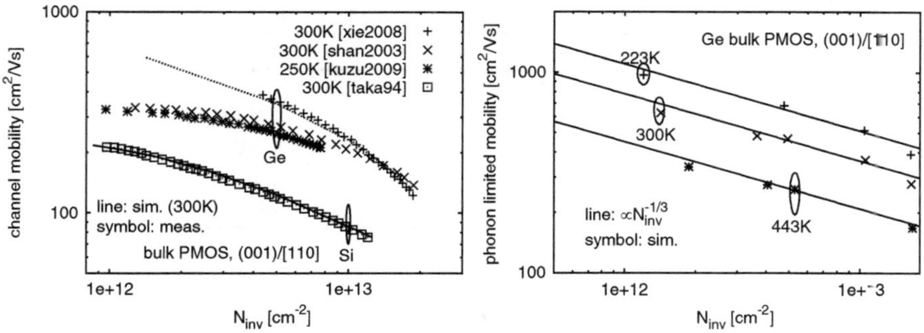

Fig. 12.4 *Left:* Simulated and measured low-field mobility of a Ge bulk PMOS compared to the low-field mobility of a Si bulk PMOS. The measurements are from [19–21] for Ge and [5] for Si. *Right:* Phonon limited mobility curves for a Ge bulk PMOS and three lattice temperatures 223, 300, and 443 K. The best fitted lines $\propto N_{\text{inv}}^{-1/3}$ are also shown

[18], GeON [19, 20], GeO_2 ($\kappa = 12.2$–14.5) with post-gate F incorporation [21]) when fabricating a high-κ/metal gate Ge PMOSFET transistor. But for modeling, this large variety causes a real problem. Many of these gate dielectrics cause additional remote scattering mechanisms, which are neither well understood nor well calibrated especially at the microscopic level [22].

In order to extract the scattering parameters that reflect the Ge material properties within the inversion layer alone a gate dielectric is needed that causes a negligible amount of additional remote scattering. For all available low-field mobility measurements in a Ge PMOS, the measured low-field mobility reported in [21] is the highest and about 3X higher than the universal mobility of a Si PMOS [5] (see Fig. 12.4 (left)). To the author's best knowledge, these are the highest reported values for the low-field mobility of a Ge PMOS (for small inversion charge, the experimental data in [20] and [19] is much lower than the one in [21] as shown in Fig. 12.4 (left)). The available low-field mobility measurements in [21] are therefore chosen for the calibration of the mobility model in this work, assuming that the influence of the gate dielectric on the low-field mobility in the work of Xie [21] is minimized and negligible.

The average optical phonon deformation potential $D_t K = 3.5 \times 10^8\,\text{eV}\,\text{cm}^{-1}$ used for bulk Ge [23, 24] before is also used for the Ge hole inversion layer. The optical phonon energy $\hbar\omega = 37.04\,\text{meV}$ is used for Ge hole inversion layer, which is consistent with [4] and nearly identical to 37 meV reported in [23,24] for bulk Ge. Only the average acoustic phonon deformation potential Ξ and the rms deviation Δ and correlation length L associated with the SR scattering mechanism are considered as adjustable parameters. The fitting of the simulated low-field mobility to the mobility measurements in [21] results in optimum values for these three parameters. The simulated results reproduce the experimental low-field mobility of Xie [21] very well (see Fig. 12.4 (left)). The calibrated scattering parameters for the Ge hole inversion layer are reported in Table 12.2. After this calibration all scattering

Table 12.2 Calibrated scattering parameters for a Ge hole inversion layer

Parameter	Present work	Literature	Unit
Ξ	13.69	7.40[a]	eV
$D_t K$	3.5	3.5[a]	10^8 eV cm^{-1}
$\hbar\omega$	37.04	37.04,[b] 37.0[a]	meV
Δ	0.5		nm
L	2.0		nm

[a][23] [b][4]

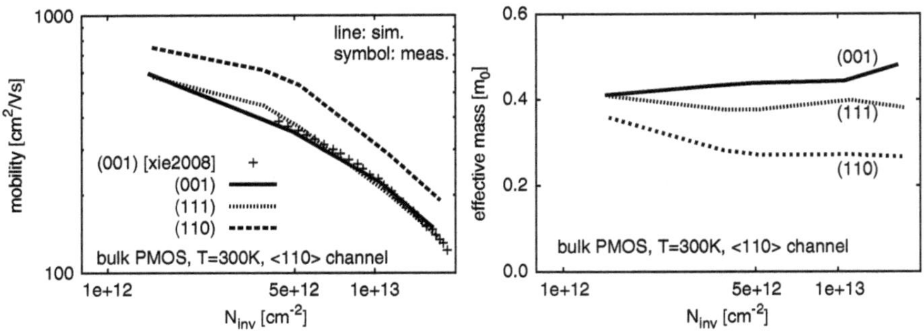

Fig. 12.5 Low-field mobility (*left*) and effective mass in the transport direction (*right*) vs. N_{inv} for (001), (110), and (111) crystallographic surface orientation in a Ge PMOS. The experimental data are from Xie [21] for the (001) surface. The channel is oriented in $\langle 110\rangle$ direction for all three surface orientations

parameters are kept fixed for all other simulations. The phonon limited mobility in a Ge PMOS is shown in Fig. 12.4 (right) for three different lattice temperatures 223, 300, and 443 K. For these three lattice temperatures the usual dependence $\mu_{phon}(N_{inv}) \propto N_{inv}^{-1/3}$ can be observed.

Figure 12.5 (left) shows mobility curves for the three surface orientations (001), (110), and (111) and the $\langle 110\rangle$ channel direction. Again, the (110) surface case yields the best effective mobility. For a high inversion charge the effective mobility for the (111) surface orientation is smaller than the mobility for the (001) surface orientation, which confirms the experimental observation from Kuzum [20]. In Fig. 12.5 (right) the dependence of the effective mass in transport direction on the inversion charge is shown. The effective mass for the (110) surface is remarkably smaller than the effective mass for the other two surfaces, which is consistent with the higher mobility for the (110) case. For the (001) and the (111) surface orientations the effective mass in a Ge PMOS (Fig. 12.5 (right)) is much smaller than the effective mass in a Si PMOS (Fig. 12.3 (right)) with the same surface and channel orientation.

Figure 12.6 shows the equi energy lines of the first subband for (001), (110), or (111) surface orientation in a Ge PMOS at $N_{inv} = 10^{13}\,\mathrm{cm}^{-2}$. The equi energy lines for the three surfaces are completely different, which is consistent with the differences in mobility and effective masses shown in Fig. 12.5. However, the

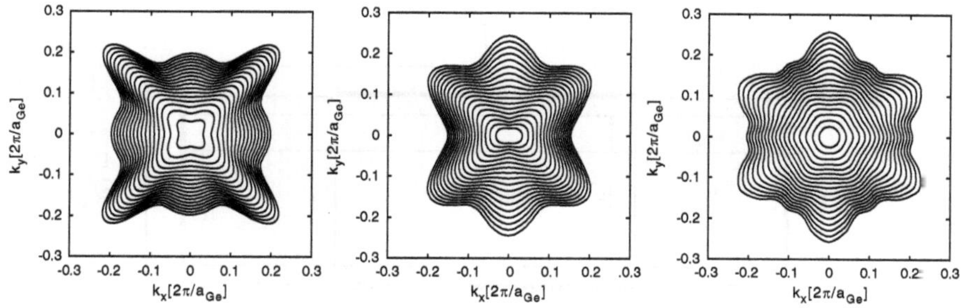

Fig. 12.6 Equi-energy lines in multiples of 20 meV of the first subband in an unstrained Ge bulk PMOS with (001) (*left*), (110) (*center*) or (111) (*right*) surface orientation. The bottom of the first subband is shifted to 0 eV. $N_{inv} = 10^{13}\,cm^{-2}$

symmetry is similar to the symmetry of the equi energy lines for the Si PMOS case as shown in Figs. 8.6 (left), 8.7 (left), and 8.8 (left). This is because both Si and Ge have the same diamond lattice structure.

12.2 Device Simulation Results

12.2.1 Numerical Aspects

A 16nm gate length Si (or Ge) double gate PMOSFET is simulated. The Si (or Ge) body and SiO_2 thicknesses of the double gate structure are 5 and 0.7 nm, respectively. The uniform acceptor doping concentration in source and drain is $10^{20}\,cm^{-3}$ and the uniform donor doping concentration in the channel region is $10^{14}\,cm^{-3}$. More details about the geometry and doping profile of the device are shown in Fig. 12.7. The (001) wafer surface and the [110] channel orientation are considered. Scattering due to phonons and surface roughness is included. A constant lattice temperature of 300 K is assumed. The gate work function difference was chosen such that the threshold voltage is about 0.18 V. The simulations are performed for both near- and non-equilibrium transport.

For $V_{GS} = -0.7\,V$ and $V_{DS} = -1\,mV$ (near equilibrium) or $V_{DS} = -0.7\,V$ (non-equilibrium), the convergence of the Gummel type loops for the SPB system are shown in Fig. 11.3. As mentioned in Chap. 11, the introduction of the Newton approach improves the convergence speed of the nonlinear relaxation loop very much for both the near and the non-equilibrium cases. In order to achieve an electrostatic potential correction of less than $10^{-7} kT/q$ near equilibrium, the new approach requires only 24 iteration steps instead of 39. The CPU time per iteration, which is dominated by the SE solver, is nearly the same for the old and the improved nonlinear relaxation scheme. The convergence behavior for a Ge PMOSFET is shown in Fig. 12.8. Here, the new method improves the convergence speed less

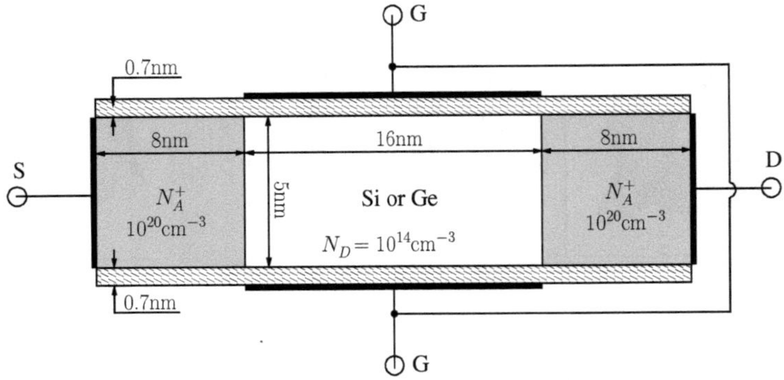

Fig. 12.7 Geometry and doping profile of the simulated DG PMOSFET

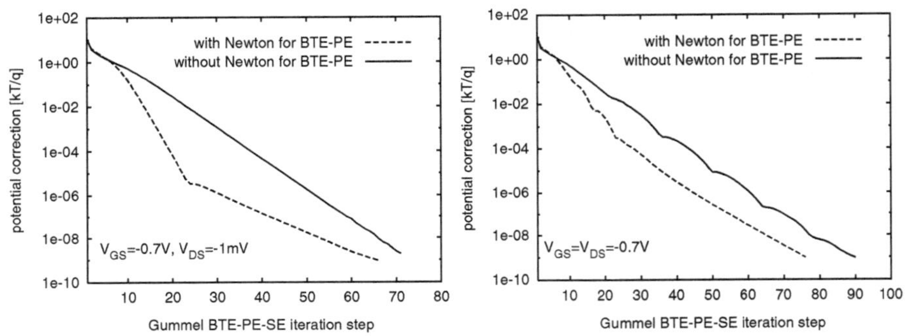

Fig. 12.8 Convergence of the BTE-PE-SE Gummel loop for $V_{GS} = -0.7$ V, $V_{DS} = -1$ mV (*left*) and $V_{GS} = V_{DS} = -0.7$ V (*right*). A 16 nm gate length Ge double gate PMOSFET with 5 nm body thickness is simulated

significant, but still there is no problem to achieve an electrostatic potential correction of less than $10^{-9}kT/q$. In order to achieve the same level of accuracy for the electrostatic potential, the new approach based on the Newton method for BTE-PE always requires less iteration steps compared to the old relaxation scheme.

The influence of the maximum order (M') of Fourier harmonics on the solution accuracy is studied in Fig. 12.9 by showing the relative difference between the drain currents for $M' = 3, 5, 7, 9, 11, 13$ and the drain current for $M' = 15$ for the Si (left) and the Ge (right) device. This difference declines below 1% with increasing M'. For $M' = 9$ the drain current difference is about 1% for Si and about 2% for the Ge device as shown in Fig. 12.9. In order to obtain a drain current difference below 1%, a Fourier expansion of the distribution function up to order 11 is required for the Si and the Ge device. This result again eliminates the concern about the efficiency of the Fourier harmonic expansion method for the solution of the BTE in a short channel PMOSFET. The average CPU time per bias point for a solution accuracy of $10^{-2}\,kT/q$ and $M' = 11$, 20 subbands, 170 energy points and 51×70 space discretization points on a state of the art LINUX Cluster with 60 processors

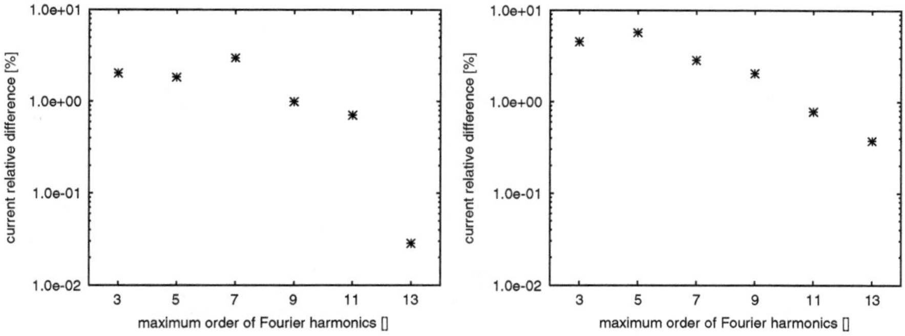

Fig. 12.9 Relative difference of I_D for various maximum order (M') of Fourier harmonics compared with I_D for $M' = 15$. Results are shown for Si (*left*) or Ge (*right*) double gate PMOSFETs with 5 nm body thickness. $V_{GS} = V_{DS} = -0.7\,V$

(2.6 GHz) and 100 GBytes of RAM is about 4 h. Within these 4 h, 30 min are used to set up the Jacobian.

12.2.2 Comparison of Transport in Strained Si or Ge Double Gate PMOSFETs with (001), (110), or (11$\bar{1}$) Surface Orientation and All Channel Directions: Homogenous Channel Low-Field Mobility, on Current, and Intrinsic Delay

The homogenous channel low-field mobility, ON current, and intrinsic delay for a double gate PMOSFET are evaluated. The geometry and doping profile of the double gate PMOSFET device considered here are as the ones assumed in the numerical aspect section (see Fig. 12.7). A constant lattice temperature of 300 K is assumed in all simulations. All effects like surface/channel orientation, stress/strain, and channel material Si/Ge are taken into account for the same geometry. Three typical surface orientations (001), (110), and (111) are simulated. Discrete channel directions are considered where the angle between the channel direction and the x'-axis of the primed coordinate system (x', y', z') is a multiple of 15°. The device is assumed to be unstrained or biaxially strained or uniaxially stressed. For uniaxial stress the stress direction is kept the same as the channel direction and a constant stress level of 1.5GPa along the device is assumed. Biaxial strain due to a lattice mismatch between the channel material and a relaxed virtual substrate is assumed. The biaxial strain tensor is determined such that the equivalent stress level is fixed at 1.5GPa as well. Both compressive and tensile stress/strain are considered. Each transport quantity is compared as a function of surface/channel orientation, stress/strain configuration and channel material. The comparison provides the

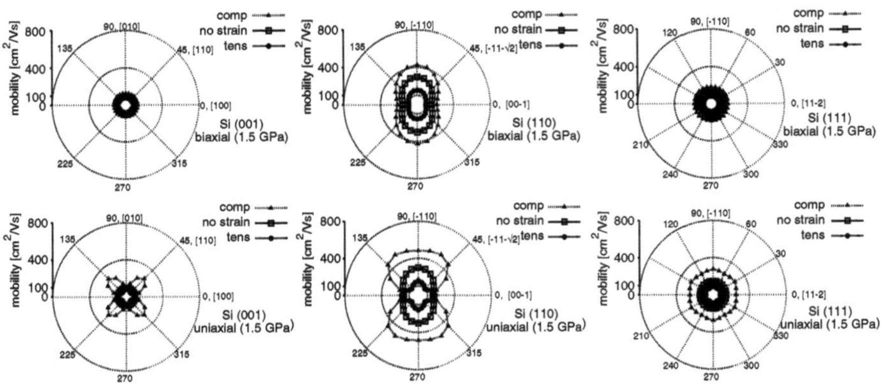

Fig. 12.10 Low-field hole mobility as a function of channel direction in a biaxially strained *(top)* or uniaxially stressed *(bottom)* Si double gate homogenous channel PMOS structure. The surface orientations (001) *(left)*, (110) *(middle)*, and (111) *(right)* are considered. For the uniaxial stress case the stress direction is the same as the channel direction. The inversion charge $N_{\text{inv}} = 1.3 \times 10^{13}\,\text{cm}^{-2}$ is kept constant

optimum selection of surface/channel orientation, stress/strain configuration, and channel material for the best transport performance.

At first, a homogenous channel situation is considered by removing source and drain regions and keeping the vertical structure of the device unchanged. The low-field mobility is evaluated for the remaining homogenous channel PMOS structures with different surface/channel orientations, stress/strain configurations, and the two channel materials Si or Ge. The mobility is calculated for a fixed inversion charge of $N_{\text{inv}} = 1.3 \times 10^{13}\,\text{cm}^{-2}$. This level of inversion charge is approximately the channel hole sheet charge for the bias $V_{\text{GS}} = -0.7\,\text{V}$ and $V_{\text{DS}} = -1\,\text{mV}$ (see Fig. 12.19). The dependence of low-field mobility on the channel direction is shown in Fig. 12.10 for Si. For the unstrained Si double gate PMOS with the traditional (001) surface and [110] channel orientation, the mobility is about 93 cm²/Vs. For each surface/channel orientation, the mobility with compressive stress/strain is always higher compared to the cases with tensile stress/strain or without stress/strain. For the same surface orientation the uniaxial stress technique is more efficient than the biaxial strain technique to boost the low-field mobility of a Si double gate PMOS. For reaching higher mobility, the (110) wafer surface orientation is better than (001) or (111). Therefore, compressive uniaxial stress combined with the (110) surface orientation is the best choice for a high mobility Si double gate PMOS and the optimum channel direction is around 60° from the $[00\bar{1}]$ direction.

The dependence of the low-field mobility on channel direction is shown in Fig. 12.11 for the homogenous Ge double gate PMOS structure. It can be clearly seen that a double gate PMOS with a Ge channel provides a much better mobility compared to Si. Nevertheless, the shape of the low-field mobility vs. channel direction characteristics for Ge is rather similar to the Si case. Again, compressive uniaxial stress with (110) surface orientation and an angle between the channel

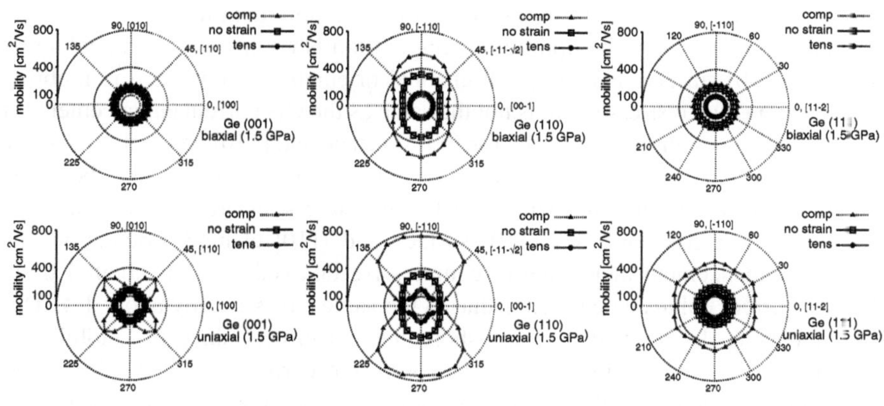

Fig. 12.11 Same as Fig. 12.10, but for a Ge double gate homogenous channel PMOS structure

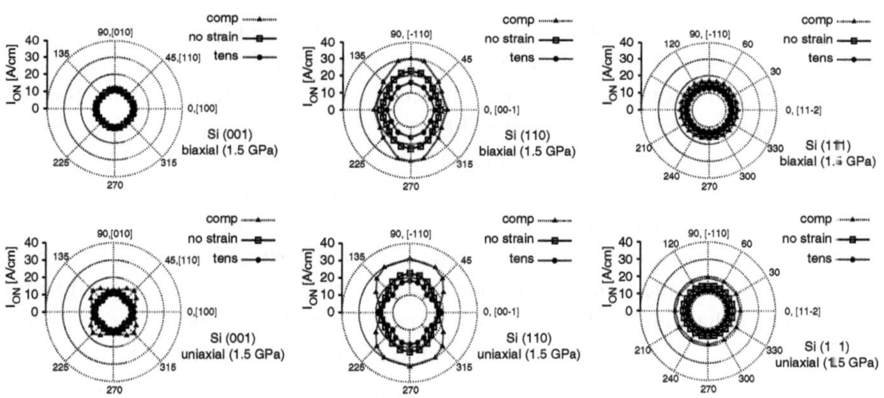

Fig. 12.12 I_{ON} as a function of channel direction for a biaxially strained *(top)* or uniaxially stressed *(bottom)* Si double gate PMOSFET. The surface orientations (001) *(left)*, (110) *(middle)*, and (111) *(right)* are considered. For the uniaxial stress case the stress direction is the same as the channel direction. $V_{GS} = V_{DS} = -0.7\,\text{V}$

direction and $[00\overline{1}]$ of about 75° is the best choice for a high mobility in a Ge double gate PMOS structure. The highest mobility is about 760 cm²/Vs. As visible on the same plot the mobility reduces slightly to 746 cm²/Vs, when the channel is oriented in $[\overline{1}10]$ direction.

For the remaining quantities like ON current device simulations are necessary. For each short channel PMOSFET, the gate work function difference is chosen such that the threshold voltage is 0 V. The ON current and intrinsic delay are evaluated for the bias $V_{GS} = V_{DS} = -0.7\,\text{V}$. The dependence of simulated ON current on channel direction ($I_{ON} = |I_D|$) is visualized in Fig. 12.12 for the Si devices. The dependence of ON current on channel direction is quite similar to the respective mobility dependence. The ON current for the (110) surface is higher than the one for (001) and (111) surfaces. Compressive stress/strain results in a higher current than tensile stress/strain. For the Si devices the maximum of ON current is about 31 A cm^{-1} and the optimum configuration is compressive uniaxial stress with the (110) surface and the $[\bar{1}10]$ channel orientation. However, the mobility enhancement due to stress/strain is not the same as the ON current enhancement for identical stress/strain. For example, for (001) surface and [110] channel orientation the mobility enhancement factor due to compressive uniaxial stress is about 2.9, but for the ON current enhancement factor the same stress results only in about a factor of 1.6. Moreover for (110) surface and $[\bar{1}10]$ channel orientation, the mobility enhancement factor due to compressive uniaxial stress is about 1.6 but the ON current enhancement factor due to the same stress is only about 1.4. This discrepancy is due to two reasons: short channel effects since the mobility is evaluated for a homogenous channel, whereas the ON current is evaluated for a short channel PMOS device, and non-linear transport effects since the mobility relates to transport near equilibrium whereas the ON current relates to transport far away from equilibrium.

Figure 12.13 shows the ON current characteristic for the Ge devices. In general, a Ge device yields a higher ON current compared to a Si device with the same stress/strain condition and the same surface/channel orientation. Again, the ON current results are similar to the mobility results. For Ge devices the ON current for the (110) surface is also higher than the one for (001) and (111) surfaces and compressive stress/strain results in a higher current than tensile stress/strain. For Ge devices the maximum ON current is about 46 A cm^{-1} and the optimum configuration is compressive uniaxial stress with a (110) surface and a $[\bar{1}10]$ channel orientation. Again, the mobility enhancement due to stress/strain is larger than the ON current enhancement for the same stress/strain. For example, for the (001) surface and the [110] channel orientation the mobility enhancement factor due to compressive uniaxial stress is about 2.3, but the ON current enhancement factor due to the same stress is only about 1.5; for the (110) surface and $[\bar{1}10]$ channel orientation the mobility enhancement factor due to compressive uniaxial stress is

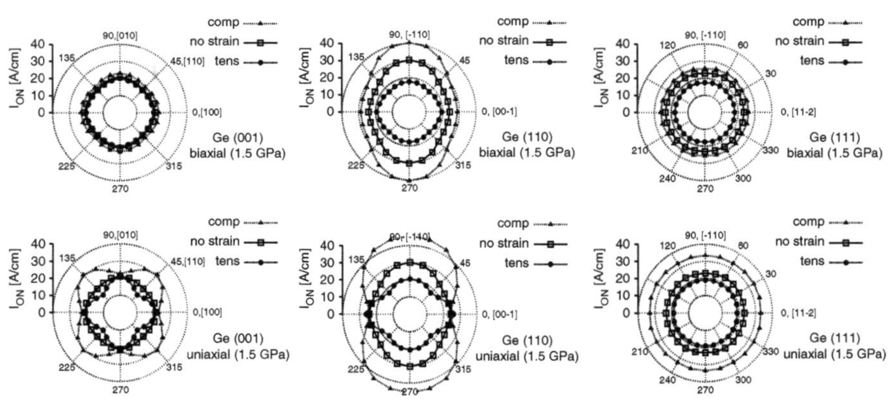

Fig. 12.13 Same as Fig. 12.12, but for a Ge double gate PMOSFET

about 1.9 whereas the ON current enhancement factor for the same stress is only about 1.4. This implies that the short channel effects and the non-linear transport effects influence the ON current in Ge devices as well.

$T_{\mathrm{delay}}(f_{\mathrm{TE}})$ is a figure of merit for the high frequency response of a MOSFET. A lower T_{delay} (higher f_{TE}) is a strong indicator for a better high frequency performance. The intrinsic delay T_{delay} is defined by

$$T_{\mathrm{delay}} = 2\pi \left.\frac{\delta Q_{\mathrm{G}}/\delta V_{\mathrm{GS}}}{\delta I_{\mathrm{ON}}/\delta V_{\mathrm{GS}}}\right|_{V_{\mathrm{DS}}=\mathrm{const.}} \tag{12.2}$$

where δQ_{G} and δI_{ON} are the small signal variations of the gate charge and the ON current caused by a small signal variation δV_{GS} under DC conditions. $T_{\mathrm{delay}} = 1/f_{\mathrm{TE}}$, where f_{TE} is the transit frequency extrapolated assuming that the frequency dependence of the small signal current gain is proportional to $1/f$. The dependence of intrinsic delay on the channel direction is shown in Fig. 12.14 for the Si devices. The intrinsic delay for the (110) surface is lower than that for the (001) or (111) surfaces and the intrinsic delay for compressive stress/strain is lower than that for tensile stress/strain. For Si devices the lowest intrinsic delay is obtained for the device with compressive biaxial strain, a (110) surface and a $[\bar{1}10]$ channel direction. Changing to the (110) surface, the $[\bar{1}10]$ channel orientation and compressive uniaxial stress for a maximum I_{ON} instead of compressive biaxial strain the intrinsic delay increases by about 22%.

The results for the intrinsic delay are shown in Fig. 12.15 for Ge devices. In general, Ge devices yield lower intrinsic delays than Si devices for the same stress/strain conditions and the same surface/channel orientations. Again, the intrinsic delay for the (110) surface is lower than that for the (001) or (111) surfaces and the

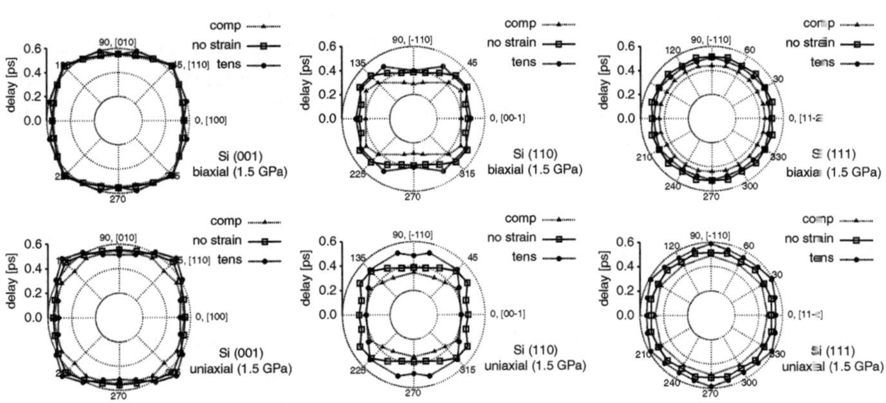

Fig. 12.14 Intrinsic delay as a function of channel direction for a biaxially strained *(top)* or uniaxially stressed *(bottom)* Si double gate PMOSFET. The surface orientations (001) *(left)*, (110) *(middle)*, and (111) *(right)* are considered. For the uniaxial stress case the stress direction is the same as the channel direction. $V_{\mathrm{GS}} = V_{\mathrm{DS}} = -0.7\,\mathrm{V}$

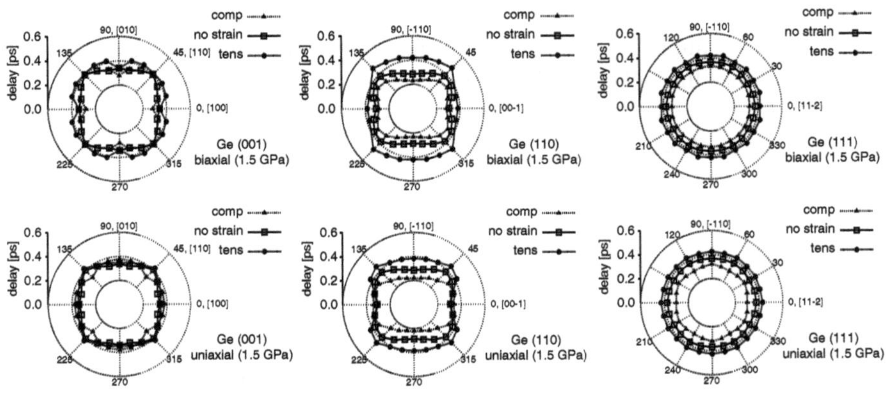

Fig. 12.15 Same as Fig. 12.14, but for a Ge double gate PMOSFET

intrinsic delay for compressive stress/strain is lower than that for tensile stress/strain. For Ge devices the lowest intrinsic delay (0.23 ps) is obtained for the device with compressive uniaxial stress, the (110) surface and the [$\bar{1}$10] channel orientation.

In summary, the optimum device with the highest ON current, a high mobility, the lowest intrinsic delay is the Ge double gate PMOSFET with compressive uniaxial stress, the (110) surface and the [$\bar{1}$10] *channel orientation. Compared to the reference Si double gate PMOSFET without strain and with the traditional (001) surface and [110] channel orientation, the optimum device increases the ON current by a factor of about 4.5, increases the mobility by a factor of about 8, and reduces the intrinsic delay by a factor of about 2.7.*

Please note that the OFF state leakage, which is very important for a high performance PMOSFET as well, has not been evaluated in this work. The bad news is that the OFF state leakage of the Ge devices discussed above is about 100 times higher than the OFF state leakage of Si devices with same stress/strain condition and surface/channel orientation. This is, for example, shown in the paper by Krishnamohan et al. [25]. The reason for this bad OFF state performance is the much lower band gap of Ge compared to Si.

12.2.3 Other Results

In the following, simulation results for other bias conditions are presented. The simulation results are for the reference Si device (unstrained Si double gate PMOSFET with the (001) surface and the [110] channel orientation) and a high performance Ge device (uniaxially stressed Ge double gate PMOSFET with the (110) surface and the [$\bar{1}$10] channel orientation, the stress direction is parallel to channel direction). Please note that the Ge device yields the highest ON state current as shown in the previous section.

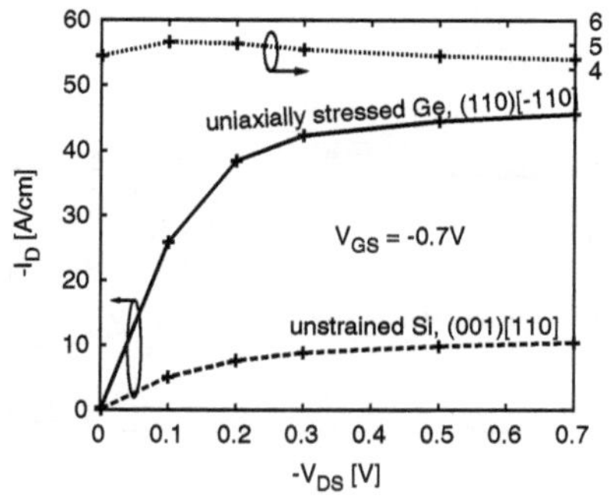

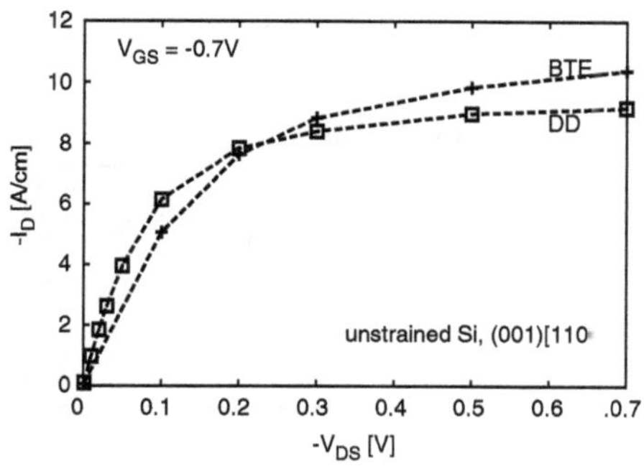

Fig. 12.16 *(Left:)* $I_D(V_{DS})$ characteristics for the unstrained Si double gate reference PMOSFET with (001) surface/[110] channel and the uniaxially stressed Ge device with (110) surface/$[\bar{1}10]$ channel (the stress direction is parallel to the channel direction). The simulations are performed for V_{DS} with *dots*. *(Right:)* $I_D(V_{DS})$ characteristic for the unstrained Si double gate reference PMOSFET ((001) surface/[110] channel) evaluated with the classical drift-diffusion model (DD) [26] and with the new device simulator based on the solution of SE-BTE-PE. $V_{GS} = -0.7\,\mathrm{V}$

Figure 12.16 (left) shows the $I_D(V_{DS})$ characteristic for a fixed $V_{GS} = -0.7\,\mathrm{V}$ of the reference Si device and the high performance Ge device. The drain current of the Ge device is always higher than the drain current of the Si reference device over the full range of V_{DS} ($0\,\mathrm{V} \geq V_{DS} \geq -0.7\,\mathrm{V}$). The ratio of the drain currents of the two devices is almost constant, even near equilibrium transport. For $V_{DS} = -1\,\mathrm{mV}$, the I_D ratio ($\approx$4.5) is much smaller than the respective homogenous channel mobility enhancement factor ($\approx$8 as mentioned before). This confirms that the short channel effects are very important for the devices with gate lengths in the range 16 nm. Figure 12.16 (right) clearly shows that the classical drift-diffusion model[1] results in an $I_D(V_{DS})$ characteristic which differs significantly from the result with the new simulator based on the solution of SE-BTE-PE. For a low $|V_{DS}|$, the DD model overestimates the drain current and for a higher $|V_{DS}|$ the DD model underestimates the drain current [27]. This is due to the fact that the DD model can not capture all details of the transport at a microscopic level in a nanoscale MOSFET [28].

The average electrostatic potential $\overline{V}(x)$ is defined as: $\overline{V}(x) = 1/N_{\mathrm{inv}}(x) \int V(x,z) \times p(x,z)\mathrm{d}z$. For the same $V_{GS} = -0.7\,\mathrm{V}$ as in the $I_D(V_{DS})$ plot (Fig. 12.16 (left)), the averaged electrostatic potential is shown in Fig. 12.17 for the two devices. For increasing $|V_{DS}|$, the potential difference $|\overline{V}(x_D) - \overline{V}(x_S)|$ increases and the

[1]The results based on the classical DD model shown here are simulated with the GALENE device simulator [26] based on the self-consistent solutions of the constitutive equation for current density, the current continuity equation, and Poisson equation. Quantum corrections are excluded from these simulations. The gate work function difference is adjusted such that the threshold voltage is the same as the one in the multisubband device simulations. A constant mobility of 93 cm^2/Vs, which is consistent with the homogenous channel low-field mobility for $N_{\mathrm{inv}} = 1.3 \times 10^{13}\,\mathrm{cm}^{-2}$, is assumed. For the mobility reduction due to a limiting drift velocity, the saturated drift velocity of holes in bulk Si ($v_{\mathrm{sat}} = 9.5 \times 10^6\,\mathrm{cm\,s}^{-1}$) is assumed. The band gap narrowing due to the heavy doping is assumed to be negligible.

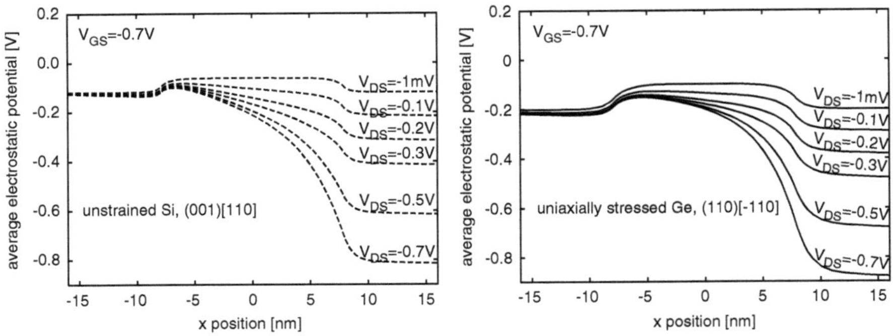

Fig. 12.17 Average electrostatic potential for the unstrained Si double gate reference PMOSFET ((001) surface/[110] channel) *(left)* and the uniaxially stressed Ge device with the (110) surface and the $[\bar{1}10]$ channel orientation (the stress direction is parallel to the channel direction)*(right)*. $V_{GS} = -0.7\,V$

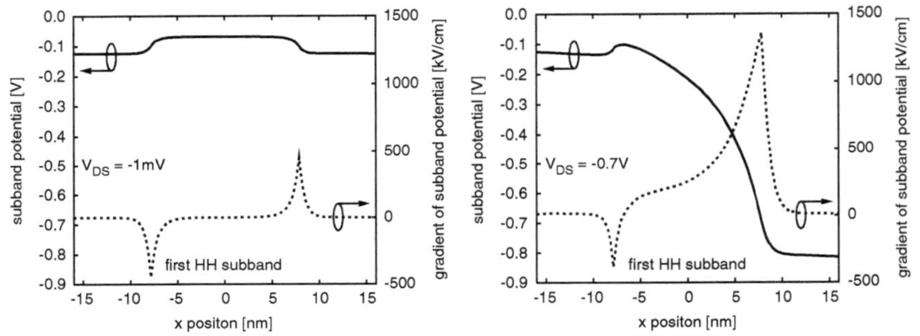

Fig. 12.18 First HH subband potential as a function of position x and its gradient for unstrained Si double gate reference PMOSFET with (001) surface/[110] channel orientation. $V_{GS} = -0.7\,V$. $V_{DS} = -1\,mV$ *(left)* and $V_{DS} = -0.7\,V$ *(right)*

gradient of $\overline{V}(x)$ in the center of the channel increases as well. The positions x_D and x_S along the transport direction x can be derived from Fig. 12.7.

The first HH subband potential which is proportional to the first HH subband potential energy $\varepsilon_{pot}^{\nu=1}(x)$ and the gradient of this potential are shown in Fig. 12.18 for $V_{GS} = -0.7\,V$ and $V_{DS} = -1\,mV$ (left) or $V_{DS} = -0.7\,V$ (right) and the Si device. The shape of the first HH subband potential is similar to the shape of $\overline{V}(x)$ shown in Fig. 12.17 (left). This is due to the high occupation of the first HH subband compared to the other subbands in the strong inversion regime with high $|V_{GS}|$. The gradients of the first HH subband potential entering the drift term of the BTE are very different for $V_{DS} = -1\,mV$ and $V_{DS} = -0.7\,V$ as shown in Fig. 12.18. Near equilibrium ($V_{DS} = -1\,mV$), the subband potential gradient of the first HH subband almost vanishes around the center of the channel ($x = 0\,nm$). This is consistent with a vanishing drain current for $V_{DS} = -1\,mV$ as shown in the $I_D(V_{DS})$ plot. For $V_{DS} = -0.7\,V$ the gradient of the first HH subband potential increases significantly

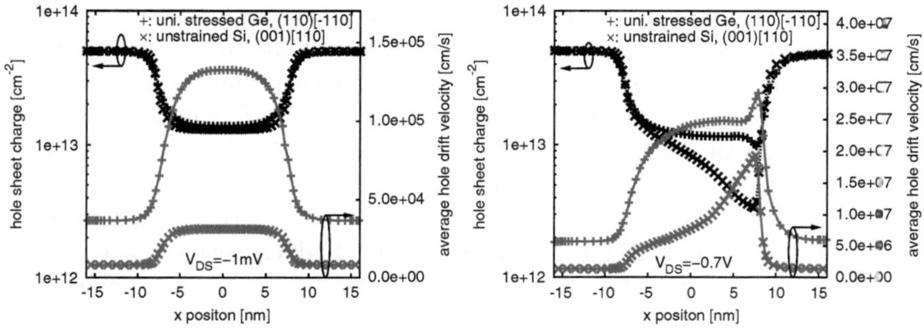

Fig. 12.19 Hole sheet charge and average drift velocity along the transport direction x of the unstrained Si double gate reference PMOSFET ((001) surface/[110] channel) and the uniaxially stressed Ge double gate PMOSFET ((110) surface/[$\bar{1}$10] channel, the stress direction is parallel to the channel direction). $V_{GS} = -0.7$ V. $V_{DS} = -1$ mV *(left)* and $V_{DS} = -0.7$ V *(right)*

around the center of the channel. The local gradient of the first HH subband potential at $x = 0$ nm reaches about 250 kV cm^{-1} and the transport situation is therefore far away from equilibrium. This results in much more drain current compared to the case with $V_{DS} = -1$ mV, consistent with the $I_D(V_{DS})$ plot. Moreover, for such a non-linear transport situation one hardly sees a range in the channel where the subband potential gradient is constant. Within the channel, the subband potential gradient increases dramatically from about -400 kV cm^{-1} at the source side to about 1.4 MV cm^{-1} at the drain side. This situation is typical for a short channel device with 16 nm gate length but hardly occurs in a long channel device.

In order to understand better why the drain current in the Ge devices is much higher, the hole sheet charge and the average hole drift velocity[2] are plotted together in Fig. 12.19 along the transport direction x in the Si and Ge devices. Near equilibrium ($V_{DS} = -1$ mV) the hole sheet charges of the two devices are nearly the same, whereas the average hole drift velocity is much higher for the Ge device. This is the reason for the large enhancement of the drain current in Ge devices near equilibrium. For $V_{DS} = -1$ mV, Fig. 12.19 (left) shows that the average hole drift velocity is almost constant around the center of the channel and very small (about 3×10^4 cm s^{-1} for Si and 1.3×10^5 cm s^{-1} for Ge). This is consistent with the constant and small subband potential gradient of the first HH subband around the center of the channel as shown in Fig. 12.18 (left) for the same V_{DS}. For a higher $|V_{DS}|$, Fig. 12.19 (right) shows that the hole sheet charge and the hole drift velocity curves change compared to the near equilibrium case and that the results for Si and Ge are very different even for the sheet charges. The hole drift velocity for the Ge device is much larger than that for the Si device. The hole

[2]The average hole drift velocity is defined as: $\langle v \rangle_x(x) = 1/N_{\text{inv}}(x) 1/(2\pi)^2 \sum_\nu \int v_x^\nu(x, \mathbf{k}) f^\nu(x, \mathbf{k}) d^2k$ where v_x^ν is the subband group velocity in the channel direction x and f^ν the subband distribution function.

sheet charge for the Ge device is also higher compared to the Si case, which can be clearly seen in the channel region. These two observations provide the underlying reasons for the much high drain current of the Ge device compared to the Si case, observable in the $I_D(V_{DS})$ plot (Fig. 12.16 (left)). Moreover, in the channel region the dependence of the hole drift velocity on x is clearly observable. For the Si device, this dependence agrees well with the dependence of the subband potential gradient on x within the channel region shown in Fig. 12.18 (right). The peak of the hole drift velocity is about $1.9 \times 10^7\,\mathrm{cm\,s^{-1}}$ for the Si device, which is much larger than the maximum hole drift velocity of about $9 \times 10^7\,\mathrm{cm\,s^{-1}}$ in an infinite long channel unstrained Si PMOS with (001) surface and the [110] channel orientation. This special velocity overshoot is due to the quasi ballistic transport effect [29]. Due to the limited flight time within such very short channels there exists a certain number of carriers that do not experience scattering during these flight time in the channel region. This establishes the quasi ballistic effect and leads to an increased carrier velocity compared to the long channel case. The simulation results shown here therefore confirm the existence of the quasi ballistic phenomenon in such short channel devices.

The $I_D(V_{GS})$ characteristics for fixed $V_{DS} = -0.7\,\mathrm{V}$ are shown in Fig. 12.20 (left) for the two devices. Again, the drain current of the Ge device is always higher than the drain current of the Si device for the full range of V_{GS}.

The average electrostatic potential for fixed $V_{DS} = -0.7\,\mathrm{V}$ is shown in Fig. 12.20 (right). For $V_{GS} = -0.1\,\mathrm{V}$, the barrier between the source and the channel is shown to be very high (about 0.2 V for the Si device and 0.3 V for the Ge device). This prevents the carriers from being injected into the channel region. This is the reason for the very small drain current with such a small $|V_{GS}|$ observed in the $I_D(V_{GS})$ plot (Fig. 12.20 (left)). For increasing $|V_{GS}|$, this barrier reduces (e.g. only about 0.02 V are left for $V_{GS} = -0.7\,\mathrm{V}$) and more carriers are injected into the channel. As a consequence, the drain current increases. This is again observable for higher $|V_{GS}|$ in the $I_D(V_{GS})$ plot.

Channel conductance near equilibrium is studied next. Near equilibrium (small $|V_{DS}|$) for a constant V_{GS} the channel conductance g_D is defined as $g_D = I_D/V_{DS}$ for a short channel PMOSFET device and as $\sigma_{D,h} = I_D/E_{||}$ for a homogenous channel PMOS structure where $E_{||}$ is the lateral electric field applied along the homogenous channel. Figure 12.21 shows the $g_D(V_{GS})$ characteristic for $V_{DS} = -0.1\,\mathrm{mV}$ and the two devices. The results show that the channel conductance of the Ge device is always higher than the channel conductance of the Si device, which is consistent with the larger drain current for the Ge device shown in Fig. 12.16 (left). For the homogenous case, the $\sigma_{D,h}$ enhancement in the Ge case is consistent with the mobility enhancement for the uniaxially stressed Ge PMOS structure as mentioned before in Sect. 12.2.2. The drift-diffusion approximation results in [28]:

$$g_D \approx g_D^{DD} = qW \left\{ \int \left[N_{\mathrm{inv}}(x)\mu(x) \right]^{-1} \mathrm{d}x \right\}^{-1}. \qquad (12.3)$$

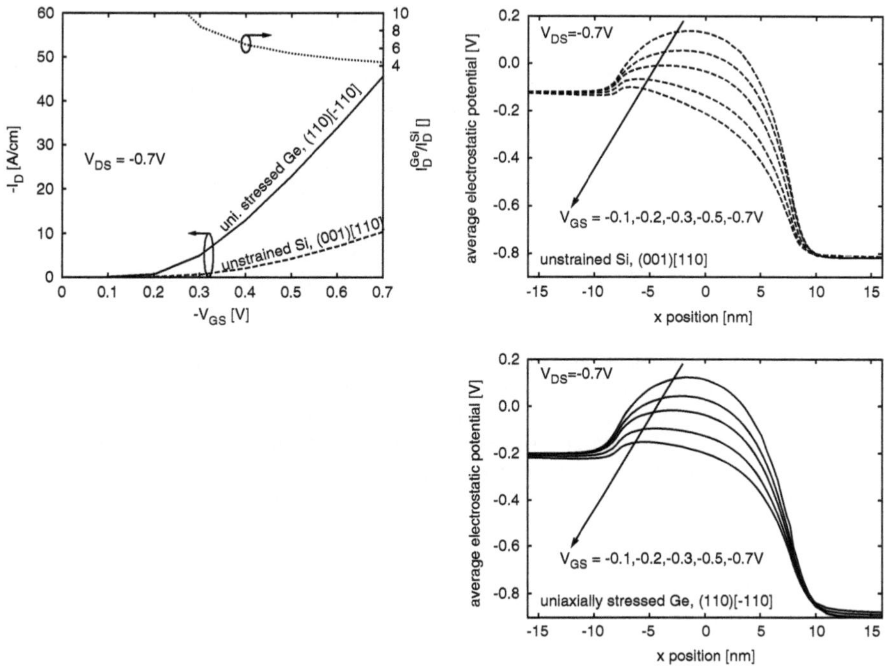

Fig. 12.20 $I_D(V_{GS})$ characteristic *(left)* and average electrostatic potential *(right)* for the unstrained Si double gate reference PMOSFET ((001) surface/[110] channel) and the uniaxially stressed Ge device ((110) surface/[$\bar{1}$10] channel, the stress direction is parallel to the channel direction.). $V_{DS} = -0.7\,\mathrm{V}$

Here, W is the channel width and $\mu(x)$ the local effective mobility along the transport direction where $\mu(x)$ for each position x is calculated using the Kubo-Greenwood formula [30] with a local Fermi level. The result holds for non-degenerate and degenerate cases. The integration (12.3) is performed along the channel from the source to the drain. Figure 12.21 (right) clearly shows that for a homogenous channel PMOS structure, the channel conductances evaluated either based on the solution of the BTE or based on the drift-diffusion approximation are the same. However, in a short channel device the drift-diffusion approximation fails (Fig. 12.21 (left)). The channel conductance evaluated by the drift-diffusion approximation always overestimates the exact channel conductance resulting from the solution of the BTE. The deviation between the two results is very large for the Ge device (e.g. about an order of magnitude for $V_{GS} = -0.2\,\mathrm{V}$). For the Si device the deviation is smaller (typical by a factor of 2 [31]). The reason for the failure of the drift-diffusion approximation which was analyzed in [28] is due to the large gradient of the built-in potential near the junctions between source or drain and the channel. This large gradient is clearly observable in Fig. 12.18 (left). The gradient of the first HH subband potential can reach $400\,\mathrm{kV\,cm^{-1}}$ at the mentioned interfaces. Short channel effects therefore play an important role even in a near equilibrium transport situation. In the homogenous channel case, however, the gradient of the subband

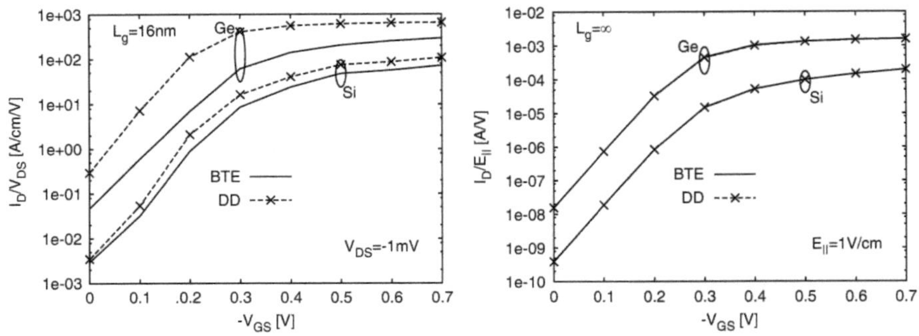

Fig. 12.21 $g_D(V_{GS})$ characteristic evaluated based on the solution of the BTE or based on the drift-diffusion (DD) model (12.3) for the two short channel double gate PMOSFETs *(left)* and for a homogenous channel double gate stack with the same vertical structure *(right)*. Two cases (unstrained Si ((001) surface/[110] channel) and uniaxially stressed Ge ((110) surface/[$\bar{1}$10] channel, the stress direction is parallel to the channel direction.)) are considered. $V_{DS} = -1\,\text{mV}$ for the short channel devices and $E_{||} = 1\,\text{V/cm}$ for homogenous channel cases

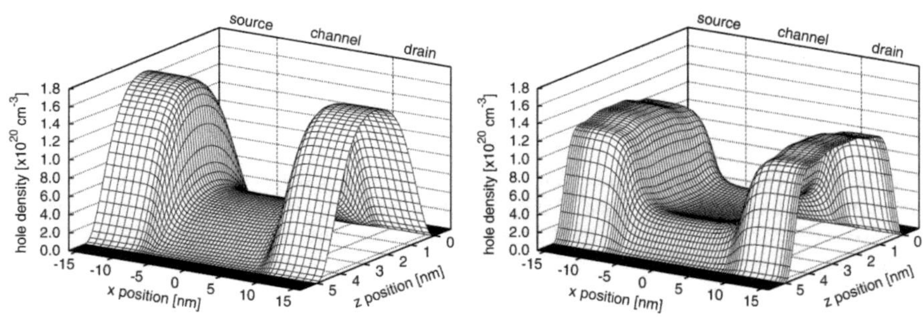

Fig. 12.22 Spatial hole density distribution within the unstrained Si double gate reference PMOSFET ((001) surface/[110] channel) *(left)* and within the uniaxially stressed Ge double gate PMOSFET ((110) surface/[$\bar{1}$10] channel, the stress direction is parallel to the channel direction) *(right)*. $V_{\text{GS}} = -0.7\,\text{V}$, $V_{\text{DS}} = -1\,\text{mV}$

potential is constant and equals to $E_{||}$. Thus, the drift-diffusion approximation works well in this case (Fig. 12.21 (right)).

The spatial hole density distribution within the Si device (Fig. 12.22 (left)) and within the Ge device (Fig. 12.22 (right)) (for $V_{\text{GS}} = -0.7\,\text{V}$ and $V_{\text{DS}} = -1\,\text{mV}$) clearly shows that size quantization is considered not only for the channel region but also for the source and drain regions. The hole density distribution for the two devices differ significantly.

12.3 Magneto Transport Effects

Magnetotransport measurements can be used to characterize the inversion layer transport in MOSFETs [32,33]. To this end, the Hall factor or the magneto resistance mobility are usually measured. In this section the magneto transport results in terms of Hall factor and magneto resistance mobility are presented for a homogeneous channel PMOS structure or for a short channel PMOSFET.

The simulation of magneto transport effects in hole inversion layers based on the self-consistent solution of the $\mathbf{k} \cdot \mathbf{p}$ SE-PE-BTE system is presented in this section. The quasi 2D hole gas subband structure is calculated by solving the 6×6 $\mathbf{k} \cdot \mathbf{p}$ SE self-consistently with the confining electrostatic potential. Based on the calculated subband structure, the transport of holes within the channel is investigated by solving the stationary BTE. A constant magnetic field of arbitrary field strength perpendicular to the substrate surface is considered. The Hall factor and the magneto resistance mobility are calculated based on the solution of BTE for small driving electric fields.

An important assumption made in this simulation approach should be noticed: *the Landau quantization effects [34–36] caused by the magnetic field can be neglected.* This assumption has been implicitly used in the simulation approach of this work because the magnetic field does not influence the SE, but it influences the BTE by the Lorentz force ((9.2)–(9.4)). This assumption is certainly questionable for a transport situation with a very low lattice temperature and a very strong magnetic field where the Landau quantization effects are measurable [37]. In this work, the magneto transport is considered for rather high lattice temperatures ($T \geq 20$ K) only. For these temperatures the neglect of the Landau quantization can be shown to be still reasonable [35].

At first the dependence of the low-field mobility tensor on the magnetic field is investigated. Based on the solution of the BTE for a small electric field $\delta\mathbf{E}$ applied along the channel direction and a constant magnetic field B perpendicular to the transport plane, the drift velocities $\langle v \rangle_1$ and $\langle v \rangle_2$ can be calculated ($1 \equiv x_\mathrm{D}$ is channel direction, $2 \equiv y_\mathrm{D}$ is the second axis in the transport plane perpendicular to 1). Please note that a non-zero B drives a current in the transverse direction 2. Therefore, $\langle v \rangle_2$ is non-zero. In the transport plane the drift velocities and electric field $\delta\mathbf{E}$ are related via a 2×2 low-field mobility tensor $\hat{\mu}$ as follows:

$$\begin{pmatrix} \langle v \rangle_1 \\ \langle v \rangle_2 \end{pmatrix} = \begin{pmatrix} \mu_{11} & \mu_{12} \\ \mu_{21} & \mu_{22} \end{pmatrix} \begin{pmatrix} \delta E \\ 0 \end{pmatrix}. \tag{12.4}$$

Here, $\mu_{11} = \langle v \rangle_1 / \delta E$ and $\mu_{21} = \langle v \rangle_2 / \delta E$. In order to determine the other two components μ_{12} and μ_{22}, the channel direction is rotated and the BTE is solved again for the new coordinate system $(1', 2')$ with $1' \equiv 2$, $2' \equiv -1$. As soon as the solution of the BTE for the new coordinate system is available, one can evaluate $\mu_{22} = \mu_{1'1'} = \langle v \rangle_{1'} / \delta E$ and $\mu_{12} = \mu_{2'1'} = \langle v \rangle_{2'} / \delta E$. For a non-zero magnetic field, all components of $\hat{\mu}$ are non-zero. Due to the fundamental symmetry of the subband structure $k^\nu(\varepsilon, \phi + \pi) = k^\nu(\varepsilon, \phi)$ which holds for all surface orientations and all stress/strain conditions as mentioned in Sect. 7.5, the mobility tensor fulfills the anti-symmetry property: $\mu_{12} = -\mu_{21}$. In general $\mu_{11} \neq \mu_{22}$. However, for a fourfold symmetry of the subband structure (e.g. (001) surface orientation without strain or with biaxial strain) the two main-diagonal components of the mobility tensor are identical.

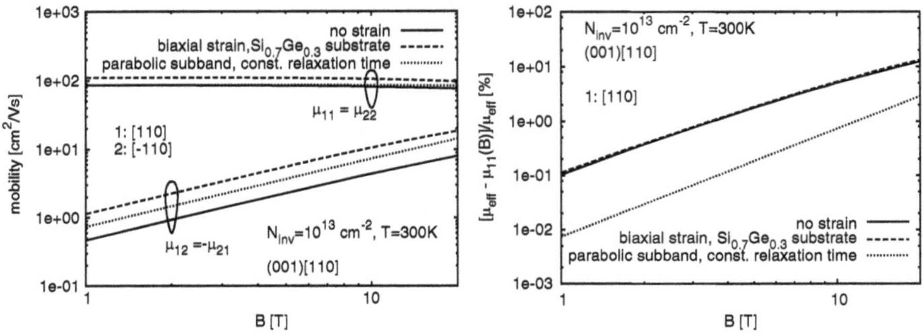

Fig. 12.23 *(Left)* Low field mobility tensor components vs. magnetic field. *(Right)* Relative reduction of the channel mobility μ_{11} vs. magnetic field. Unstrained or biaxially strained Si homogenous channel bulk PMOS structures with (001) surface and [110] channel orientation are simulated. Results obtained by using the analytical theory based on the assumption of a parabolic subband are included. $N_{\mathrm{inv}} = 10^{13}\,\mathrm{cm}^{-2}$. $T = 300\,\mathrm{K}$

Figure 12.23 (left) shows the mobility tensor components as a function of the magnetic field for an unstrained or biaxially strained Si homogenous channel bulk PMOS structure with (001) surface and [110] channel orientation. The biaxial strain is assumed to be consistent with a pseudomorphic growth of Si on a relaxed $Si_{0.7}Ge_{0.3}$ virtual substrate. The simulations are performed for an inversion charge of $10^{13}\,\mathrm{cm}^{-2}$ at room temperature. The main diagonal components of mobility tensor are the same in this case due to the fourfold symmetry of the subband structure. μ_{11} ($= \mu_{22}$) is shown to reduce with increasing magnetic field. $\mu_{11}(B)$ for the biaxially strained PMOS structure is higher than $\mu_{11}(B)$ for the unstrained PMOS structure. This is consistent with the fact that μ_{eff} in a biaxially strained Si hole inversion layer is higher than the μ_{eff} in an unstrained Si PMOS structure for zero magnetic field. Here, $\mu_{\mathrm{eff}} = \mu_{11}(B = 0)$ denotes the low-field channel mobility without a magnetic field. The off-diagonal mobility component $\mu_{12}(= -\mu_{21})$ shows the opposite trend: μ_{12} increases with increasing magnetic field. Moreover, $\mu_{12}(B)$ for biaxial strain structure is much higher than $\mu_{12}(B)$ for the unstrained case. Simulation results assuming only one analytical parabolic subband and the validity of the relaxation time approximation with a constant relaxation time are included in Fig. 12.23 (left). To get these results the subband structure is analytically expressed with an effective mass and the relaxation time is evaluated under the condition that μ_{eff} reproduces the effective mobility resulting from the solution of the BTE with the realistic subband and the full scattering integral of the unstrained Si PMOS structure. The results show that the $\mu_{11}(B)$ curve resulting from the simplifying assumption is rather similar to the exact $\mu_{11}(B)$ characteristic for the unstrained Si PMOS structure. However, the $\mu_{12}(B)$ curve resulting from the simplifying assumption is higher than the exact one for the unstrained Si PMOS structure. For all three cases, oscillations do not occur in the $\mu_{ij}(B)$ curves even for B up to 20 T or higher because the Landau quantization is excluded from the simulations.

The relative reduction of μ_{11} compared to μ_{11} without B is shown in Fig. 12.23 (right) for different levels of B up to 20 T. The $\Delta\mu_{11}(B)$ characteristics for unstrained or biaxially strained Si bulk PMOS are quite similar. However, the $\Delta\mu_{11}(B)$ curve resulting from the simplifying assumption is much lower. For all three cases, $\Delta\mu_{11}$ increases quadratically with increasing B for $B < 4$ T. For a higher B the quadratic behavior of the $\Delta\mu_{11}(B)$ characteristic can not be observed (visible in Fig. 12.23 (right) for unstrained or biaxially strained Si PMOS and $B > 10$ T).

The magneto resistance mobility μ_{MR} is defined based on $\Delta\mu_{11}(B)$ as follows [31–33]:

$$\mu_{MR}(B) := \sqrt{\frac{\mu_{eff} - \mu_{11}(B)}{\mu_{eff}}} \cdot \frac{1}{B}, \qquad \mu_{MR,0} := \lim_{B\to 0} \mu_{MR}(B). \tag{12.5}$$

In literature the approximation $\mu_{MR,0} \approx \mu_{eff}$ is often used [32, 33]. This is the fundamental assumption for the low-field mobility extraction with the magneto resistance technique [32, 33]. However, it can be shown that the equality $\mu_{MR,0} = \mu_{eff}$ is only correct under very special conditions: (1) the subband structure is parabolic and isotropic and at the same time (2) for the scattering integral of the BTE a relaxation time approximation with a constant relaxation time must hold. For a realistic PMOS structure the subband structure is non-parabolic and fully anisotropic (Figs. 8.4–8.8) and the scattering integral of the BTE can in general not be expressed using a constant relaxation time [24], therefore *both (1) and (2) are violated.* Consequently, for a realistic PMOS structure $\mu_{MR,0}$ and μ_{eff} should be different.

Figure 12.24 (left) shows μ_{MR} compared to μ_{eff} for two realistic PMOS structures and the special case with parabolic subband and relaxation time approximation with a constant relaxation time. For this special case with parabolic subband and constant relaxation time approximation, the simulation results clearly show that

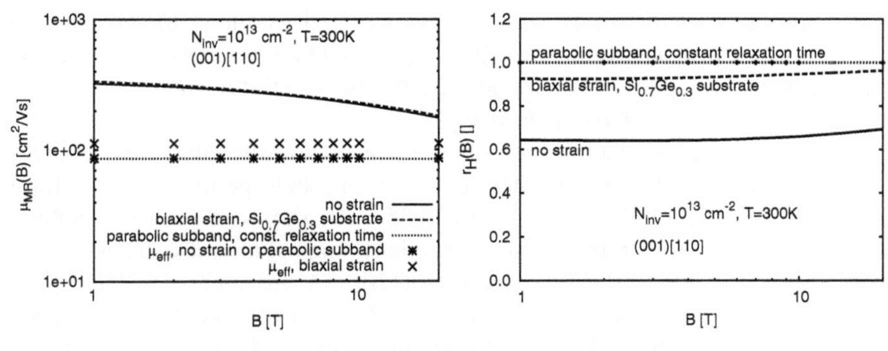

Fig. 12.24 *(Right)* Magneto resistance mobility extracted with different levels of B compared to the effective channel mobility μ_{eff}. *(Left)* Hall factor extracted with different levels of B. Simulation conditions are the same as in Fig. 12.23

$\mu_{\mathrm{MR},0} = \mu_{\mathrm{eff}}$, which confirms the theoretical analysis. For the two realistic PMOS structures, the simulation results show that μ_{MR} substantially overestimates the effective channel mobility μ_{eff}. μ_{MR} for unstrained or biaxially strained Si PMOS structures is rather similar due to the small deviation of the respective $\Delta\mu_{11}(B)$ curves shown in Fig. 12.23 (right).

For the Hall factor calculation, the Hall field $\mathbf{E}_{\mathrm{H}}$ must be determined. Due to the presence of B perpendicular to the wafer plane, $\mathbf{E}_{\mathrm{H}}$ is parallel to the transport plane and is oriented in the direction 2 ($\mathbf{e}_{y_{\mathrm{D}}}$). The Hall field is determined based on the assumption that the current in the direction 2 vanishes. This implies $\langle v \rangle_2 = 0$ and leads with (12.4) to the following equation for the evaluation of E_{H}:

$$\begin{pmatrix} \langle v \rangle_1^{\mathrm{H}} \\ 0 \end{pmatrix} = \begin{pmatrix} \mu_{11} & \mu_{12} \\ \mu_{21} & \mu_{22} \end{pmatrix} \begin{pmatrix} \delta E \\ E_{\mathrm{H}} \end{pmatrix}. \tag{12.6}$$

This leads to $E_{\mathrm{H}} = -(\mu_{21}/\mu_{22})\delta E$. Moreover, the drift velocity component $\langle v \rangle_1^{\mathrm{H}}$ can also be evaluated: $\langle v \rangle_1^{\mathrm{H}} = [(\mu_{11}\mu_{22} - \mu_{12}\mu_{21})/\mu_{22}]\delta E$. The Hall factor is defined in terms of E_{H} and $\langle v \rangle_1^{\mathrm{H}}$ as follows:

$$r_{\mathrm{H}}(B) := \frac{E_{\mathrm{H}}}{\langle v \rangle_1^{\mathrm{H}} B} = -\frac{\mu_{21}}{\mu_{11}\mu_{22} - \mu_{12}\mu_{21}} \cdot \frac{1}{B}, \qquad r_{\mathrm{H},0} = \lim_{B \to 0} r_{\mathrm{H}}(B). \tag{12.7}$$

The calculation of r_{H} using (12.7) involves all components of the mobility tensor, which is more complicated than the calculation of μ_{MR} with (12.5). It should be mentioned that for the special case where only one parabolic, isotropic subband and the validity of the relaxation time approximation with a constant relaxation time is assumed (see (1) and (2) above), $r_{\mathrm{H},0} = 1$ holds. The analytical result is therefore confirmed by the simulation result based on the solution of the BTE as shown in Fig. 12.24 (right). This agreement is an important validation for the magneto transport model based on the solution of the BTE presented in this work. Fig. 12.24 (right) also shows the Hall factor for two realistic PMOS structures. These are different from 1.0. The Hall factor for biaxially strained structure is about 0.92 and only 0.63 for the structure without strain. This allows to conclude that the Hall factor depends strongly on the warped structure of the subbands (see the subband structures of the hole inversion layers with biaxial strain and without strain in Fig. 8.6 (center) and (left), respectively).

Magneto transport quantities are often measured for low lattice temperatures. To study this typical situation, the simulations are now performed for a lower lattice temperature of $T = 20\,\mathrm{K}$. Figure 12.25 (left) shows the channel mobility $\mu_{11}(B)$ characteristic for an unstrained, biaxially strained or uniaxially stressed Si bulk PMOS with the (001) surface and the [110] channel orientation and $N_{\mathrm{inv}} = 10^{13}\,\mathrm{cm}^{-2}$. The biaxial strain is consistent with the one in the simulations for 300 K done before (tensile strain, $\mathrm{Si}_{0.7}\mathrm{Ge}_{0.3}$ virtual substrate). For the uniaxial stress case, a uniform compressive 1.5 GPa stress is assumed along the transport direction [110]. The typical increase of the mobility for low lattice temperatures compared to the

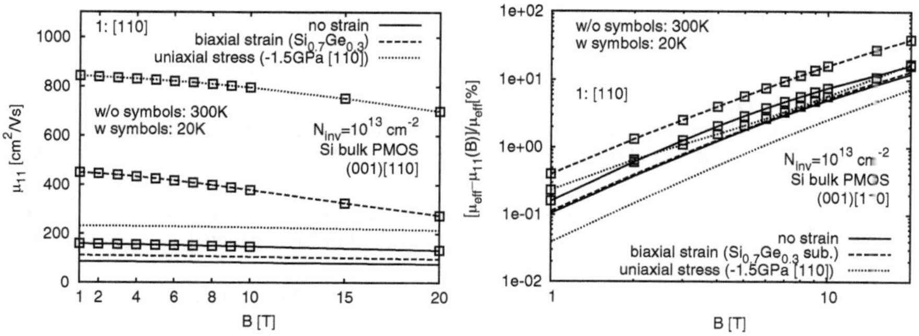

Fig. 12.25 *(Left)* Channel mobility μ_{11} vs. magnetic field. *(Right)* Relative reduction of channel mobility μ_{11} vs. magnetic field. Unstrained, biaxially strained or uniaxially stressed Si homogenous channel bulk PMOS structures are simulated. The (001) surface and the [110] channel orientation are considered. $N_{\text{inv}} = 10^{13}\,\text{cm}^{-2}$. $T = 300\,\text{K}$ or $T = 20\,\text{K}$ is considered

mobility at room temperature is visible for all three cases. Especially the hole mobility without B for the uniaxial stress case reaches $800\,\text{cm}^2/\text{Vs}$ for $T = 20\,\text{K}$ which is about an order of magnitude larger than the μ_{eff} of an unstrained Si structure at room temperature. For the same lattice temperature of either 20 or 300 K, the following relation holds: $\mu_{11}^{\text{no strain}}(B) < \mu_{11}^{\text{biaxial strain}}(B) < \mu_{11}^{\text{uniaxial stress}}(B)$. However, the structure with the highest μ_{eff} does not always result in the highest (relative) $\Delta\mu_{11}(B)$. This situation is illustrated in Fig. 12.25 (right). For $T = 300\,\text{K}$, the results show that $\Delta\mu_{11}(B)$ of the uniaxially stressed structure is much smaller compared to the other two cases with biaxial strain or without strain. For $T = 20\,\text{K}$, the highest $\Delta\mu_{11}(B)$ results for the biaxially strained structure. For this structure $\Delta\mu_{11}(B)$ is about 17% for $T = 20\,\text{K}$ and $B = 10\,\text{T}$ and $\Delta\mu_{11}(B) \approx 7\%$ for $T = 300\,\text{K}$ and the same B. Therefore, the best sensitivity of the current w.r.t the magnetic field is achieved for the biaxially strained case.

References

1. Wang, E.X., Matagne, P., Shifren, L., Obradovic, B., Kotlyar, R., Cea, S., Stettler, M. Giles, M.D.: Physics of hole transport in strained silicon MOSFET inversion layers. IEEE Trans. Electron Dev. **53**(8), 1840–1851 (2006)
2. Fischetti, M.V., Ren, Z., Solomon, P.M., Yang, M., Rim, K.: Six-band $k \cdot p$ calculation of the hole mobility in silicon inversion layers: Dependence on surface orientation, strain, and silicon thickness. J. Appl. Phys. **94**, 1079–1095 (2003)
3. Oberhuber, R., Zandler, G., Vogl, P.: Subband structure and mobility of two-dimensional holes in strained Si/SiGe MOSFETs. *Phys. Rev. B*, **58**, 9941–9948 (1998)
4. Fischetti, M.V., Laux, S.E.: Band structure, deformation potentials, and carrier mobility in strained Si, Ge and SiGe alloys. J. Appl. Phys., **80**, 2234–2252 (1996)
5. Takagi, S., Toriumi, A., Iwase, M., Tango, H.: On the universality of inversion layer mobility in Si MOSFET's: Part I–Effects of substrate impurity concentration. IEEE Trans. Electron Dev. **41**, 2357–2362 (1994)

6. Levenberg, K.: "A method for the solution of certain problems in least squares". Quart. Appl. Math. **2**, 164–168 (1944)
7. Marquardt, D.: An algorithm for least-squares estimation of nonlinear parameters. SIAM J. Appl. Math. **11**, 431–441 (1963)
8. Tsutsui, G., Hiramoto, T.: Mobility and threshold-voltage comaprision between (110)- and (100)-oriented ultrathin-body Silicon MOSFETs. IEEE Trans. Electron Dev. **53**(10), 2582–2588 (2006)
9. Chleirigh, C., Theodore, N., Fukuyama, H., Mure, S., Ehrke, H.-U., Domenicucci, A., Hoyt, J.L.: Thickness dependence of hole mobility in ultrathin SiGe-channel p-MOSFETs. IEEE Trans. Electron Dev. **55**(10), 2687–2694 (2008)
10. Uchida, K., Takagi, S.: Carrier scattering induced by thickness fluctuation of silicon-on-insulator film in ultrathin-body metal-oxide-semiconductor field-effect transistors. Appl. Phys. Lett. **82**(17), 2916–2918 (2003)
11. Jin, S., Fischetti, M.V., Tang, T.: Modeling of surface-roughness scattering in ultrathin-body SOI MOSFETs. *IEEE Trans. Electron Dev.* **54**(9), 2191–2203 (2007)
12. Sakaki, H., Noda, T., Hirakawa, K., Tanaka, M., Matsusue, T.: Interface roughness scattering in GaAs/AlAs quantum wells. Appl. Phys. Lett. **51**(23), 1934–1936 (1987)
13. Irie, H., Kita, K., Kyuno, K., Toriumi, A.: In-plane mobility anisotropy and universality under uni-axial strains in n- and p-MOS inversion layers on (100), (110), and (111) Si. In: IEDM Tech. Dig. 225–228 (2004)
14. Saitoh, M., Kobayashi, S., Uchida, K.: Physical understanding of fundamental properties of Si (110) pMOSFETs inversion-layer capacitance, mobility universality, and uniaxial stress effects . In: IEDM Tech. Dig. 711–714 (2007)
15. Gusev, E.P., Shang, H., Copel, M., Gribelyuk, M., D'Emic, C., Kozlowski, P., Zabel, T.: Microstructure and thermal stability of HfO_2 gate dielectric deposited on Ge(100). Appl. Phys. Lett. **85**(12), 2334–2336 (2004)
16. Chui, C.O., Ramanathan, S., Triplett, B..B, McIntyre, P.C., Saraswat, K.C.: Germanium MOS capacitors incorporating ultrathin high-κ gate dielectric. IEEE Electron Dev. Lett. **23**(8), 473–475 (2002)
17. Iwauchi, S., Tanaka, T.: Interface properties of Al_2O_3-Ge structure and characteristics of Al_2O_3-Ge MOS transistor. Jpn. J. Appl. Phys. **10**, 260–265 (1971)
18. Yu, D.S., Chiang, K.C., Cheng, C.F., Chin, A., Chunxiang, Z., Li, M.F., Kwong, D.-L.: Fully silicided NiSi:Hf-$LaAlO_3$/SG-GOI n-MOSFETs with high electron mobility. IEEE Electron Dev. Lett. **25**(8), 559–561 (2004)
19. Shang, H., Okorn-Schimdt, H., Ott, J., Kozlowski, P., Steen, S., Jones, E.C., Wong, H.-S.P., Hanesch, W.: Electrical characterization of Germanium p-channel MOSFETs. IEEE Electron Dev. Lett. **24**(4), 242–244 (2003)
20. Kuzum, D., Pethe, A.J., Krishnamohan, T., Saraswat, K.C.: Ge (100) and (111) N- and P-FETs with high mobility and low-T mobility characterization. IEEE Trans. Electron Dev. **56**(4),
648–655 (2009)
21. Xie, R., Phung, T.H., He, W., Sun, Z., Yu, M., Cheng, Z., Zhu, C.: High mobility high-κ/Ge pMOSFETs with 1 nm EOT – New concept on interface engineering and interface characterization. In: IEDM Tech. Dig. pp. 393–396 (2008)
22. Zhang, Y., Fischetti, M.V.: Calculation of hole mobility in Ge and III-V p-channels. In: Proceedings of IWCE, pp. 41–44 (2009)
23. Bufler, F.M.: Monte–carlo-simulation von verspannten Si/SiGe–strukturen. Vortrag bei Siemens, Bereich Halbleiter, München (1997)
24. Jungemann, C., Meinerzhagen, B.: Hierarchical Device Simulation: The Monte-Carlo Perspective. Computational Microelectronics. Springer, New York (2003)
25. Krishnamohan, T., Kim, D., Dinh, T.V., Pham, A.T., Meinerzhagen, B., Jungemann, C., Saraswat, K.: Comparison of (001), (110) and (111) Uniaxial- and Biaxial-Strained-Ge and Strained-Si PMOS DGFETs for all channel orientations: Mobility enhancement, drive current, delay and off-state leakage. In: IEDM Tech. Dig. pp. 899–902 (2008)

26. Engl, W.L., Dirks, H.K., Meinerzhagen, B.: Device modeling. In: Proceedings of IEEE, **71**, 10–33 (1983)
27. Pham, A.T., Jungemann, C., Nguyen, C.D., Meinerzhagen, B.: A semiempirical surface scattering model for quantum corrected monte-carlo simulation of unstrained Si and strained Si/SiGe PMOSFETs. Mater. Sci. Eng. B **135**, 224–227 (2006)
28. Jungemann, C., Grasser, T., Neinhüs, B., Meinerzhagen, B.: Failure of moments-based transport models in nanoscale devices near equilibrium. IEEE Trans. Electron Dev. **52**(11), 2404–2408 (2005)
29. Lundstrom, M., Ren, Z.: Essential physics of carrier transport in nanoscale MOSFETs. IEEE Trans. Electron Dev. **49**(1), 133–141 (2002)
30. Pham, A.T., Jungemann, C., Meinerzhagen, B.: Physics-based modeling of hole inversion layer mobility in strained SiGe on insulator. IEEE Trans. Electron Dev. **54**(9), 2174–2182 (2007)
31. Pham, A.T., Jungemann, C., Meinerzhagen, B.: Deterministic multisubband device simulations for strained double gate PMOSFETs including magnetotransport. In: IEDM Tech. Dig. 895–898 (2008)
32. Chaisantikulwat, W., Mouisa, M., Ghibaudoa, G., Gallonb, C., Fenouillet-Berangerc, C., Mauded, D.K., Skotnickib, T., Cristoloveanu, S.: Differential magnetoresistance technique for mobility extraction in ultra-short channel FDSOI transistors. Solid State Electron. **50**(4), 637–643 (2006)
33. Meziani, Y.M., Lusakowski, J., Teppe, F., Dyakonova, N., Knap, W., Romanjek, K., Ferrier, M., Clerc, R., Ghibaudo, G., Boeuf, F., Skotnicki, T.: Magnetoresistance mobility measurements in sub $0.1\mu m$ Si MOSFETs. In: Proceedings of ESSDERC, pp. 157–160 (2004)
34. Landau, L.D., Lifschitz, E.M.: Quantum Mechanics (Non-Relativistic Theory), 3rd edn. Pergamon Press, Oxford (1977)
35. Pham, A.T., Jungemann, C., Meinerzhagen, B.: Simulation of Landau quantization effects due to strong magnetic fields in (110) Si hole inversion layers. In: Proceedings of IWCE, 335–338 (2010)
36. Latussek, V., Bangert, E., Landwehr, G.: Self-consistent calculations of landau levels for symmetric p-type inversion layers. Annalen der Physik **503**(6), 394–414 (1991)
37. Takahashi, T., Yamahata, G., Ogi, J., Kodera, T., Oda, S., Uchida, K.: Direct observation of subband structures in (110) pMOSFETs under high magnetic field: Impact of energy split between bands and effective masses on hole mobility. IEDM Tech. Dig. pp. 470–480 (2009)

Appendix A

A.1 Discretization of the EFA SE in the Quantization Direction

This appendix provides the details about the discretization of the EFA SE in z' with the box integration method. At grid point j, the second, first and zero order derivatives in (7.48) are discretized by box integration:

$$\int_{j-\frac{1}{2}}^{j+\frac{1}{2}} -\hat{\mathscr{M}}_{\mathbf{kp}}^{(2)}(z') \frac{\partial^2 \boldsymbol{\zeta}(z')}{\partial z'^2} \mathrm{d}z' \rightarrow \int_{j-\frac{1}{2}}^{j+\frac{1}{2}} -\frac{\partial}{\partial z'} \left(\hat{\mathscr{M}}_{\mathbf{kp}}^{(2)}(z') \frac{\partial \boldsymbol{\zeta}(z')}{\partial z'} \right) \mathrm{d}z'$$

$$= -\frac{\hat{\mathscr{M}}_{\mathbf{kp}}^{(2)}{}_{,j-1} + \hat{\mathscr{M}}_{\mathbf{kp}}^{(2)}{}_{,j}}{2(z'_j - z'_{j-1})} \boldsymbol{\zeta}_{j-1} - \frac{\hat{\mathscr{M}}_{\mathbf{kp}}^{(2)}{}_{,j+1} + \hat{\mathscr{M}}_{\mathbf{kp}}^{(2)}{}_{,j}}{2(z'_{j+1} - z'_j)} \boldsymbol{\zeta}_{j+1}$$

$$+ \left[\frac{\hat{\mathscr{M}}_{\mathbf{kp}}^{(2)}{}_{,j-1} + \hat{\mathscr{M}}_{\mathbf{kp}}^{(2)}{}_{,j}}{2(z'_j - z'_{j-1})} + \frac{\hat{\mathscr{M}}_{\mathbf{kp}}^{(2)}{}_{,j+1} + \hat{\mathscr{M}}_{\mathbf{kp}}^{(2)}{}_{,j}}{2(z'_{j+1} - z'_j)} \right] \boldsymbol{\zeta}_j, \tag{A.1}$$

$$\int_{j-\frac{1}{2}}^{j+\frac{1}{2}} -i \hat{\mathscr{M}}_{\mathbf{kp}}^{(1)}(z') \frac{\partial \boldsymbol{\zeta}(z')}{\partial z'} \mathrm{d}z'$$

$$\rightarrow -\frac{i}{2} \hat{\mathscr{M}}_{\mathbf{kp}}^{(1)}{}_{,j} \int_{j-\frac{1}{2}}^{j+\frac{1}{2}} \frac{\partial \boldsymbol{\zeta}(z')}{\partial z'} \mathrm{d}z' - \frac{i}{2} \int_{j-\frac{1}{2}}^{j+\frac{1}{2}} \frac{\partial \hat{\mathscr{M}}_{\mathbf{kp}}^{(1)}(z') \boldsymbol{\zeta}(z')}{\partial z'} \mathrm{d}z'$$

$$= \frac{i}{4} \left[\hat{\mathscr{M}}_{\mathbf{kp}}^{(1)}{}_{,j} + \hat{\mathscr{M}}_{\mathbf{kp}}^{(1)}{}_{,j-1} \right] \boldsymbol{\zeta}_{j-1} - \frac{i}{4} \left[\hat{\mathscr{M}}_{\mathbf{kp}}^{(1)}{}_{,j} + \hat{\mathscr{M}}_{\mathbf{kp}}^{(1)}{}_{,j+1} \right] \boldsymbol{\zeta}_{j+1}, \tag{A.2}$$

S.-M. Hong et al., *Deterministic Solvers for the Boltzmann Transport Equation*, Computational Microelectronics, DOI 10.1007/978-3-7091-0778-2,

$$\int_{j-\frac{1}{2}}^{j+\frac{1}{2}} \hat{\mathbf{H}}'^{(0)}(z')\boldsymbol{\zeta}(z')\mathrm{d}z' \rightarrow \int_{j-\frac{1}{2}}^{j} \hat{\mathbf{H}}'^{(0)}(z')\boldsymbol{\zeta}(z')\mathrm{d}z' + \int_{j}^{j+\frac{1}{2}} \hat{\mathbf{H}}'^{(0)}(z')\boldsymbol{\zeta}(z')\mathrm{d}z'$$

$$= \frac{1}{4}\left[(z'_j - z'_{j-1})\left(\hat{\mathbf{H}}'^{(0)}{}_{j-1} + \hat{\mathbf{H}}'^{(0)}{}_{j}\right) + (z'_{j+1} - z'_j)\left(\hat{\mathbf{H}}'^{(0)}{}_{j+1} + \hat{\mathbf{H}}'^{(0)}{}_{j}\right)\right]\boldsymbol{\zeta}_j \tag{A.3}$$

respectively. The discretized equations of all grid points can be casted into matrix form (7.50), where

$$\hat{\mathbf{D}}_j = \frac{\hat{\mathscr{M}}^{(2)}_{\mathbf{kp}}{}_{,j} + \hat{\mathscr{M}}^{(2)}_{\mathbf{kp}}{}_{,j-1}}{2(z'_j - z'_{j-1})} + \frac{\hat{\mathscr{M}}^{(2)}_{\mathbf{kp}}{}_{,j} + \hat{\mathscr{M}}^{(2)}_{\mathbf{kp}}{}_{,j+1}}{2(z'_{j+1} - z'_j)}$$
$$+ \frac{(z'_j - z'_{j-1})\left(\hat{\mathbf{H}}'^{(0)}{}_{,j} + \hat{\mathbf{H}}'^{(0)}{}_{,j-1}\right)}{4} + \frac{(z'_{j+1} - z'_j)\left(\hat{\mathbf{H}}'^{(0)}{}_{,j} + \hat{\mathbf{H}}'^{(0)}{}_{,j+1}\right)}{4} \tag{A.4}$$

and

$$\hat{\mathbf{D}}^-_j = -\frac{\hat{\mathscr{M}}^{(2)}_{\mathbf{kp}}{}_{,j} + \hat{\mathscr{M}}^{(2)}_{\mathbf{kp}}{}_{,j-1}}{2(z'_j - z'_{j-1})} + \frac{i}{4}\left(\hat{\mathscr{M}}^{(1)}_{\mathbf{kp}}{}_{,j} + \hat{\mathscr{M}}^{(1)}_{\mathbf{kp}}{}_{,j-1}\right) \tag{A.5}$$

and

$$\hat{\mathbf{D}}^+_j = -\frac{\hat{\mathscr{M}}^{(2)}_{\mathbf{kp}}{}_{,j} + \hat{\mathscr{M}}^{(2)}_{\mathbf{kp}}{}_{,j+1}}{2(z'_{j+1} - z'_j)} - \frac{i}{4}\left(\hat{\mathscr{M}}^{(1)}_{\mathbf{kp}}{}_{,j} + \hat{\mathscr{M}}^{(1)}_{\mathbf{kp}}{}_{,j+1}\right) \tag{A.6}$$

and $\hat{\mathbf{C}}$ is a positive-definite and diagonal matrix with $c_{jj} = \frac{z'_{j+1} - z'_{j-1}}{2}$. Because $\hat{\mathscr{M}}^{(2)}_{\mathbf{kp}}$ and $\hat{\mathbf{H}}'^{(0)}$ are hermitian, it is obviously that $\hat{\mathbf{D}}^\dagger_j = \hat{\mathbf{D}}_j$. Moreover:

$$(\hat{\mathbf{D}}^-_j)^\dagger = -\frac{\hat{\mathscr{M}}^{\dagger(2)}_{\mathbf{kp}}{}_{,j} + \hat{\mathscr{M}}^{\dagger(2)}_{\mathbf{kp}}{}_{,j-1}}{2(z'_j - z'_{j-1})} + \frac{i^*}{4}\left(\hat{\mathscr{M}}^{\dagger(1)}_{\mathbf{kp}}{}_{,j} + \hat{\mathscr{M}}^{\dagger(1)}_{\mathbf{kp}}{}_{,j-1}\right)$$
$$= -\frac{\hat{\mathscr{M}}^{(2)}_{\mathbf{kp}}{}_{,j} + \hat{\mathscr{M}}^{(2)}_{\mathbf{kp}}{}_{,j-1}}{2(z'_j - z'_{j-1})} - \frac{i}{4}\left(\hat{\mathscr{M}}^{(1)}_{\mathbf{kp}}{}_{,j-1} + \hat{\mathscr{M}}^{(1)}_{\mathbf{kp}}{}_{,j}\right)$$
$$= \hat{\mathbf{D}}^+_{j-1} \tag{A.7}$$

because $\hat{\mathscr{M}}^{(1)}_{\mathbf{kp}}$ and $\hat{\mathscr{M}}^{(2)}_{\mathbf{kp}}$ are hermitian.

A.2 Definition of Fourier Harmonics

The Fourier harmonic $Y_m(\phi)$ is defined as

$$Y_m(\phi) = c_m \cos(m\phi + \varphi_m) \tag{A.8}$$

with the normalization factor c_m and phase φ_m given by

$$c_m = \sqrt{\frac{1}{(1+\delta_{m,0})\pi}} \quad \text{and} \quad \varphi_m = \begin{cases} 0, & m \geq 0 \\ \frac{\pi}{2}, & m < 0 \end{cases}. \tag{A.9}$$

It can be easily proven that $\int_0^{2\pi} Y_m(\phi) Y_n(\phi) \mathrm{d}\phi = \delta_{m,n}$ and $\mathrm{d}Y_m(\phi)/\mathrm{d}\phi = -mY_{-m}(\phi)$.

A.3 Integrals of Products of Fourier Harmonics

This appendix proves the following formula for the integral of products of Fourier harmonics

$$\int_0^{2\pi} \prod_{i=1}^{N} Y_{m_i}(\phi)\mathrm{d}\phi = \frac{2\pi}{2^{N-1}} \left(\prod_{i=1}^{N} c_{m_i}\right) \sum_{j_1 \in \{1\}} \sum_{j_2 \in \{-1,1\}} \cdots \sum_{j_N \in \{-1,1\}} \Delta\left(\sum_{i=1}^{N} j_i m_i\right) \cos\left(\sum_{i=1}^{N} j_i \varphi_{m_i}\right), \tag{A.10}$$

where $\Delta(n)$ denotes the Kronecker symbol $\delta_{n,0}$.

Applying the basic formula $\cos\alpha\cos\beta = \frac{1}{2}[\cos(\alpha+\beta) + \cos(\alpha-\beta)]$ N times, the product of N harmonic functions reads

$$\prod_{i=1}^{N} Y_{m_i}(\phi) = \frac{1}{2^{N-1}} \left(\prod_{i=1}^{N} c_{m_i}\right) \sum_{j_1 \in \{1\}} \sum_{j_2 \in \{-1,1\}} \cdots \sum_{j_N \in \{-1,1\}} \cos\left(\sum_{i=1}^{N} j_i \left(\phi m_i + \varphi_{m_i}\right)\right). \tag{A.11}$$

The integral of the product of N harmonic functions reads

$$\int_0^{2\pi} \prod_{i=1}^{N} Y_{m_i}(\phi)\mathrm{d}\phi = \frac{1}{2^{N-1}} \left(\prod_{i=1}^{N} c_{m_i} \right) \sum_{j_1 \in \{1\}} \sum_{j_2 \in \{-1,1\}} \cdots \sum_{j_N \in \{-1,1\}} \int_0^{2\pi} \cos\left(\sum_{i=1}^{N} j_i \left(\phi m_i + \varphi_{m_i}\right) \right) \mathrm{d}\phi. \quad \text{(A.12)}$$

Because $\int_0^{2\pi} \cos\left(\sum_{i=1}^{N} j_i \left(\phi m_i + \varphi_{m_i}\right) \right) \mathrm{d}\phi = 2\pi \Delta \left(\sum_{i=1}^{N} j_i m_i \right) \cos\left(\sum_{i=1}^{N} j_i \varphi_{m_i} \right)$, formula (A.10) is obtained.

In order to implement (A.10) into the program source code, the sums over $j_1 \ldots j_N$ within (A.10) are replaced by a single sum over κ

$$\int_0^{2\pi} \prod_{i=1}^{N} Y_{m_i}(\phi)\mathrm{d}\phi = \frac{2\pi}{2^{N-1}} \left(\prod_{i=1}^{N} c_{m_i} \right) \times \sum_{\kappa=1}^{2^{N-1}} \Delta \left(\sum_{i=1}^{N} (-1 + 2b_{\kappa,i})\, m_i \right) \cos\left(\sum_{i=1}^{N} (-1 + 2b_{\kappa,i})\, \varphi_{m_i} \right), \quad \text{(A.13)}$$

where $b_{\kappa,i}$ is either 0 or 1. It is defined as follows: $b_{\kappa,1} = 1$; for $i = 2, \ldots, N$, $b_{\kappa,i}$ represents the $(i-1)$th bit of the binary number κ containing $(N-1)$ bits.

As a consequence of (A.10),

$$m_1 \pm m_2 \pm \cdots \pm m_N \neq 0 \rightarrow \int_0^{2\pi} \prod_{i=1}^{N} Y_{m_i}(\phi)\mathrm{d}\phi = 0 \quad \text{(A.14)}$$

because

$$m_1 \pm m_2 \pm \cdots \pm m_N \neq 0 \rightarrow \sum_{j_1 \in \{1\}} \sum_{j_2 \in \{-1,1\}} \cdots \sum_{j_N \in \{-1,1\}} \Delta \left(\sum_{i=1}^{N} j_i m_i \right) = 0. \quad \text{(A.15)}$$

Applying (A.10) for $N = 2$ gives the orthonormality of Y functions $\int_0^{2\pi} Y_{m_1}(\phi) Y_{m_2}(\phi)\mathrm{d}\phi = \delta_{m_1,m_2}$.

The symbol $\Upsilon_{m,m',m''} = \int_0^{2\pi} Y_m(\phi) Y_{m'}(\phi) Y_{m''}(\phi)\mathrm{d}\phi$ can be calculated using (A.10) with $N = 3$ and $m_1 = m$, $m_2 = m'$, $m_3 = m''$.

The symbol $\{\Upsilon C\}_{m,m',m''} = \int_0^{2\pi} Y_{-m}(\phi) Y_{m'}(\phi) Y_{m''}(\phi) \cos\phi \mathrm{d}\phi$ is expressed in the form

$$\{\Upsilon C\}_{m,m',m''} = \sqrt{\pi} \int_0^{2\pi} Y_{-m}(\phi) Y_{m'}(\phi) Y_{m''}(\phi) Y_1(\phi)\mathrm{d}\phi \quad \text{(A.16)}$$

and therefore can be calculated easily using (A.10) with $N = 4$ and $m_1 = -m$, $m_2 = m'$, $m_3 = m''$, $m_4 = 1$.

Similarly the symbol $\{\Upsilon S\}_{m,m',m''} = \int_0^{2\pi} Y_{-m}(\phi)Y_{m'}(\phi)Y_{m''}(\phi)\sin\phi \mathrm{d}\phi$ can be expressed as

$$\{\Upsilon S\}_{m,m',m''} = \sqrt{\pi}\int_0^{2\pi} Y_{-m}(\phi)Y_{m'}(\phi)Y_{m''}(\phi)Y_{-1}(\phi)\mathrm{d}\phi, \tag{A.17}$$

which can be calculated using (A.10) with $N = 4$ and $m_1 = -m$, $m_2 = m'$, $m_3 = m''$, $m_4 = -1$.

A.4 Integrals with Delta Function

Some important results for integrals with a delta function are described in this appendix. At first, the integral of a product of a function $h(x)$ and a delta function or the derivative of delta function can be derived:

$$\int_{-\infty}^{+\infty} h(x)\delta(x - x_0)\mathrm{d}x = h(x_0), \tag{A.18}$$

$$\int_{-\infty}^{+\infty} h(x)\frac{\partial\delta(x - x_0)}{\partial x}\mathrm{d}x = -\frac{\partial h(x_0)}{\partial x}. \tag{A.19}$$

Here, $\delta(x)$ is defined as

$$\delta(x) = \begin{cases} 0, & x \neq 0 \\ \infty, & x = 0 \end{cases} \quad \text{and} \quad \int_{-\infty}^{+\infty} \delta(x)\mathrm{d}x = 1. \tag{A.20}$$

Applying (A.18) one can simplify the integral over the 2D $\mathbf{k}$-space of a product of a microscopic subband quantity $h^\nu(\mathbf{k})$, subband distribution $f^\nu(\mathbf{k})$ and delta function

$$\begin{aligned} &\frac{1}{(2\pi)^2}\int h^\nu(\mathbf{k}) f^\nu(\mathbf{k})\delta(\tilde{\varepsilon} - \tilde{\varepsilon}^\nu(\mathbf{k}))\mathrm{d}^2k \\ &= \int_0^{2\pi}\int h^\nu(\tilde{\varepsilon}^\nu(\mathbf{k},\phi),\phi)\underbrace{\frac{1}{(2\pi)^2} f^\nu(\mathbf{k})k\frac{\partial k}{\partial\tilde{\varepsilon}^\nu}}_{g^\nu(\tilde{\varepsilon}^\nu(\mathbf{k},\phi),\phi)}\delta(\tilde{\varepsilon} - \tilde{\varepsilon}^\nu(\mathbf{k}))\mathrm{d}\tilde{\varepsilon}^\nu\mathrm{d}\phi \\ &= \int_0^{2\pi} h^\nu(\tilde{\varepsilon},\phi) g^\nu(\tilde{\varepsilon},\phi)\mathrm{d}\phi. \end{aligned} \tag{A.21}$$

Applying (A.19) one can solve the integral of the product of $h^\nu(\mathbf{k})$ and derivative of delta function

$$\int h^\nu(\mathbf{k}) \frac{\partial \delta(\tilde{\varepsilon} - \tilde{\varepsilon}^\nu(\mathbf{k}))}{\partial \tilde{\varepsilon}^\nu} \mathrm{d}\tilde{\varepsilon}^\nu = \int h^\nu(\tilde{\varepsilon}^\nu(\mathbf{k}), \phi) \frac{\partial \delta(\tilde{\varepsilon} - \tilde{\varepsilon}^\nu(\mathbf{k}))}{\partial \tilde{\varepsilon}^\nu} \mathrm{d}\tilde{\varepsilon}^\nu$$
$$= - \left. \frac{\partial h^\nu(\tilde{\varepsilon}^\nu, \phi)}{\partial \tilde{\varepsilon}^\nu} \right|_{\tilde{\varepsilon}^\nu = \tilde{\varepsilon}} = - \frac{\partial h^\nu(\tilde{\varepsilon}, \phi)}{\partial \tilde{\varepsilon}} \quad \text{(A.22)}$$

A.5 Projection of Drift and Diffusion Terms

In this appendix the projection of the LHS of the BTE is shown. Integration by parts yields for the projection of the LHS (assuming that the surface integral vanishes for infinitely large energies):

$$\begin{aligned}
D &\equiv \frac{1}{(2\pi)^2} \int \{\mathrm{LHS}\} \delta(\tilde{\varepsilon} - \tilde{\varepsilon}^\nu(x, \mathbf{k})) Y_m(\phi) \mathrm{d}^2 k \\
&= \frac{1}{(2\pi)^2} \int \frac{1}{\hbar} \mathbf{F}^{\nu\mathrm{T}} (\nabla_\mathbf{k} f^\nu) \, \delta(\tilde{\varepsilon} - \tilde{\varepsilon}^\nu(x, \mathbf{k})) Y_m(\phi) \mathrm{d}^2 k \\
&\quad + \frac{1}{(2\pi)^2} \int \mathbf{v}^{\nu\mathrm{T}} (\nabla_\mathbf{r} f^\nu) \, \delta(\tilde{\varepsilon} - \tilde{\varepsilon}^\nu(x, \mathbf{k})) Y_m(\phi) \mathrm{d}^2 k \\
&= - \frac{1}{(2\pi)^2} \int \frac{1}{\hbar} f^\nu \mathbf{F}^{\nu\mathrm{T}} \nabla_\mathbf{k} \{\delta(\tilde{\varepsilon} - \tilde{\varepsilon}^\nu(x, \mathbf{k})) Y_m(\phi)\} \, \mathrm{d}^2 k \\
&\quad - \frac{1}{(2\pi)^2} \int \frac{1}{\hbar} f^\nu \delta(\tilde{\varepsilon} - \tilde{\varepsilon}^\nu(x, \mathbf{k})) Y_m(\phi) \nabla_\mathbf{k}^\mathrm{T} \mathbf{F}^\nu \mathrm{d}^2 k \\
&\quad + \frac{1}{(2\pi)^2} \int \mathbf{v}^{\nu\mathrm{T}} (\nabla_\mathbf{r} f^\nu) \, \delta(\tilde{\varepsilon} - \tilde{\varepsilon}^\nu(x, \mathbf{k})) Y_m(\phi) \mathrm{d}^2 k \\
&= D_\mathbf{k} + D_\mathbf{r},
\end{aligned} \quad \text{(A.23)}$$

where

$$\begin{aligned}
D_\mathbf{k} &= - \frac{1}{(2\pi)^2} \int \frac{1}{\hbar} f^\nu \mathbf{F}^{\nu\mathrm{T}} \nabla_\mathbf{k} \{\delta(\tilde{\varepsilon} - \tilde{\varepsilon}^\nu(x, \mathbf{k})) Y_m(\phi)\} \, \mathrm{d}^2 k \\
&= - \frac{1}{(2\pi)^2} \int \frac{1}{\hbar} f^\nu \mathbf{F}^{\nu\mathrm{T}} Y_m(\phi) \nabla_\mathbf{k} \delta(\tilde{\varepsilon} - \tilde{\varepsilon}^\nu(x, \mathbf{k})) \mathrm{d}^2 k \\
&\quad - \frac{1}{(2\pi)^2} \int \frac{1}{\hbar} f^\nu \delta(\tilde{\varepsilon} - \tilde{\varepsilon}^\nu(x, \mathbf{k})) \mathbf{F}^{\nu\mathrm{T}} \nabla_\mathbf{k} Y_m(\phi) \mathrm{d}^2 k
\end{aligned} \quad \text{(A.24)}$$

and

$$D_{\mathbf{r}} = -\frac{1}{(2\pi)^2}\int \frac{1}{\hbar} f^{\nu}\delta(\tilde{\varepsilon}-\tilde{\varepsilon}^{\nu}(x,\mathbf{k}))Y_m(\phi)\nabla_{\mathbf{k}}^{\mathrm{T}}\mathbf{F}^{\nu}\mathrm{d}^2k$$

$$+\frac{1}{(2\pi)^2}\int \mathbf{v}^{\nu\mathrm{T}}\left(\nabla_{\mathbf{r}} f^{\nu}\right)\delta(\tilde{\varepsilon}-\tilde{\varepsilon}^{\nu}(x,\mathbf{k}))Y_m(\phi)\mathrm{d}^2k$$

$$= \frac{1}{(2\pi)^2}\int \delta(\tilde{\varepsilon}-\tilde{\varepsilon}^{\nu}(x,\mathbf{k}))Y_m(\phi)\nabla_{\mathbf{r}}^{\mathrm{T}}\left(\mathbf{v}^{\nu} f^{\nu}\right)\mathrm{d}^2k \tag{A.25}$$

because

$$\frac{1}{\hbar}\nabla_{\mathbf{k}}^{\mathrm{T}}\mathbf{F}^{\nu} = \frac{1}{\hbar}\nabla_{\mathbf{k}}^{\mathrm{T}}\mathbf{F}_{\mathrm{E}}^{\nu} + \frac{q}{\hbar}\nabla_{\mathbf{k}}^{\mathrm{T}}(\mathbf{v}^{\nu}\times\mathbf{B})$$

$$= -\nabla_{\mathbf{r}}^{\mathrm{T}}\frac{1}{\hbar}\nabla_{\mathbf{k}}\varepsilon^{\nu} + B\frac{q}{\hbar}\underbrace{\left(\frac{\partial v_y^{\nu}}{\partial k_x}-\frac{\partial v_x^{\nu}}{\partial k_y}\right)}_{0} = -\nabla_{\mathbf{r}}^{\mathrm{T}}\mathbf{v}^{\nu} \tag{A.26}$$

Taking into account the fact that $\mathbf{v}^{\nu} = \frac{1}{\hbar}\nabla_{\mathbf{k}}\varepsilon^{\nu} = \frac{1}{\hbar}\nabla_{\mathbf{k}}\tilde{\varepsilon}^{\nu}$ and $\mathbf{F}^{\nu\mathrm{T}}\mathbf{v}^{\nu} = \mathbf{F}_{\mathrm{E}}^{\nu\mathrm{T}}\mathbf{v}^{\nu} + \mathbf{F}_{\mathrm{B}}^{\nu\mathrm{T}}\mathbf{v}^{\nu} = F_{\mathrm{E}}^{\nu}v_x^{\nu}$, the $D_{\mathbf{k}}$-term can be expressed in terms of the Fourier expansion coefficients of g:

$$D_{\mathbf{k}} = -\int_0^{2\pi}\int_0^{\infty} g^{\nu}\mathbf{F}^{\nu\mathrm{T}}Y_m(\phi)\underbrace{\left(\frac{1}{\hbar}\nabla_{\mathbf{k}}\tilde{\varepsilon}^{\nu}\right)}_{\mathbf{v}^{\nu}}\frac{\partial\delta(\tilde{\varepsilon}^{\nu}-\tilde{\varepsilon})}{\partial\tilde{\varepsilon}^{\nu}}\mathrm{d}\tilde{\varepsilon}^{\nu}\mathrm{d}\phi$$

$$-\frac{1}{(2\pi)^2}\int \frac{1}{\hbar} f^{\nu}\delta(\tilde{\varepsilon}-\tilde{\varepsilon}^{\nu}(x,\mathbf{k}))\mathbf{F}^{\nu\mathrm{T}}\nabla_{\mathbf{k}}Y_m(\phi)\mathrm{d}^2k$$

$$= \sum_{m'=-M'}^{M'}\frac{\partial}{\partial\tilde{\varepsilon}}\left\{ g_{m'}^{\nu}(x,\tilde{\varepsilon})\int_0^{2\pi} Y_m(\phi)Y_{m'}(\phi)F_{\mathrm{E}}^{\nu}(x,\tilde{\varepsilon},\phi)v_x^{\nu}(x,\tilde{\varepsilon},\phi)\mathrm{d}\phi\right\}$$

$$-\frac{1}{(2\pi)^2}\int \frac{1}{\hbar} f^{\nu}\delta(\tilde{\varepsilon}-\tilde{\varepsilon}^{\nu}(x,\mathbf{k}))\mathbf{F}^{\nu\mathrm{T}}\nabla_{\mathbf{k}}Y_m(\phi)\mathrm{d}^2k \tag{A.27}$$

using (A.22) for the simplification of the integral over $\tilde{\varepsilon}^{\nu}$ within the first term. Using (A.21) the second term occurring in $D_{\mathbf{k}}$ can be also simplified and expressed in terms of $g_{m'}$

$$\frac{1}{(2\pi)^2}\int \frac{1}{\hbar} f^{\nu}\delta(\tilde{\varepsilon}-\tilde{\varepsilon}^{\nu}(x,\mathbf{k}))\mathbf{F}^{\nu\mathrm{T}}\nabla_{\mathbf{k}}Y_m(\phi)\mathrm{d}^2k$$

$$= \int_0^{2\pi}\frac{g^{\nu}(x,\tilde{\varepsilon},\phi)}{\hbar k^{\nu}(x,\tilde{\varepsilon},\phi)}F_{\mathrm{E}}^{\nu}(x,\tilde{\varepsilon},\phi)\mathbf{e}_x^{\mathrm{T}}(-\sin\phi\mathbf{e}_x+\cos\phi\mathbf{e}_y)(-m)Y_{-m}(\phi)\mathrm{d}\phi$$

$$+\int_0^{2\pi}\frac{g^\nu(x,\tilde{\varepsilon},\phi)}{\hbar k^\nu(x,\tilde{\varepsilon},\phi)}qBv_k^\nu(x,\tilde{\varepsilon},\phi)(-1)\mathbf{e}_{\phi'}^{\mathrm{T}}\mathbf{e}_\phi(-m)Y_{-m}(\phi)\mathrm{d}\phi$$

$$=\sum_{m'=-M'}^{M'}g_{m'}^\nu(x,\tilde{\varepsilon})\left\{m\int_0^{2\pi}Y_{-m}(\phi)Y_{m'}(\phi)\frac{F_{\mathrm{E}}^\nu(x,\tilde{\varepsilon},\phi)}{\hbar k^\nu(x,\tilde{\varepsilon},\phi)}\sin\phi\mathrm{d}\phi\right.$$

$$\left.+Bm\int_0^{2\pi}Y_{-m}(\phi)Y_{m'}(\phi)\frac{qv_k^\nu(x,\tilde{\varepsilon},\phi).}{\hbar k^\nu(x,\tilde{\varepsilon},\phi)}\mathrm{d}\phi\right\} \tag{A.28}$$

due to $\mathbf{F}_{\mathrm{B}}=qB\mathbf{v}\times\mathbf{e}_z=qB\left(v_k\mathbf{e}_k+v_\phi\mathbf{e}_\phi\right)\times\mathbf{e}_z=-qBv_k\mathbf{e}_\phi+qBv_\phi\mathbf{e}_k$.

The identity $\partial\tilde{\varepsilon}^\nu/\partial x=-\tilde{F}_{\mathrm{E}}^\nu$ leads for $D_{\mathbf{r}}$ to

$$D_{\mathbf{r}}=\frac{1}{(2\pi)^2}\int\delta(\tilde{\varepsilon}-\tilde{\varepsilon}^\nu(x,\mathbf{k}))Y_m(\phi)\nabla_{\mathbf{r}}^{\mathrm{T}}\left(\mathbf{v}^\nu f^\nu\right)\mathrm{d}^2k$$

$$=\frac{1}{(2\pi)^2}\int\delta(\tilde{\varepsilon}-\tilde{\varepsilon}^\nu(x,\mathbf{k}))Y_m(\phi)\frac{\partial\left(v_x^\nu f^\nu\right)}{\partial x}\mathrm{d}^2k$$

$$=\frac{\partial}{\partial x}\frac{1}{(2\pi)^2}\int Y_m(\phi)v_x^\nu f^\nu\delta(\tilde{\varepsilon}-\tilde{\varepsilon}^\nu(x,\mathbf{k}))\mathrm{d}^2k$$

$$-\frac{1}{(2\pi)^2}\int Y_m(\phi)v_x^\nu f^\nu\frac{\partial\delta(\tilde{\varepsilon}-\tilde{\varepsilon}^\nu(x,\mathbf{k}))}{\partial x}\mathrm{d}^2k$$

$$=\frac{\partial}{\partial x}\int_0^{2\pi}Y_m(\phi)v_x^\nu(x,\tilde{\varepsilon},\phi)g^\nu(x,\tilde{\varepsilon},\phi)\mathrm{d}\phi$$

$$-\int_0^{2\pi}\int_0^\infty Y_m(\phi)v_x^\nu g^\nu\underbrace{\left(\frac{\partial\tilde{\varepsilon}^\nu}{\partial x}\right)}_{-\tilde{F}_{\mathrm{E}}^\nu}\frac{\partial\delta(\tilde{\varepsilon}^\nu-\tilde{\varepsilon})}{\partial\tilde{\varepsilon}^\nu}\mathrm{d}\tilde{\varepsilon}^\nu\mathrm{d}\phi \tag{A.29}$$

Finally, $D_{\mathbf{r}}$ reads

$$D_{\mathbf{r}}=\sum_{m'=-M'}^{M'}\frac{\partial}{\partial x}\left\{g_{m'}^\nu(x,\tilde{\varepsilon})\int_0^{2\pi}Y_m(\phi)Y_{m'}(\phi)v_x^\nu(x,\tilde{\varepsilon},\phi)\mathrm{d}\phi\right\}$$

$$-\frac{\partial}{\partial\tilde{\varepsilon}}\left\{g_{m'}^\nu(x,\tilde{\varepsilon})\int_0^{2\pi}Y_m(\phi)Y_{m'}(\phi)\tilde{F}_{\mathrm{E}}^\nu(x,\tilde{\varepsilon},\phi)v_x^\nu(x,\tilde{\varepsilon},\phi)\mathrm{d}\phi\right\} \tag{A.30}$$

using (A.21) and (A.22). Putting $D_{\mathbf{k}}$ and $D_{\mathbf{r}}$ together gives the final form for D:

$$D=D_{\mathbf{k}}+D_{\mathbf{r}}$$

$$=\sum_{m'=-M'}^{M'}\frac{\partial}{\partial\tilde{\varepsilon}}\left\{g_{m'}^\nu(x,\tilde{\varepsilon})F_{\mathrm{E,pot}}^\nu(x)\int_0^{2\pi}Y_m(\phi)Y_{m'}(\phi)v_x^\nu(x,\tilde{\varepsilon},\phi)\mathrm{d}\phi\right\}$$

$$+\frac{\partial}{\partial x}\left\{g^{\nu}_{m'}(x,\tilde{\varepsilon})\int_0^{2\pi} Y_m(\phi)Y_{m'}(\phi)v^{\nu}_x(x,\tilde{\varepsilon},\phi)\mathrm{d}\phi\right\}$$

$$-g^{\nu}_{m'}(x,\tilde{\varepsilon})\left\{m\int_0^{2\pi} Y_{-m}(\phi)Y_{m'}(\phi)\frac{F^{\nu}_{\mathrm{E}}(x,\tilde{\varepsilon},\phi)}{\hbar k^{\nu}(x,\tilde{\varepsilon},\phi)}\sin\phi\mathrm{d}\phi\right.$$

$$\left.+Bm\int_0^{2\pi} Y_{-m}(\phi)Y_{m'}(\phi)\frac{qv^{\nu}_k(x,\tilde{\varepsilon},\phi)}{\hbar k^{\nu}(x,\tilde{\varepsilon},\phi)}\mathrm{d}\phi\right\} \tag{A.31}$$

due to $F^{\nu}_{\mathrm{E}}(x,\mathbf{k}) - \tilde{F}^{\nu}_{\mathrm{E}}(x,\mathbf{k}) = F^{\nu}_{\mathrm{E,pot}}(x) \equiv -\partial\varepsilon^{\nu}_{\mathrm{pot}}(x)/\partial x$.

Within the primed coordinate system (x',y',z') the polar angle $\overline{\phi}$ of the wave vector $\mathbf{k}$ is given by $\overline{\phi} = \phi + \varphi_{\mathrm{D}}$. Therefore, the projection D finally reads:

$$D = \sum_{m'=-M'}^{M'} \frac{\partial}{\partial\tilde{\varepsilon}}\left\{F^{\nu}_{\mathrm{E,pot}}(x)d^{\nu}_{m,m'}(x,\tilde{\varepsilon})g^{\nu}_{m'}(x,\tilde{\varepsilon})\right\} + \frac{\partial}{\partial x}\left\{d^{\nu}_{m,m'}(x,\tilde{\varepsilon})g^{\nu}_{m'}(x,\tilde{\varepsilon})\right\}$$

$$-\left[b^{\mathrm{E}\nu}_{m,m'}(x,\tilde{\varepsilon}) + b^{\mathrm{B}\nu}_{m,m'}(x,\tilde{\varepsilon})\right]g^{\nu}_{m'}(x,\tilde{\varepsilon}), \tag{A.32}$$

where the coefficients $d^{\nu}_{m,m'}$, $b^{\mathrm{E}\nu}_{m,m'}$ and $b^{\mathrm{B}\nu}_{m,m'}$ are given by:

$$d^{\nu}_{m,m'}(x,\tilde{\varepsilon}) = \int_0^{2\pi} Y_m(\overline{\phi})Y_{m'}(\overline{\phi})v^{\nu}_x(x,\tilde{\varepsilon},\overline{\phi})\mathrm{d}\overline{\phi} \tag{A.33}$$

$$b^{\mathrm{E}\nu}_{m,m'}(x,\tilde{\varepsilon}) = m\int_0^{2\pi} Y_{-m}(\overline{\phi})Y_{m'}(\overline{\phi})\frac{F^{\nu}_{\mathrm{E}}(x,\tilde{\varepsilon},\overline{\phi})}{\hbar k^{\nu}(x,\tilde{\varepsilon},\overline{\phi})}\sin(\overline{\phi}-\varphi_{\mathrm{D}})\mathrm{d}\overline{\phi} \tag{A.34}$$

$$b^{\mathrm{B}\nu}_{m,m'}(x,\tilde{\varepsilon}) = mB\int_0^{2\pi} Y_{-m}(\overline{\phi})Y_{m'}(\overline{\phi})\frac{qv^{\nu}_k(x,\tilde{\varepsilon},\overline{\phi})}{\hbar k^{\nu}(x,\tilde{\varepsilon},\overline{\phi})}\mathrm{d}\overline{\phi}. \tag{A.35}$$

(A.32) is equivalent to the LHS of (9.22).

A.6 Projection of the Scattering Integral

This appendix provides the derivation of (9.36). The total scattering integral (9.33) is split into in- and out-scattering integrals:

$$\hat{S}^{\nu}\{f\} = \hat{S}^{\nu}_{\mathrm{in}}\{f\} - \hat{S}^{\nu}_{\mathrm{out}}\{f\}, \tag{A.36}$$

where

$$\hat{S}^{\nu}_{\mathrm{in}}\{f\} = \frac{1}{(2\pi)^2}\sum_{\nu'}\int(1-f^{\nu}(x,\mathbf{k}))S^{\nu,\nu'}(x,\mathbf{k}|\mathbf{k}')f^{\nu'}(x,\mathbf{k}')\mathrm{d}^2k' \tag{A.37}$$

and

$$\hat{S}^{\nu}_{\text{out}}\{f\} = \frac{1}{(2\pi)^2}\sum_{\nu'}\int (1 - f^{\nu'}(x,\mathbf{k}'))S^{\nu',\nu}(x,\mathbf{k}'|\mathbf{k})f^{\nu}(x,\mathbf{k})\mathrm{d}^2k' \tag{A.38}$$

Using (A.21) the projection of in-scattering reads

$$\begin{aligned}
&\frac{1}{(2\pi)^2}\int \hat{S}^{\nu}_{\text{in}}\{f\}\delta(\tilde{\varepsilon} - \tilde{\varepsilon}^{\nu}(x,\mathbf{k}))Y_m(\phi)\mathrm{d}^2k \\
&= \frac{1}{(2\pi)^4}\int (1 - f^{\nu}(x,\mathbf{k}))\delta(\tilde{\varepsilon} - \tilde{\varepsilon}^{\nu}(x,\mathbf{k}))Y_m(\phi) \\
&\quad \times \int \sum_{\nu',\eta,\mp}\sum_{l,l'=-L}^{L} c^{\nu,\nu'}_{\eta,\mp,l,l'}\left(x,\varepsilon^{\nu}(x,\mathbf{k}),\varepsilon^{\nu'}(x,\mathbf{k}')\right)Y_l(\phi)Y_{l'}(\phi') \\
&\quad \times \delta\left(\tilde{\varepsilon}^{\nu}(x,\mathbf{k}) + \varepsilon^{\nu}_{\text{pot}}(x) - \varepsilon^{\nu'}(x,\mathbf{k}') \mp \hbar\omega_{\eta}\right) f^{\nu'}(x,\mathbf{k}')\mathrm{d}^2k'\mathrm{d}^2k \\
&= \sum_{l,l'=-L}^{L}\sum_{\nu',\eta,\mp} c^{\nu,\nu'}_{\eta,\mp,l,l'}\left(x,\tilde{\varepsilon},\tilde{\varepsilon}'\right)\int_0^{2\pi}\left[Z^{\nu}(x,\tilde{\varepsilon},\phi) - g^{\nu}(x,\tilde{\varepsilon},\phi)\right]Y_m(\phi)Y_l(\phi)\mathrm{d}\phi \\
&\quad \times \int_0^{2\pi} g^{\nu'}\left(x,\tilde{\varepsilon}',\phi'\right)Y_{l'}(\phi')\mathrm{d}\phi',
\end{aligned} \tag{A.39}$$

where $\tilde{\varepsilon}'$ is the subband kinetic energy of the initial state given by (9.38) due to Fermi's Golden Rule (9.37).

Introducing the Fourier expansion coefficients of Z and g gives:

$$\begin{aligned}
&\frac{1}{(2\pi)^2}\int \hat{S}^{\nu}_{\text{in}}\{f\}\delta(\tilde{\varepsilon} - \tilde{\varepsilon}^{\nu}(x,\mathbf{k}))Y_m(\phi)\mathrm{d}^2k \\
&= \sum_{m'=-M'}^{M'}\sum_{l=-L}^{L}\Upsilon_{m,m',l}\sum_{l'=-L}^{L}\sum_{\nu',\eta,\mp}\left[Z^{\nu}_{m'}(x,\tilde{\varepsilon}) - g^{\nu}_{m'}(x,\tilde{\varepsilon})\right] \\
&\quad c^{\nu,\nu'}_{\eta,\mp,l,l'}(x,\tilde{\varepsilon},\tilde{\varepsilon}')g^{\nu'}_{l'}(x,\tilde{\varepsilon}').
\end{aligned} \tag{A.40}$$

The project of out-scattering can be derived analogously:

$$\begin{aligned}
&\frac{1}{(2\pi)^2}\int \hat{S}^{\nu}_{\text{out}}\{f\}\delta(\tilde{\varepsilon} - \tilde{\varepsilon}^{\nu}(x,\mathbf{k}))Y_m(\phi)\mathrm{d}^2k \\
&= \frac{1}{(2\pi)^4}\int f^{\nu}(x,\mathbf{k})\delta(\tilde{\varepsilon} - \tilde{\varepsilon}^{\nu}(x,\mathbf{k}))Y_m(\phi) \\
&\quad \times \int \sum_{\nu',\eta,\mp}\sum_{l,l'=-L}^{L} c^{\nu',\nu}_{\eta,\pm,l',l}\left(x,\varepsilon^{\nu'}(x,\mathbf{k}'),\varepsilon^{\nu}(x,\mathbf{k})\right)Y_l(\phi)Y_{l'}(\phi')
\end{aligned}$$

$$\times\,\delta\left(\varepsilon^{\nu'}(x,\mathbf{k}')-\varepsilon^{\nu}_{\text{pot}}(x)-\tilde{\varepsilon}^{\nu}(x,\mathbf{k})\mp\hbar\omega_{\eta}\right)\left(1-f^{\nu'}(x,\mathbf{k}')\right)\mathrm{d}^2k'\mathrm{d}^2k$$

$$=\sum_{l,l'=-L}^{L}\sum_{\nu',\eta,\mp}c^{\nu',\nu}_{\eta,\pm,l,l'}\left(x,\tilde{\varepsilon}',\tilde{\varepsilon}\right)\int_0^{2\pi}g^{\nu}(x,\tilde{\varepsilon},\phi)Y_m(\phi)Y_l(\phi)\mathrm{d}\phi$$

$$\times\int_0^{2\pi}\left[Z^{\nu'}(x,\tilde{\varepsilon}',\phi')-g^{\nu'}\left(x,\tilde{\varepsilon}',\phi'\right)\right]Y_{l'}(\phi')\mathrm{d}\phi' \qquad \text{(A 41)}$$

using (A.21). Therefore, it yields

$$\frac{1}{(2\pi)^2}\int\hat{S}^{\nu}_{\text{out}}\{f\}\delta(\tilde{\varepsilon}-\tilde{\varepsilon}^{\nu}(x,\mathbf{k}))Y_m(\phi)\mathrm{d}^2k$$

$$=\sum_{m'=-M'}^{M'}\sum_{l=-L}^{L}\Upsilon_{m,m',l}\sum_{l'=-L}^{L}\sum_{\nu',\eta,\mp}\left[Z^{\nu'}_{l'}(x,\tilde{\varepsilon}')-g^{\nu'}_{l'}(x,\tilde{\varepsilon}')\right]$$

$$c^{\nu',\nu}_{\eta,\pm,l',l}(x,\tilde{\varepsilon}',\tilde{\varepsilon})g^{\nu}_{m'}(x,\tilde{\varepsilon}). \qquad \text{(A.42)}$$

Taking the difference of the in and out-terms yields (9.36).

A.7 Scattering Transition Rates

A.7.1 Phonon Scattering

The scattering model for a heterostructure SiGe PMOSFET is considered in this appendix. Due to the non-uniform material of the channel, the isotropic overlap factor of phonon scattering, which is approximated as in [2], is calculated separately for each layer within the channel:

$$\mathcal{G}^{\nu',\nu}_{\iota}(x)=\int_{\iota\text{th layer}}\left|\boldsymbol{\zeta}^{\nu'\dagger}(x,z,\mathbf{k})\boldsymbol{\zeta}^{\nu}(x,z,\mathbf{k})\right|^2_{\mathbf{k}=\mathbf{0}}\mathrm{d}z, \qquad \text{(A.43)}$$

where $\boldsymbol{\zeta}$ denotes the 6-component envelope vector (i.e. the eigen vector of the 6×6 $\mathbf{k}\cdot\mathbf{p}$ SE). For each SiGe layer the corresponding bulk phonon modes of unstrained Si and Ge are assumed and their scattering rates are linearly interpolated according to the Ge content [1]. Any interaction of phonons between layers is neglected. Therefore, in terms of the overlap factor for each layer, the transition rate of acoustic phonon scattering can be modeled as:

$$S^{\nu',\nu}_{\text{ac}}(x,\mathbf{k}'|\mathbf{k})=\delta(\varepsilon^{\nu'}(x,\mathbf{k}')-\varepsilon^{\nu}(x,\mathbf{k}))\times\sum_{\iota}\mathcal{G}^{\nu',\nu}_{\iota}(x)\left\{[1-\beta_{\iota}(x)]\frac{2\pi k_{\text{B}}T(\Xi^{\text{Si}})^2}{\hbar\rho^{\text{Si}}(u^{\text{Si}}_{l})^2}\right.$$

$$\left.+\beta_{\iota}(x)\frac{2\pi k_{\text{B}}T(\Xi^{\text{Ge}})^2}{\hbar\rho^{\text{Ge}}(u^{\text{Ge}}_{l})^2}\right\} \qquad \text{(A.44)}$$

and for optical phonon scattering:

$$S_{\text{opt}}^{\nu',\nu}(x,\mathbf{k}'|\mathbf{k}) = \sum_{\iota} \mathcal{G}_{\iota}^{\nu',\nu}(x)$$

$$\times \Bigg\{ [1-\beta_{\iota}(x)] \frac{\pi (D_t K^{\text{Si}})^2}{\rho^{\text{Si}} \omega^{\text{Si}}} \left[n(\hbar\omega^{\text{Si}}) + \frac{1}{2} \mp \frac{1}{2} \right] \delta\left(\varepsilon^{\nu'}(x,\mathbf{k}') - \varepsilon^{\nu}(x,\mathbf{k}) \mp \hbar\omega^{\text{Si}} \right)$$

$$+ \beta_{\iota}(x) \frac{\pi (D_t K^{\text{Ge}})^2}{\rho^{\text{Ge}} \omega^{\text{Ge}}} \left[n(\hbar\omega^{\text{Ge}}) + \frac{1}{2} \mp \frac{1}{2} \right] \delta\left(\varepsilon^{\nu'}(x,\mathbf{k}') - \varepsilon^{\nu}(x,\mathbf{k}) \mp \hbar\omega^{\text{Ge}} \right) \Bigg\}, \tag{A.45}$$

where β_{ι} is the Ge-content of layer ι, $\hbar$ the Planck constant, Ξ the average acoustic deformation potential, ρ the mass density, u_l the longitudinal sound velocity, $D_t K$ the average optical deformation potential, ω the optical phonon frequency, and $n(\hbar\omega) = \left(e^{\frac{\hbar\omega}{k_B T}} - 1 \right)^{-1}$ the Bose-Einstein distribution. The mass density and longitudinal sound velocity of Si and Ge are taken from [7]. The deformation potentials and optical phonon frequencies are adjustable parameters which are extracted by matching the simulated channel mobility to specially selected measurements as described in Sect. 12.1.

A.7.2 Alloy Scattering

When the channel contains SiGe alloy material, the alloy scattering needs to be taken into account. Based on the Harrison and Hauser model for the bulk materials [4], the transition rate of alloy scattering in a 2D hole gas is modeled as:

$$S_{\text{alloy}}^{\nu',\nu}(x,\mathbf{k}'|\mathbf{k}) = \delta(\varepsilon^{\nu'}(x,\mathbf{k}') - \varepsilon^{\nu}(x,\mathbf{k})) \sum_{\iota} \mathcal{G}_{\iota}^{\nu',\nu}(x) \beta_{\iota}(x)[1-\beta_{\iota}(x)] \frac{\pi a_{\iota}^3 U^2}{4\hbar}. \tag{A.46}$$

a is the lattice constant which is calculated following [7]. U is the alloy scattering potential. A wide range of U between 0.2 and 1.0 eV has been reported in the literature. An alloy scattering potential $U = 0.484$ eV is used in this work.

A.7.3 Surface Roughness Scattering

The surface roughness (SR) scattering mechanism is one of the most important scattering mechanisms for transport in a MOS transistor. Surface roughness scattering describes the scattering of particles at the side walls of the quantum well which are located at the oxide/semiconductor interfaces. The transition rate of SR scattering can be determined based on the Fermi's Golden Rule

$$S_{\mathrm{SR}}^{\nu',\nu}(x,\mathbf{k}'|\mathbf{k}) = \delta_{\nu',\nu}\frac{2\pi}{\hbar}|M_{\mathrm{GPN}}^{\nu',\nu}(x,\mathbf{k}',\mathbf{k})|^2\delta(\varepsilon^{\nu'}(x,\mathbf{k}') - \varepsilon^{\nu}(x,\mathbf{k})) \quad \text{(A.47)}$$

The Kronecker symbol $\delta_{\nu',\nu}$ in the $S_{\mathrm{SR}}^{\nu',\nu}(x,\mathbf{k}'|\mathbf{k})$ implies that only intra-subband surface roughness scattering is considered in this work. Similar to [6] the inter-subband surface roughness scattering is neglected in this work as well. Here the square of scattering matrix elements $|M_{\mathrm{GPN}}^{\nu',\nu}(x,\mathbf{k}',\mathbf{k})|^2$ are approximated such that

$$|M_{\mathrm{GPN}}^{\nu',\nu}(\mathbf{k}',\mathbf{k})|^2 \approx \left\langle|\Delta_{\mathbf{q}}|^2\right\rangle|\Gamma_{\mathrm{GPN}}^{\nu',\nu}(x,\mathbf{k}',\mathbf{k})|^2. \quad \text{(A.48)}$$

The power spectrum $\left\langle|\Delta_{\mathbf{q}}|^2\right\rangle$ with an exponential shape [2, 3, 5] is employed in this work:

$$\left\langle|\Delta_{\mathbf{q}}|^2\right\rangle = \pi\Delta^2L^2(1+q^2L^2/2)^{-2}. \quad \text{(A.49)}$$

Δ and L are rms height and correlation length of the roughness, respectively, which are determined by fitting the simulated low-field mobility with the experimental data (see Sect. 12.1 of Chap. 12).

The scattering matrix elements $\Gamma_{\mathrm{GPN}}^{\nu',\nu}(x,\mathbf{k}',\mathbf{k})$ are approximated in a manner similar to the work of Fischetti [2]:

$$\Gamma_{\mathrm{GPN}}^{\nu',\nu}(x,\mathbf{k}',\mathbf{k}) \approx \left.\frac{\partial\zeta^{\nu'\dagger}(x,0,\mathbf{k})}{\partial z}\hat{\mathscr{M}}^{(2)}\frac{\partial\zeta^{\nu}(x,0,\mathbf{k})}{\partial z}\right|_{\mathbf{k}=0}, \quad \text{(A.50)}$$

where $z = 0$ identifies the location of the oxide/semiconductor interface. Matrix $\hat{\mathscr{M}}^{(2)}$, which is independent of $\mathbf{k}$, is consistent with the one in the polynomial representation (7.40) of the $\mathbf{k}\cdot\mathbf{p}$ Hamiltonian.

References

1. Bufler, F.M., Graf, P., Meinerzhagen, B., Adeline, B., Rieger, M.M., Kibbel, H., Fischer, G.: Low- and high-field electron-transport parameters for unstrained and strained $Si_{1-x}Ge_x$. IEEE Electron Dev. Lett. **18**(6), 264–266 (1997)
2. Fischetti, M.V., Ren, Z., Solomon, P.M., Yang, M., Rim, K.: Six-band $k \cdot p$ calculation of the hole mobility in silicon inversion layers: Dependence on surface orientation, strain, and silicon thickness. J. Appl. Phys. **94**, 1079–1095 (2003)
3. Goodnick, S.M., Ferry, D.K., Wilmsen, C.W., Liliental, Z., Fathy, D., Krivanek, O.L.: Surface roughness at the Si(100)-SiO_2 interface. Phys. Rev. B **32**, 8171–8186 (1985)
4. Harrison, J.W., Hauser, J.R.: Alloy scattering in ternary III-V compounds. Phys. Rev. B **13**, 5347–5350 (1976)
5. Jin, S., Fischetti, M.V., Tang, T.: Modeling of surface-roughness scattering in ultrathin-body SOI MOSFETs. IEEE Trans. Electron Dev. **54**(9), 2191–2203 (2007)
6. Jungemann, C., Emunds, A., Engl, W.L.: Simulation of linear and nonlinear electron transport in homogeneous silicon inversion layers. Solid State Electron. **36**, 1529–1540 (1993)
7. Rieger, M.M., Vogl, P.: Electronic-band parameters in strained $Si_{1-x}Ge_x$ alloys on $Si_{1-y}Ge_y$ substrates. Phys. Rev. B **48**, 14276–14287 (1993)

Index

S.-M. Hong et al., *Deterministic Solvers for the Boltzmann Transport Equation*, Computational Microelectronics, DOI 10.1007/978-3-7091-0778-2,

MIX
Papier aus verantwortungsvollen Quellen
Paper from responsible sources
FSC® C105338

If you have any concerns about our products,
you can contact us on
ProductSafety@springernature.com

In case Publisher is established outside the EU,
the EU authorized representative is:
Springer Nature Customer Service Center GmbH
Europaplatz 3, 69115 Heidelberg, Germany

Printed by Libri Plureos GmbH
in Hamburg, Germany